Poultry: Breeding, Health, Nutrition, and Management

Poultry: Breeding, Health, Nutrition, and Management

Editor

István Komlósi

MDPI • Basel • Beijing • Wuhan • Barcelona • Belgrade • Manchester • Tokyo • Cluj • Tianjin

Editor
István Komlósi
Institute of Animal Science,
Biotechnology and Nature
Conservation, Faculty of
Agricultural and Food
Sciences and Environmental
Management, University of
Debrecen
Hungary

Editorial Office
MDPI
St. Alban-Anlage 66
4052 Basel, Switzerland

This is a reprint of articles from the Special Issue published online in the open access journal *Agriculture* (ISSN 2077-0472) (available at: https://www.mdpi.com/journal/agriculture/special_issues/Poultry_Breeding_Health_Nutrition_Management).

For citation purposes, cite each article independently as indicated on the article page online and as indicated below:

LastName, A.A.; LastName, B.B.; LastName, C.C. Article Title. *Journal Name* **Year**, *Volume Number*, Page Range.

ISBN 978-3-0365-4055-9 (Hbk)
ISBN 978-3-0365-4056-6 (PDF)

Contents

About the Editor

István Komlósi graduated from the University of Agriculture Debrecen. After studying biostatistics and quantitative genetics at the UNSW, Kensington, and the UNE, Armidale, Australia, as a postgraduate he started researching breeding programs. He completed his PhD, entitled "Application of video image analysis to predict carcass conformation in sheep", at the University of Wales, Bangor, UK. He also works as an advisor for the National Sheep Breeding Association and the Hungarian Simmental Breeder's Association on breeding programs. He leads research in animal welfare in cattle and poultry. In 2013, he completed his Doctor of Science degree at the Hungarian Academy of Science, and he is now a Professor in animal science at Debrecen University. He has published 293 scientific papers and has 551 independent citations. István Komlósi is the Head of the Institute of Animal Science, Biotechnology and Biodiversity, and the Head of the Animal Science Doctoral School. His teaching covers the principles of livestock production, quantitative genetics, and experimental design. He has lectured in applied business statistics at Curtin University, Perth, Australia.

 *agriculture*

MDPI

Editorial

Recent Advancements in Poultry Health, Nutrition and Sustainability

István Komlósi

Faculty of Agricultural and Food Science and Environmental Management, University of Debrecen, 4032 Debrecen, Hungary; komlosi@agr.unideb.hu

Citation: Komlósi, I. Recent Advancements in Poultry Health, Nutrition and Sustainability. *Agriculture* **2022**, *12*, 516. https://doi.org/10.3390/agriculture12040516

Received: 20 March 2022
Accepted: 25 March 2022
Published: 5 April 2022

Publisher's Note: MDPI stays neutral with regard to jurisdictional claims in published maps and institutional affiliations.

As the largest animal protein producer, the poultry industry is within the focus of mixed-diet consumers, as well as the livestock industry in general. The poultry industry is also extremely fast to uptake new technologies such as biotechnology, mechanization, robotics, and climate and nutrient control in order to be economically efficient and sustainable. Reducing the use of chemical fertilizers in agriculture is one of the EU Green Deal's priorities. The NPK supply of a 100 ha field by pelletized poultry litter was found to possess a smaller environmental impact compared to several combinations of chemical fertilizers [1]. Off-site coupled industrial chicken manure recycling technology (Hosoya compost) fundamentally affects the agricultural value of organic-based products [2]. Achieving the reduction in the N emissions in the poultry industry is vital. Feeding poultry with low-protein diets is an option, but their effects on the performance parameters and excreta composition of broiler chickens need to be investigated. It was found that the urinary N content of broiler chicken's excreta is lower than can be found in the literature, which should be considered in the ammonia inventory calculations [3]. The amount of phosphorus in the diet is also under debate. However, a P-deficient diet caused rickets in commercial chicks within three days [4]. There is constant pressure from pathogens and new threats, such as avian flu, that require new treatments and biosecurity measures. There are many novel approaches and answers to these challenges. Footpad dermatitis and hepatic lipidosis are health problems in fattening turkeys, where the positive influence of a higher methionine content in feed can be found on foot health and antioxidative capacity of the liver, although protein reduction might reduce growth performance [5]. The One Health approach, which requires a holistic approach, where genetics, nutrition, health treatment, and management need to be considered together, has gained ground in the poultry industry. *Citrullus colocynthis* (CC) has been known as a natural medicinal plant with wide-ranging biological activities, including antioxidant, anti-inflammatory, and antilipidemic effects. Dietary supplementation with 2.0 g/kg of ECCs could be considered a successful nutritional approach to producing healthier, lower-cholesterol eggs for consumers, in addition to enhancing the physiological and productive performance of laying hens by alleviating the stress of intensive commercial production [6]. Ascaridiosis in poultry results in a reduction in body weight gain, egg production, as well as microelement levels. Infected poultry have higher demands on feed with the addition of essential elements including zinc. The enrichment of the diet with inorganic zinc has a positive effect on the relative percentage of CD4+ lamina propria lymphocytes in the jejunum and on heterophil counts in the blood. In addition, inorganic zinc has an anti-inflammatory effect and activates IgA-producing cells in the jejunum of chicks infected with *A. galli* [7]. The microbiome of animals, both in the digestive tract and in the skin, plays an important role in protecting the host. The skin is one of the largest surface organs for animals, and therefore, the destabilization of the microbiota on its surface can increase the risk of diseases that may adversely affect animals' health and production rates. The housing environment in which animals live may affect the microbiota of their skin [8]. The gut microbiome seems to be a good indicator of the balanced health of an animal. The intestinal health of poultry is of great importance for the birds' growth and development. Probiotics-driven shifts in the gut microbiome can exert considerable indirect effect on the birds' welfare and production performance. The fecal bacteriobiome can be very useful for the global meta-analysis in order to gain a better insight into bacterial functioning and interactions with gut microbiota to improve poultry health, welfare and production performance [9]. For the adjustment of breeding programs for local, commercial, and exotic breeds, and to implement molecular breeding, a proper comprehension of phenotypic and genotypic variation is a *sine qua non* for sustainable breeding [10]. The status of animal welfare needs to be constantly monitored and improved. The use and effect of a new beak-abrasive material not yet examined on mortality of non-beak-trimmed laying hens of different genotypes housed in an alternative pen has been examined [11], and increased the behavioral repertoire of hens. Alternative food sources that support healthy human nutrition are in heavy demand. Ostrich meat, as lean meat with low intramuscular fat (0.5%) and cholesterol content, is suitable for this purpose [12].

Conflicts of Interest: The authors declare no conflict of interest.

References

1. Kiss, N.É.; Tamás, J.; Szőllősi, N.; Gorliczay, E.; Nagy, A. Assessment of composted pelletized poultry litter as an alternative to chemical fertilizers based on the environmental impact of their production. *Agriculture* **2021**, *11*, 1130. [CrossRef]
2. Gorliczay, E.; Boczonádi, I.; Kiss, N.É.; Tóth, F.A.; Pabar, S.A.; Biró, B.; Kovács, L.R.; Tamás, J. Microbiological effectivity evaluation of new poultry farming organic waste recycling. *Agriculture* **2021**, *11*, 683. [CrossRef]
3. Such, N.; Pál, L.; Strifler, P.; Horváth, B.; Koltay, I.A.; Rawash, M.A.; Farkas, V.; Mezőlaki, Á.; Wágner, L.; Dublecz, K. Effect of feeding low protein diets on the production traits and the nitrogen composition of excreta of broiler chickens. *Agriculture* **2021**, *11*, 781. [CrossRef]
4. Xu, L.; Li, N.; Farnell, Y.Z.; Wan, X.; Yang, H.; Zhong, X.; Farnell, M.B. Effect of feeding a high calcium: Phosphorus ratio, phosphorous deficient diet on Hypophosphatemic rickets onset in broilers. *Agriculture* **2021**, *11*, 955. [CrossRef]
5. Lingens, J.B.; Abd El-Wahab, A.; de Paula Dorigam, J.C.; Lemme, A.; Brehm, R.; Langeheine, M.; Visscher, C. Evaluation of methionine sources in protein reduced diets for turkeys in the late finishing period regarding performance, footpad health and liver health. *Agriculture* **2021**, *11*, 901. [CrossRef]
6. Alzarah, M.I.; Alaqil, A.A.; Abbas, A.O.; Nassar, F.S.; Mehaisen, G.M.K.; Gouda, G.F.; Abd El-Atty, H.K.; Moustafa, E.S. Inclusion of *Citrullus colocynthis* seed extract into diets induced a hypolipidemic effect and improved layer performance. *Agriculture* **2021**, *11*, 808. [CrossRef]
7. Karaffová, V.; Revajová, V.; Dvorožňáková, E.; Grešáková, Ľ.; Levkut, M.; Ševčíková, Z.; Herich, R.; Levkut, M. Effect of inorganic zinc on selected immune parameters in chicken blood and jejunum after *A. galli* infection. *Agriculture* **2021**, *11*, 551. [CrossRef]
8. Cholewińska, P.; Michalak, M.; Wojnarowski, K.; Skowera, S.; Smoliński, J.; Czyż, K. Levels of firmicutes, actinobacteria phyla and lactobacillaceae family on the skin surface of broiler chickens (Ross 308) depending on the nutritional supplement and the housing conditions. *Agriculture* **2021**, *11*, 287. [CrossRef]
9. Naumova, N.B.; Alikina, T.Y.; Zolotova, N.S.; Konev, A.V.; Pleshakova, V.I.; Lescheva, N.A.; Kabilov, M.R. *Bacillus*-Based Probiotic treatment modified bacteriobiome diversity in duck feces. *Agriculture* **2021**, *11*, 406. [CrossRef]
10. Larkina, T.A.; Barkova, O.Y.; Peglivanyan, G.K.; Mitrofanova, O.V.; Dementieva, N.V.; Stanishevskaya, O.I.; Vakhrameev, A.B.; Makarova, A.V.; Shcherbakov, Y.S.; Pozovnikova, M.V.; et al. Evolutionary subdivision of domestic chickens: Implications for local breeds as assessed by phenotype and genotype in comparison to commercial and fancy breeds. *Agriculture* **2021**, *11*, 914. [CrossRef]
11. Farkas, T.P.; Orbán, A.; Szasz, S.; Rapai, A.; Garamvölgyi, E.; Sütő, Z. Examination of the usage of a new beak-abrasive material in different laying hen genotypes (preliminary results). *Agriculture* **2021**, *11*, 947. [CrossRef]
12. Brassó, L.D.; Szabó, V.; Komlósi, I.; Pusztahelyi, T.; Várszegi, Z. Preliminary study of slaughter value and meat characteristics of 18 months ostrich reared in hungary. *Agriculture* **2021**, *11*, 885. [CrossRef]

Article

Levels of Firmicutes, Actinobacteria Phyla and Lactobacillaceae Family on the Skin Surface of Broiler Chickens (Ross 308) Depending on the Nutritional Supplement and the Housing Conditions

Paulina Cholewińska [1,*], Marta Michalak [2], Konrad Wojnarowski [1], Szymon Skowera [1], Jakub Smoliński [1] and Katarzyna Czyż [1]

1 Institute of Animal Breeding, Wroclaw University of Environmental and Life Sciences, 51-630 Wroclaw, Poland; konrad.wojnarowski@upwr.edu.pl (K.W.); 119633@student.upwr.edu.pl (S.S.); 107848@student.upwr.edu.pl (J.S.); katarzyna.czyz@upwr.edu.pl (K.C.)
2 Department of Animal Nutrition and Feed Management, Wroclaw University of Environmental and Life Sciences, 51-630 Wroclaw, Poland; marta.michalak@upwr.edu.pl
* Correspondence: paulina.cholewinska@upwr.edu.pl

Abstract: The microbiome of animals, both in the digestive tract and in the skin, plays an important role in protecting the host. The skin is one of the largest surface organs for animals; therefore, the destabilization of the microbiota on its surface can increase the risk of diseases that may adversely affect animals' health and production rates, including poultry. The aim of this study was to evaluate the effect of nutritional supplementation in the form of fermented rapeseed meal and housing conditions on the level of selected bacteria phyla (Firmicutes, Actinobacteria, and family Lactobacillaceae). The study was performed on 30 specimens of broiler chickens (Ross 308), individually kept in metabolic cages for 36 days. They were divided into 5 groups depending on the feed received. On day 36, skin swabs were individually collected. Temperature and humidity were measured in the room. The temperature was measured every 2 days (18 measurements × 6 points). The results of Real-Time PCR analysis have shown a significant effect of the feed additive on the level of Firmicutes phylum on the skin. On the other hand, a variable level of the tested bacteria was shown depending on the location of the cages. The Firmicutes phylum and Lactobacillaceae family achieved the highest level in the top-window zone. However, in the case of the Actinobacteria phylum, the highest level was found at the top-door and middle-door zones. The obtained results suggest that the conditions in which animals live may affect the microbiota of their skin.

Keywords: skin microbiota; poultry; welfare; diet

Citation: Cholewińska, P.; Michalak, M.; Wojnarowski, K.; Skowera, S.; Smoliński, J.; Czyż, K. Levels of Firmicutes, Actinobacteria Phyla and Lactobacillaceae Family on the Skin Surface of Broiler Chickens (Ross 308) Depending on the Nutritional Supplement and the Housing Conditions. *Agriculture* **2021**, *11*, 287. https://doi.org/10.3390/agriculture11040287

Academic Editor: István Komlósi

Received: 22 February 2021
Accepted: 23 March 2021
Published: 26 March 2021

Publisher's Note: MDPI stays neutral with regard to jurisdictional claims in published maps and institutional affiliations.

1. Introduction

Poultry, and in particular domestic chickens (*Gallus gallus domesticus*), is one of the main sources of animal protein due to one of the best feed/meat conversion ratios [1]. Due to humanitarian reasons and the production volume itself, it is important to maintain adequate welfare when keeping animals [2]. One of the welfare aspects is avoiding animal exposure to mechanical injuries possibility [3], which often result in disruption of the skin continuity, which may lead to many bacterial diseases [4]. It is not, however, the only circumstance related to the skin that we should pay attention to when keeping animals.

The skin of vertebrates is one of the largest organs in terms of surface, and for decades it has been attributed mainly to the role of an insulator separating the internal system of animals from the external environment. Over time, however, our knowledge about the skin and the roles and mechanisms occurring within or on its surface has grown significantly [5]. We already know that the protective barrier function is maintained not only by ectoderm cells but also by bacteria living on the surface of the skin, creating a specific ecosystem—a microbiota that, in order to fulfill its proper function, requires a balance both in terms of the species composition itself and the prevailing physicochemical conditions [6,7]. Factors such as nutrition, temperature, humidity and sunlight may affect the structure of the microbiota, which may directly affect animal health, and therefore, generate higher breeding costs [6]. The microbiota consists of bacteria, protozoa, fungi and archaea. However, bacteria are the most abundant both inside the body and on the skin. In mammals, the most numerous groups of bacteria belong to the Firmicutes and Bacteroidetes phyla. Yet, on the skin itself, bacteria from the Firmicutes, Bacteroidetes, Actinobacteria and Proteobacteria phyla are largely present; however, it depends on the taxon and even the species of the animal or the place where it lives [8]. Unfortunately, in the case of broiler chickens, no tests were carried out to systematize the correct microbiological composition, which in the future may provide additional information about their health status and thus their welfare.

The use of feed additives, including fermented feed, may affect the microbiological composition of the digestive system, increasing the amount of, e.g., bacteria from the Lactobacillaceae family. Changes in the gut can have an impact on overall health status [9] and thus also on the skin [10]. However, this relationship has not yet been thoroughly investigated for broiler chickens and should be undertaken in the future, and was also the reason for conducting the research presented in this paper.

This research focuses on the effects of nutrition and the housing environment on the levels of selected bacteria phyla on the skin of caged broiler chickens.

2. Materials and Methods

2.1. Animals

The research was conducted on 30 broiler chickens (Ross 308, male) kept in metabolic cages (1 broiler chicken per cage). The animals were kept in cages from the first day of life. They were divided into 5 food groups, 6 chickens per group (Table 1):

- A—control (no rapeseed meal);
- B—addition of 5% of standard rapeseed meal;
- C—addition of 10% of standard rapeseed meal;
- D—addition of 5% standard rapeseed meal and 5% fermented rapeseed meal;
- E—addition of 10% standard rapeseed meal and 10% fermented rapeseed meal.

Table 1. Nutritional composition of experimental diets.

Mixture	Group	Dry Matter (%)	Ash (%)	Protein (%)	Fat (%)	Fiber (%)	Gross Energy (kcal)
Starter (Day: 1–7)	A	87.94	4.46	21.90	7.63	7.24	4275.29
	B	87.90	3.99	20.77	7.45	7.39	4252.56
	C	87.70	4.17	22.78	7.81	7.67	4306.74
	D	87.63	4.41	22.37	7.72	7.53	4293.92
	E	87.99	4.83	24.07	7.41	7.10	4243.74
Grower 1 (Day: 8–28)	A	89.17	6.33	19.98	8.11	7.62	4268.95
	B	89.03	5.95	19.25	7.80	6.92	4247.90
	C	89.37	6.41	19.83	8.27	6.61	4246.39
	D	89.16	5.90	20.67	8.60	6.55	4294.93
	E	89.59	5.92	20.23	7.87	7.41	4308.50
Grower 2 (Day: 28–35)	A	89.76	6.58	20.85	8.90	6.01	4356.83
	B	89.90	5.98	20.77	9.89	6.17	4398.55
	C	89.58	6.02	20.22	9.62	8.11	4364.70
	D	89.74	6.40	20.61	9.02	6.31	4358.05
	E	89.57	6.62	20.90	8.86	8.61	4327.62

They received feed and water ad libitum.

The composition of the additive fermented rapeseed meal: protein 29%, fiber 10%, lactic acid 6%, LAB(lactic acid bacteria) log 6–7, glucosinolate reduction.

The room temperature from day 1 was about 32 °C; then it was gradually lowered by about 0.5 to 1 °C until the 20th day to obtain the temperature of 21 °C, at which it was kept until the sampling and weighing day. Humidity in the rooms during their rearing ranged from 60–70%. During the day, they had no access to external light. The only source of light was automatic lighting from 6.00 to 23.00 throughout the entire period of their maintenance. The lighting on both sides of the cages ranged between 24 to 20 lux from top to bottom; LED lighting was used.

The condition of the tested animals was good, and they did not show any symptoms, such as diarrhea, skin problems, increased body temperature, lack of appetite, and apathy, etc.

The authors confirm that the ethical policies of the journal were adhered to. All animals that qualified for the study were subjected to standard procedures without any harm or discomfort. The study did not require the consent of the Local Ethical Committee for Animal Experiments (Act of 15 January 2015 on protection animals used for scientific or educational purposes), OJ 2015, 266, implementing the Directive 2010/63/EU of the European Parliament and the Council of 22 September 2010 on the protection of animals used for scientific purposes.

2.2. Sampling

Samples were collected on the 36th day of age individually from each animal in the area around the neck. Feathers were trimmed on a small area so the swabs would not have contact with them. Skin swabs were collected with sterile brushes, placed in test tubes with sterile water, and then frozen at −26 °C until analysis (5 days).

2.3. Assessment of the Housing Environment

Throughout the experimental period, the maintenance conditions were assessed, the differences in temperature between the arrangement of the cages (top, middle and bottom of the cages, as well as on the door and window side) were examined (Table 2). The temperature was measured every 2 days (18 measurements×6 points). The arrangement of the cages in the room is shown in the diagram below (Figure 1). Additionally, the room humidity was measured at the same points, but its range was constant-SD (Standard Deviation) = 5.25.

Table 2. Average differences in room temperature depending on the arrangement of cages (°C) in relation to the general room temperature.

Cage Position	Top/ Window	Top/ Door	Middle/ Door	Middle/ Window	Bottom/ Window	Bottom/ Door
Mean	+1.05	+/−0.42	+/−0.31	+0.61	−0.65	−0.44
SD	0.30	0.28	0.34	0.15	0.15	0.35

SD-Standard devation.

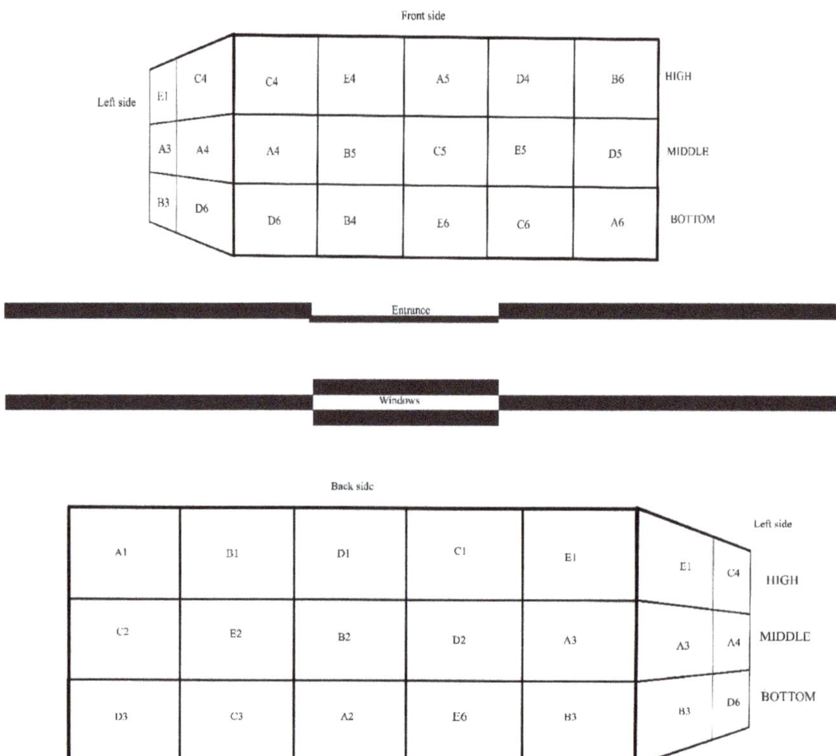

Figure 1. Diagram presenting arrangement of individual digestibility cages.

2.4. DNA (Deoxyrybonucleic acid) Isolation

A genomic bacteria AX Mini (A&A Biotechnology, Gdansk, Poland) was used for DNA isolation. The quality of DNA obtained as a result of isolation was verified using the NanoDrop 2000 spectrophotometer (Thermo Scientific, Wilmington, NC, USA). The average DNA content was 90–110 µg/µL. The level of impurities in the samples was as follows: for parameter 260/230:2.0–2.2 and for parameter 260/280:1.8–2.0 (in accordance with the guidelines from Thermo Scientific). When high levels of impurity or low-quality DNA were found, the samples were isolated again or purified using a clean-up concentrator (A&A Biotechnology, Gdansk, Poland)

2.5. Real-Time PCR Analysis

Bio-Rad CFX Connect 96 Touch apparatus was used to perform real-time PCR (Polymerase Chain Reaction) analysis. Bio-Rad SsoAdvanced™ Universal SYBR® Green Supermix kit (Bio-Rad Laboratories, Inc., Hercules CA, USA) at a volume of 10 µL was applied in 3 technical repetitions (Table 3). A NTC test (no template control) was additionally performed for each gene. The strategy of RT–PCR analysis involved the amplification of genes specific for the examined phyla in the presence of the reference gene for all bacteria (Table 4) [10,11].

Table 3. PCR mix components.

Component	Volume in 10 µL of Reaction
SsoAdvanced™ Universal SYBR® Green Supermix	5 µL
Starter (Foward + Reverse)	1 µL (0.8 µM)
Matrix DNA	2 µL ($0.04 - 0.015 \times 10^{-4}$)
Sterile water	2 µL

PCR-Polymerase Chain Reaction.

Table 4. Primers used during RT–PCR.

Name	Forward (5′–3′)	Reverse (5′–3′)	Source
Universal Eubacterial genes	530F (5′-GTC CCA GCM GCN GCG G)	1100R (5′-GGG TTN CGN TCG TTG)	[12]
Firmicutes	928F-Firm (5′-TGA AAC TYA AAG GAA TTG ACG)	1040FirmR (5′-ACC ATG CAC CTG TC)	[13]
Actinobacteria	Act1159R TCCGAGT-TRACCCCGGC	Eub338F ACGGGCG-GTGTGTACA	[14]
Lactobacillaceae	lac1 forward (5′-AGC AGT AGG GAA TCT TCC A)	Lac2Seq (5′-ATTTCACCGCTACACATG)	[15]

RT-PCR-Real Time Polymerase Chain Recation.

In order to determine the performance of individual genes, a standard curve was drawn for the genes under the study. A sample dilution of 10^{-3} was selected for analysis (from the 10^{-1} to 10^{-7} series). The analysis was performed in accordance with a 40 cycles protocol: polymerase activation and DNA denaturation 95 °C (3 min), denaturation 95 °C (15 s), annealing 60.5 °C (15 s), extension and plate reading at 72 °C (40 s). The melting curves analysis for the samples was performed at temperatures ranging from 65 °C (5 s) to 95 °C (0.5 °C increments in 2 s).

The data obtained were then processed using the CFX Maestro software v. 1.1 (Bio-Rad Laboratories, Inc., Hercules CA, USA). The sample with a DNA level of 100 µg/µL and impurities at a level consistent with the standards was an arbitrary calibrator. The efficiency of individual primers was normal (in accordance with the standards established by Bio-Rad) and amounted to 89.4% for Firmicutes, 93.6% for Actinobacteria, 97.7% for Lactobacillaceae; Universal—94.4. CFX Maestro calculated the results from the number of the reference gene matrix and the differences at the relative normalized expression (RNE = $\Delta\Delta C_q$) phylum's level, taking into account the amplification efficiency of individual genes [10].

2.6. Statistical Analysis

The results were analyzed with the use of the Statistica software (v. 13.3, StatSoft Inc., Tulsa, OK, USA). The data distribution was checked with the Shapiro–Wilk test. The data relating to the bodyweight of the chickens were analyzed using the ANOVA (analysis of variance) test, while the remaining data (no normal distribution) using the Wilcoxon pair test and the Friedman ANOVA ($p > 0.05$). The differences in the case of the ANOVA test were determined using Tukey's test.

3. Results

The weighing performed on the 36th day of life showed that the group with the addition of 5% standard rapeseed meal and 5% fermented rapeseed meal (D) had the highest body weight. On the other hand, the lowest body weight was found in individuals from the group with the addition of 5% of standard rapeseed meal (Table 5). Additionally, the performed statistical analysis in terms of the arrangement of cages (level and side-window/door) did not show any significant statistical differences.

Table 5. Bodyweight of broiler chickens at 36 days of age (g).

Group	A	B	C	D	E
Mean	1729.92 [a]	1615.83 [A]	1707.29 [A]	1920.76 [B,b]	1642.57 [A]
SD	79.17	66.05	129.68	83.98	33.18

$p > 0.05$—a,b; $p > 0.01$—A,B.

The analysis of the RT–PCR results showed that the group fed with 10% standard rapeseed meal (C) had a significantly higher RNE level of Firmicutes compared to the group fed with 10% fermented rapeseed meal and 10% standard (E). In addition, this level was generally the highest also compared to the other nutritional groups; similar trends are also visible in the comparison of the level of Actinobacteria and the Lactobacillaceae family. The lowest level of the Actinobacteria phylum was characteristic for the control group (A), and the highest, the aforementioned group C. The lowest level of the Lactobacillaceae family was characteristic for group B—5% addition of standard rapeseed meal (Figure 2).

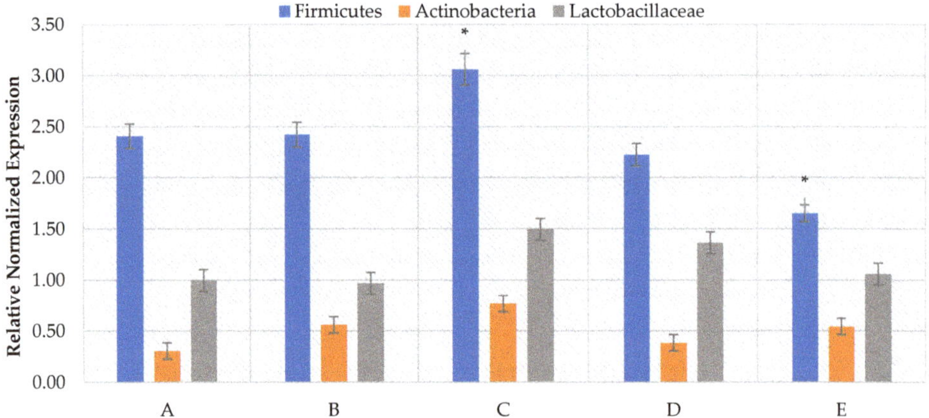

Figure 2. Level of relative normalized expression (RNE) of Firmicutes, Actinobacteria phyla and *Lactobacillaceae* family depending on the nutritional supplement. * $p = 0.048$. (**A**)—control (no rapeseed meal); (**B**)—addition of 5% of standard rapeseed meal; (**C**)—addition of 10% of standard rapeseed meal; (**D**)—addition of 5% standard rapeseed meal and 5% fermented rapeseed meal; (**E**)—addition of 10% standard rapeseed meal and 10% fermented rapeseed meal.

The analysis of the level of selected bacterial phyla also showed differences depending on the arrangement of the cages. The animals placed in the highest cages differed in the level of the Firmicutes phylum depending on the side of the location—doors or windows. In the window-up zone, animals were characterized by a much higher level (about 3 times) of this phylum compared to the up-door zone (Figure 3). Additionally, the temperature in the window-top sector was on average higher by 1.05 °C, and on the door side by 0.4 °C. Similar results were also obtained in the case of the Lactobacillaceae family, where the difference between the RNE level was more than twofold (Figure 4). Significant differences were also observed for the Actinobacteria phylum (Figure 5). The most significant differences in the level of the phyla were between the top-window and top-door zones. The highest RNE level was found in the middle cages in the middle-door zone, while the lowest in the upper part in the top-window zone.

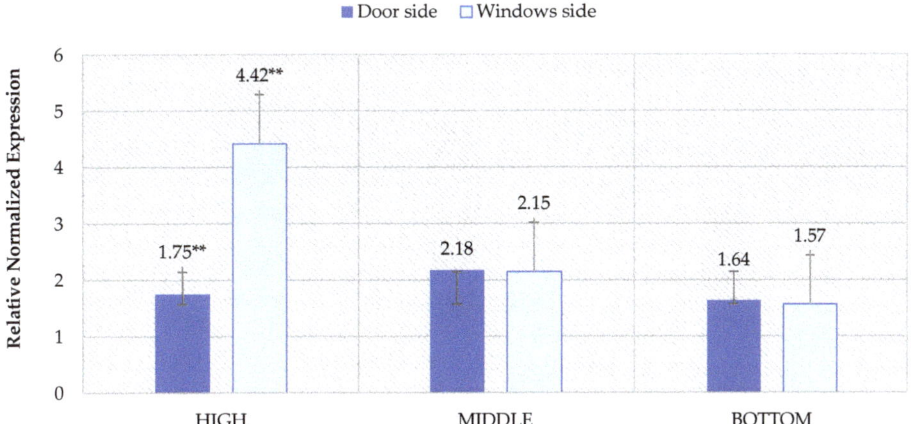

Figure 3. Level of relative normalized expression (RNE) of Firmicutes depending on the cage setting. ** $p = 0.00427$.

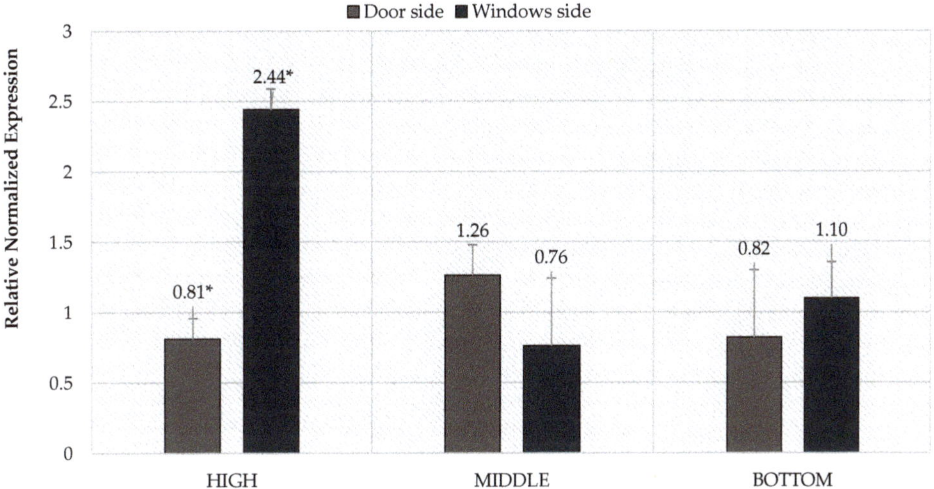

Figure 4. Level of relative normalized expression (RNE) of Lactobacillaceae family depending on the cage setting. * $p = 0.0428$.

However, when analyzing only the position of the cages in terms of height, significant differences were shown only in the level of the Firmicutes phylum (Figure 6). The highest level of this phylum was characterized by the groups staying in the upper cages and the lowest in the lowest cages. There is also a downward trend in RNE levels in the Lactobacillaceae family. On the other hand, in the case of the Actinobacteria phylum, there is a slight upward trend between individuals housed in the upper and lower cages. However, the highest level occurred in the middle frames. On the other hand, when analyzing the level of the tested bacteria broken down into the window and door zones only, no significant differences in the RNE level were found.

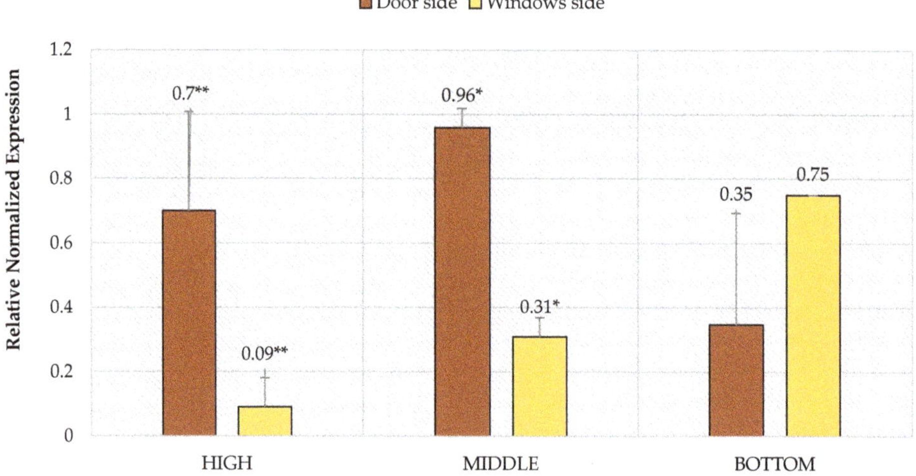

Figure 5. Level of relative normalized expression (RNE) of Actinobacteria phylum depending on the cage setting. ** p = 0.00427; * p = 0.046.

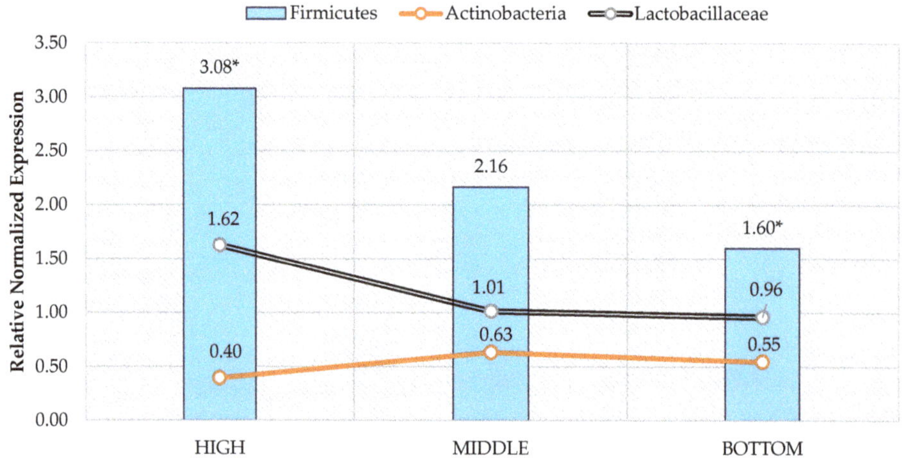

Figure 6. The level of studied phyla and families depending on the height of the cages. * p = 0.038.

4. Discussion

The microbial composition of the skin and the digestive tract is important because of its role in regulating the immune system and overall health status maintenance [16,17]. The skin microbiota changes with age but is somewhat stable in adult animals. In the case of animals, the main factors influencing the composition of their skin microbiome are environment, diet, livelihood, and geographical area [6,18,19]. In the experiment carried out on adult animals, it was shown that depending on the feed supplement obtained; there were differences in the level of the Firmicutes phylum on the skin surface. The group with the addition of 10% rapeseed meal (C) had the highest level of this phylum, while the group with the addition of 10% standard rapeseed meal and 10% fermented rapeseed meal had the lowest level of this phylum. Statistical differences were observed only in the case of Firmicutes phylum. Such a phenomenon, diet affecting microbiota of the skin, has been observed previously for felines and humans [20–22] but not for poultry.

The housing conditions and animal welfare of livestock, including poultry, can influence the microbiological composition of the organism. Research and analyses conducted by, among others, Hoffman et al. [8] or Ross et al. [6] indicate that the environment in which animals are kept may affect the skin microbiota. In the conducted study, there were observed statistical differences between groups on the basis of the cage arrangement. The highest level of the Firmicutes phylum has been observed in the case of individuals located the highest from the side of the windows; similar results were obtained in the case of the Lactobacillaceae family; in addition, the highest temperature was recorded in this place. Taking into account the sensitivity of bacteria to changing environmental conditions, the bacteria on the skin surface are the most sensitive to changes, because even despite the constant internal temperature of the host, the temperature of the skin surface also depends on the environment in which it lives [23–25]. However, for this phylum and family, no significant differences were found in the lower and mid-level cages, which may have been due to a more stable holding temperature.

On the other hand, the greatest differentiation occurred at the level of the Actinobacteria phylum. Its highest level occurred in the cages arranged in the middle part from the side of the door, and an equally high-level occurred in the upper part of the cages on the same side. The upper part from the window side had the highest temperature in the room, and the middle part from the same side was also slightly higher than the general temperature, which could have influenced the higher level of this phylum from the door side, where the temperature was usually lower. Similar relationships also occurred at the lower level; in this part, the temperature was generally slightly lower on the window side and slightly higher on the door side. This indicates that this phylum is better kept at a slightly lower temperature than the aforementioned bacteria and allows the conclusion that its level, as in the case of the others, is also dependent on the temperature on the skin surface [26–28]. As can be seen, despite the presence of the animals in the same room, the studied phyla appeared at different levels, depending on the arrangement of the cages. This indicates that the housing system in the case of poultry can significantly affect the microbiological composition of its skin. Both Firmicutes and Actinobacteria phyla are characterized by bacteria that have a positive effect on the condition of the skin and are relatively pathogenic. In the Actinobacteria phylum, these include, but are not limited to, *C. amicolatum*, *C. striatum*, *C. jeikeium*, *C. urealyticum*, and *C. xerosis*. In the case of the Firmicutes phyla, these are, for example, *Staphylococcus epidermidis*. Disturbance of homeostasis in the composition of the microbiota may result in an increased risk of skin diseases [16,17,27,28].

Due to the lack of similar studies, it is not possible to unequivocally assess the impact of individual factors included in the experiment. Most of the studies performed on broilers were focusing only on microbiota associated with the digestive tract. There is a clearly visible lack of studies focused on chicken skin microbiota; thus given issue requires further research and analysis. In addition, in the future, this type of research should be performed using sequencing to detect as many bacteria on the skin as possible, as was the case with humans.

5. Conclusions

The conducted experiment showed that the diet had a significant effect on broilers' skin microbiota in terms of Firmicutes phylum. Additionally, environmental factors related to the keeping of animals—the location of cages—also had a significant impact on levels of selected bacteria. It has been shown that, depending on their arrangement, there may be changes in the level of Firmicutes (including the Lactobacillaceae family) and Actinobacteria phyla, depending on the ambient temperature. These changes indicate that normalizing the conditions in the room where animals are located may positively impact their skin microbiota and thus their health.

Author Contributions: Conceptualization, M.M., K.W. and P.C.; methodology M.M., P.C.; formal analysis, P.C., K.W. and M.M.; investigation, S.S., J.S. and P.C.; data curation, P.C., K.W.; writing—original draft preparation, P.C. and M.M.; writing—review and editing, J.S. and K.W.; visualization, P.C., J.S., S.S.; supervision, K.C.; project administration, K.C. All authors have read and agreed to the published version of the manuscript.

Funding: This research received no external funding.

Institutional Review Board Statement: All animals that qualified for the study were subjected to standard procedures without any harm or discomfort, and therefore, the study did not require the consent of the Local Ethical Commission for Animal Experiments at the Institute of Immunology and Experimental Therapy of the Polish Academy of Sciences in Wroclaw, Poland (Act of 15 January 2015 on protection animals used for scientific or educational purposes).

Informed Consent Statement: Not applicable.

Data Availability Statement: The data presented in this study are available on request from the corresponding author. The data are not publicly available due to privacy.

Acknowledgments: The authors would like to thank Magdalena Wołoszyńska (Department of Genetics, Wroclaw University of Environmental and Life Sciences, 51-630 Wroclaw, Poland) for the access to the equipment.

Conflicts of Interest: All authors declare no conflict of interest.

References

1. FAO. 2017. Available online: http://www.fao.org/poultry-production-products/production/en/ (accessed on 9 January 2021).
2. Cholewińska, P.; Górniak, W.; Wojnarowski, K. Impact of selected environmental factors on microbiome of the digestive tract of ruminants. *BMC Vet. Res.* **2021**, *17*, 1–10. [CrossRef]
3. Weeks, C.A.; Lambton, S.L.; Williams, A.G. Implications for welfare, productivity and sustainability of the variation in reported levels of mortality for laying hen flocks kept in different housing systems: A meta-analysis of ten studies. *PLoS ONE* **2016**, *11*, e0146394. [CrossRef]
4. Atterbury, R.J.; Gigante, A.M.; Lozano, M.D.L.S.R.; Medina, R.D.M.; Robinson, G.; Alloush, H.; Allen, V.M. Reduction of salmonella contamination on the surface of chicken skin using bacteriophage. *Virol. J.* **2020**, *17*, 1–8. [CrossRef]
5. Jablonski, N.G. *Skin: A Natural History*; University of California Press: Berkeley, CA, USA, 2008.
6. Ross, A.A.; Rodrigues Hoffmann, A.; Neufeld, J.D. The skin microbiome of vertebrates. *Microbiome* **2019**, *7*, 79. [CrossRef]
7. Nagase, S.; Ogai, K.; Urai, T.; Shibata, K.; Matsubara, E.; Mukai, K.; Okamoto, S. Distinct skin microbiome and skin physiological functions between bedridden older patients and healthy people: A single-center study in Japan. *Front. Med.* **2020**, *7*, 101. [CrossRef]
8. Hoffmann, A.R. The cutaneous ecosystem: The roles of the skin microbiome in health and its association with inflammatory skin conditions in humans and animals. *Adv. Vet. Dermatol.* **2017**, *8*, 71–83.
9. Naji, S.A.H.; Al-Zanili, I.F.B.; Al-Gharawi, J. The effect of feed wetting and fermentation on the intestinal flora, humoral and cellular immunity of broiler chicks. *Int. J. Adv. Res.* **2015**, *3*, 87–94.
10. Brandwein, M.; Katz, I.; Katz, A.; Kohen, R. Beyond the gut: Skin microbiome compositional changes are associated with BMI. *Hum. Microbiome J.* **2019**, *13*, 100063. [CrossRef]
11. Górniak, W.; Cholewińska, P.; Szeligowska, N.; Wołoszyńska, M.; Soroko, M.; Czyż, K. Effect of intense exercise on the level of bacteroidetes and *Firmicutes phyla* in the digestive system of thoroughbred racehorses. *Animals* **2021**, *11*, 290. [CrossRef] [PubMed]
12. Dowd, S.E.; Callaway, T.R.; Wolcott, R.D.; Sun, Y.; McKeehan, T.; Hagevoort, R.G.; Edrington, T.S. Evaluation of the bacterial diversity in the feces of cattle using 16S rDNA bacterial tag-encoded FLX amplicon pyrosequencing (bTEFAP). *BMC Microbiol.* **2008**, *8*, 1–8. [CrossRef]
13. De Gregoris, T.B.; Aldred, N.; Clare, A.S.; Burgess, J.G. Improvement of phylum-and class-specific primers for real-time PCR quantification of bacterial taxa. *J. Microbiol. Methods* **2011**, *86*, 351–356. [CrossRef]
14. Blackwood, C.B.; Oaks, A.; Buyer, J.S. Phylum-And class-specific PCR primers for general microbial community analysis. *Appl. Environ. Microbiol.* **2005**, *71*, 6193–6198. [CrossRef]
15. Walter, J.; Hertel, C.; Tannock, G.W.; Lis, C.M.; Munro, K.; Hammes, W.P. Detection of lactobacillus, pediococcus, leuconostoc, and weissella species in human feces by using group-specific PCR primers and denaturing gradient gel electrophoresis. *Appl. Environ. Microbiol.* **2001**, *67*, 2578–2585. [CrossRef]
16. Yan, D.; Issa, N.; Afifi, L. The Role of the skin and gut microbiome in psoriatic disease. *Curr. Dermatol. Rep.* **2017**, *6*, 94–103. [CrossRef]
17. Xiao, Y.; Yun, X.; Weidong, Z.; Jinggang, C.; Kaifeng, L.; Hua, Y. Microbial community mapping in intestinal tract of broiler chicken. *Poult. Sci.* **2017**, *96*, 1387–1393. [CrossRef]

18. Henderson, G.; Cox, F.; Ganesh, S.; Jonker, A.; Young, W.; Global Rumen Census Collaborators; Janssen, P.H. Rumen microbial community composition varies with diet and host, but a core microbiome is found across a wide geographical range. *Sci. Rep.* **2015**, *5*, 14567. [CrossRef] [PubMed]
19. Oakley, B.B.; Morales, C.A.; Line, J.; Berrang, M.E.; Meinersmann, R.J.; Tillman, G.E.; Wise, M.G.; Siragusa, G.R.; Hiett, K.L.; Seal, B.S. The poultry-associated microbiome: Network analysis and farm-to-fork characterizations. *PLoS ONE* **2013**, *8*, e57190. [CrossRef] [PubMed]
20. Older, C.E.; Diesel, A.B.; Lawhon, S.D.; Queiroz, C.R.; Henker, L.C.; Rodrigues Hoffmann, A. The feline cutaneous and oral microbiota are influenced by breed and environment. *PLoS ONE* **2019**, *14*, e0220463. [CrossRef] [PubMed]
21. Morvan, P.Y.; Vallee, R. Evaluation of the effects of stressful life on human skin microbiota. *Appl. Microbiol. Open Access* **2018**, *4*, 1–8.
22. Lehtimäki, J.; Karkman, A.; Laatikainen, T.; Paalanen, L.; Von Hertzen, L.; Haahtela, T.; Ruokolainen, L. Patterns in the skin microbiota differ in children and teenagers between rural and urban environments. *Sci. Rep.* **2017**, *7*, 1–11. [CrossRef] [PubMed]
23. Shawkey, M.D.; Mills, K.L.; Dale, C. Microbial diversity of wild bird feathers revealed through culture-based and culture-independent techniques. *Microb. Ecol.* **2005**, *50*, 40–47. [CrossRef] [PubMed]
24. Romanovsky, A.A. Skin temperature: Its role in thermoregulation. *Acta Physiol.* **2014**, *210*, 498–507. [CrossRef] [PubMed]
25. Theurer, M.E.; White, B.J.; Anderson, D.E.; Miesner, M.D.; Mosier, D.A.; Coetzee, J.F.; Amrine, D.E. Effect of transportation during periods of high ambitne temperature on physiologic ang behavioral indices of beef heifers. *Am. J. Vet. Res.* **2013**, *74*, 481–490. [CrossRef] [PubMed]
26. Jutte, L.S.; Merrick, M.A.; Ingersoll, C.D.; Edwards, J.E. The relationship between intramuscular temperature, skin temperature, and adipose thickness during cryotherapy and rewarming. *Arch. Phys. Med. Rehabil.* **2001**, *82*, 845–850. [CrossRef] [PubMed]
27. Li, W.; Ma, Z. FBA Ecological guild: Trio of firmicutes-bacteroidetes alliance against actinobacteria in human oral microbiome. *Sci. Rep.* **2020**, *10*, 287. [CrossRef]
28. Zhao, Y.; Qian, L.; Yuquan, W.; Hongyang, C.; Xu, Z.; Xueqin, W.; Si, S.; Zimin, W. Effect of actinobacteria agent inoculation methods on cellulose degradation during composting based on redundancy analysis. *Bioresour. Technol.* **2016**, *219*, 196–203. [CrossRef] [PubMed]

*agriculture*

MDPI

Article

Bacillus-Based Probiotic Treatment Modified Bacteriobiome Diversity in Duck Feces

Natalia B. Naumova [1],*, Tatiana Y. Alikina [1], Natalia S. Zolotova [2], Alexey V. Konev [2], Valentina I. Pleshakova [2], Nadezhda A. Lescheva [2] and Marsel R. Kabilov [1]

[1] Institute of Chemical Biology and Fundamental Medicine, Siberian Branch of the Russian Academy of Sciences, Lavrentieva 8, 630090 Novosibirsk, Russia; alikina@niboch.nsc.ru (T.Y.A.); kabilov@niboch.nsc.ru (M.R.K.)
[2] The Faculty of Veterinary Medicine, Stolypin Omsk State Agrarian University, Institutskaya pl. 1, 644008 Omsk, Russia; ns.zolotova@omgau.org (N.S.Z.); av.konev@omgau.org (A.V.K.); vi.pleshakova@omgau.org (V.I.P.); lescheva@list.ru (N.A.L.)
* Correspondence: naumova@niboch.nsc.ru

Abstract: The intestinal health of poultry is of great importance for birds' growth and development; probiotics-driven shifts in gut microbiome can exert considerable indirect effect on birds' welfare and production performance. The information about gut microbiota of ducks is scarce; by using high throughput metagenomic sequencing with Illumina Miseq we examined fecal bacterial diversity of Peking ducks grown on conventional and *Bacillus*-probiotic-enriched feed. The probiotic supplementation drastically decreased the presence of the opportunistic pathogen *Escherichia/Shigella*, which was the major and sole common dominant in all samples. Seventy other bacterial species in the ducks' fecal assemblages were found to have probiotic-related differences, which were interpreted as beneficial for ducks' health as was confirmed by the increased production performance of the probiotic-fed ducks. Bacterial α-biodiversity indices increased in the probiotic-fed group. The presented inventory of the duck fecal bacteriobiome can be very useful for the global meta-analysis of similar data in order to gain a better insight into bacterial functioning and interactions with other gut microbiota to improve poultry health, welfare and production performance.

Keywords: 16S rRNA gene; amplicon sequencing; ducks; probiotic; gut microbiome

Citation: Naumova, N.B.; Alikina, T.Y.; Zolotova, N.S.; Konev, A.V.; Pleshakova, V.I.; Lescheva, N.A.; Kabilov, M.R. *Bacillus*-Based Probiotic Treatment Modified Bacteriobiome Diversity in Duck Feces. *Agriculture* **2021**, *11*, 406. https://doi.org/10.3390/agriculture11050406

Academic Editor: István Komlósi

Received: 7 April 2021
Accepted: 26 April 2021
Published: 1 May 2021

Publisher's Note: MDPI stays neutral with regard to jurisdictional claims in published maps and institutional affiliations.

1. Introduction

Over 21 million ducks are raised for human consumption each year in the Russian Federation [1], yet so far no research has been conducted on the gut microbiome of the Pekin duck breed, maintained in the country. The intestinal health of poultry is of great importance for birds' welfare and hence their production performance, food safety and environmental consequences [2]. Industrial poultry production still relies on antibiotics as growth promoters, although probiotics nowadays are becoming an increasingly indispensable pharmacological component for production of high quality food [3]. Probiotic preparations can be based on different microorganisms, including the spore-producing ones like *Bacillus* [4,5], which are Gram-positive, aerobic, spore-forming bacteria ubiquitous in the environment. Importantly, they have high stability under adverse environmental conditions, which is indispensable for probiotic cells to survive processing and storage of feed, its passage through the gastrointestinal tract and subsequent chemical digestion processes. The antagonistic effect of such probiotics on the pathogenic gut microflora of humans and animals has been known since long ago [6]. However, most research about the effect of dietary administration of probiotic *Bacillus* strains on growth performance has been conducted in chicken, mouse, and pig [7–10], and yet only recently it was experimentally shown that certain strains of *B. subtilis* can provide beneficial effects on the growth of young broiler chickens and have the potential to replace antibiotic growth promoters [11] or improve egg quality [12]. Similar studies on ducks have been fewer [13,14], and we failed to find reports on gut microbiome research using probiotic *Bacillus* strains for Pekin ducks.

Knowledge of the microbiome profiles in regional agricultural populations could help in drawing a global picture of duck gut microbiota, leading to a better insight into the regional effects of production technologies such as the use of probiotics, prebiotics, synbiotics, enzymes and antibiotics. As there is still a gap in knowledge concerning the effectiveness of probiotic supplementation in shaping gastrointestinal taxonomic profiles in ducks, the objective of the study was to examine composition and structure of ducks' gut bacterial assemblages by estimating diversity of phylogenetically significant fragments of 16S rRNA genes from the feces of ducks grown on conventional and probiotic-enriched feed by using high throughput metagenomic sequencing.

2. Materials and Methods

2.1. Duck Breed and Experimental Design

All experimental procedures involving ducks met the guidelines approved by the institutional animal care and use committee and were performed in accordance with the Russian National Law concerning the care of animals for research purposes, as well as in compliance with the European Commission Directive 2010/63/EU on the protection of animals used for scientific purposes [15]. Ducks *Anas platyrhynchos* of the Peking breed Agidel variety were raised and grown at a poultry farm in the Omsk region, Russia, and female ducks were used in the study. Until 10 days of age the ducklings were kept in stainless steel cages at +28–30 °C and 65–70% relative humidity on a small-mesh-flooring; after that they were put into a bigger house where they could freely roam on a deep pine-shaving-based litter at +25 °C. Then, the 30-d-old birds were placed into the premises with similar deep litter, air temperature of +14–20 °C, drinking water ad libitum and access to artificial ponds. From the first day of life to three weeks of age, the ducks were fed ad libitum with a starter diet (wheat, soya beans, oil free sunflower seed, sunflower seed cake, fish flour, methionine, threonine, lysine, sodium chloride, premix), providing 3100 kcal/kg of feed and 23% of crude protein. Then, from four to five weeks of age, the birds were fed with a grower diet with threonine substituted with cysteine and providing 3150 kcal/kg of feed and 21% of crude protein. Additionally, from 6 weeks of age until the end of the performance the ducks were fed with a similar diet but providing 3200 kcal/kg of feed and 20% of crude protein.

The ducks were assembled in two groups of sixteen birds in each. One group received conventional feed as described above supplemented with a probiotic (probiotic-fed) during the entire growth period of 60 d as per manufacturer's instructions, i.e., 0.4 kg/t during the first 15 d followed by 1.0 kg/t till the end of the growth. The other group received only conventional feed (control).

The commercially distributed probiotic preparation Olin®, produced for Probiotic-Plus LLC (Russia) [16], was used in the study. According to the manufacturer, the preparation contains dried biomass of antagonistically active strains of *Bacillus subtilis* and *Bacillus licheniformis*, registered in the Russian Collection of Industrial Microorganisms under accession numbers 10172 and 10135, respectively, with plate counts of at least 2×10^9 CFU per 1 g of the preparation [17].

2.2. Sample Collection

At sixty days of age, all birds were weighed, and five apparently healthy ducks were selected at random from each group, caught, not fed for 8 h, but could drink ad libitum, and then euthanized by cervical dislocation in compliance with the European Commission Directive 2010/63/EU on the protection of animals used for scientific purposes [15]. Within two hours the recta were opened using sterile scissors, and the contents were collected into sterile vials and frozen at −196 °C. In the laboratory the samples were stored at −80 °C prior to the DNA extraction.

2.3. Extraction of Total Nucleic Acid from Feces

Total DNA was extracted from 250 mg of feces using the DNeasy Powersoil Kit (Qiagen, Germany) as per the manufacturer's instructions [18] to lyse microbial cells and obtain high-quality DNA solutions free from PCR inhibitors. The bead-beating was performed using a TissueLyser II (Qiagen, Germany), for 10 min at 30 Hz. No further purification of the DNA was needed. The quality of the DNA was assessed using agarose gel electrophoresis.

2.4. 16S rRNA Gene Amplification and Sequencing

The 16S DNA region was amplified with the primer pair F343 (5′-TACGGRAGGCAG CAG-3′) and R803 (5′-CTACCAGGGTATCTAATCC-3′) combined with Illumina adapter sequences [19]. PCR amplification was performed as described earlier [20]. A total of 200 ng PCR product from each sample was pooled together and purified through a MinElute Gel Extraction Kit (Qiagen, Germany). The obtained libraries were sequenced with 2×300 bp paired-ends reagents on MiSeq (Illumina, San Diego, CA, USA) in the SB RAS Genomics Core Facility (ICBFM SB RAS, Novosibirsk, Russia). The read data reported in this study were submitted to the GenBank under the study accession PRJNA523560.

2.5. Bioinformatic and Statistical Analyses

Raw sequences were analyzed with the UPARSE pipeline [21] using Usearch v11.0. The UPARSE pipeline included merging of paired reads; read quality filtering; length trimming; merging of identical reads (dereplication); discarding singleton reads; removing chimeras and operational taxonomic unit (OTU) clustering using the UPARSE-OTU algorithm. The OTU sequences were assigned a taxonomy using the SINTAX [22] and 16S RDP training set v.16 [23].

Taxonomic structure of thus obtained sequence assemblages, i.e., a collection of different species at one site at one time [24], was estimated by the ratio of the number of taxon-specific sequence reads to the total number of sequence reads, i.e., by the relative abundance of taxa, expressed as a percentage.

Statistical analyses of the data were perfumed using Statistica v.13.3 software (Statsoft, Tulsa, OK, USA). Comparison of relative abundances of different bacterial taxa in fecal samples of the control and probiotic-fed group was carried out using the Mann–Whitney nonparametric test, whereas comparison of ducks' production characteristics ANOVA and Fisher's least significant difference test were carried out. The rarefaction curves were obtained using iNEXT 2.0.15 in R-package [25] and biodiversity indices calculated with the help of PAST 2.17 software [26].

3. Results

3.1. Taxonomic Richness and Structure of Duck Fecal Bacterial Assemblages

After 16S gene amplicon sequencing, quality filtering and chimera removal a total of 666,588 high-quality DNA sequences were obtained from feces of the 10 ducks. High-quality reads were clustered using >97% sequence identity into 568 bacterial operational taxonomic units (OTUs). The obtained sets of sequences for each sample were analyzed by plotting the number of OTUs against the total number of sequence reads (Figure 1). The resulting rarefaction curves demonstrated sufficient out coverage to describe the bacterial composition and compare assemblages of different groups [27].

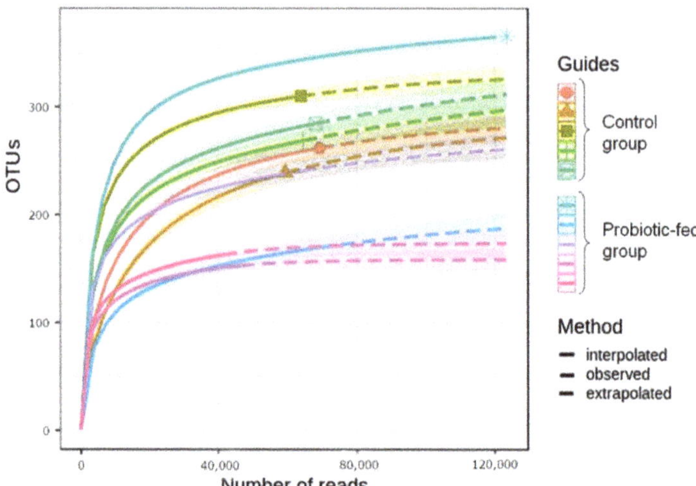

Figure 1. Rarefaction curves for the OTU number in fecal bacterial assemblages of the ducks.

The total number of different-level taxa identified in the study is shown in Table 1. The mooutOTU-rich phyla were *Proteobacteria* (147 OTUs, or 26% of the total number of identified OTUs) and *Firmicutes* with (134 OTUs, or 24%), followed by *Actinobacteria* (131 OTUs, or 23%) and *Bacteroideteout*25 OTU, or 4%). Taxonomic richness in the studied samples was found to drastically decrease if only the dominant members, i.e., the ones contributing at least 1% into the total number of sequence reads, of the bacterial assemblages, were taken into account (Table 1).

Table 1. Taxonomic richness of fecal bacterial assemblages of ducks.

Taxon Level	Taxonomic Attribution			
	All OTUs	Dominant [a] OTUs		
		Both Groups	Control Group	Probiotic-Fed Group
Phylum	15	4	3	3
Class	36	6	4	6
Order	63	6	4	6
Family	137	9	4	9
Genus	251	12	5	12
OTU	567	13	5	13

[a] OTUs were considered dominant if their relative abundance was more than 1%.

Two bacterial phyla—*Firmicutes* and *Proteobacteria*—collectively accounted for more than 90% of the total sequence reads in fecal assemblages (Figure 2a). The overwhelming majority of sequences represented three classes (Figure 2b), three orders (Figure 2c), just six families (Figure 2d) and six genera (*Escherichia/Shigella*, *Terrisporobacter*, *Streptococcus*, *Enterococcus*, *Romboutsia* and an unclassified representative of *Clostridiaceae*, Figure 3). The commercial probiotic preparation, fed to the ducks in the study, was found to contain 54 OTUs, with one OTU (*Bacillus* sp.) accounting for 58% of the total number of sequence reads, detected in the preparation. Other dominant components of the probiotic preparation were *Pseudomonas* spp. (three OTUs), *Comamonas* sp. (one OTU) and unclassified *Enterobacteriaceae* (two OTUs). These bacteria were practically absent in fecal bacteriobiomes of both groups. Overall *Bacillus* class was represented by seven OTUs in fecal assemblages of ducks (Figure 2b), collectively accounting for a tiny portion of the total number of sequence reads (0.015% and 0.002% in the control and probiotic-fed groups, respectively).

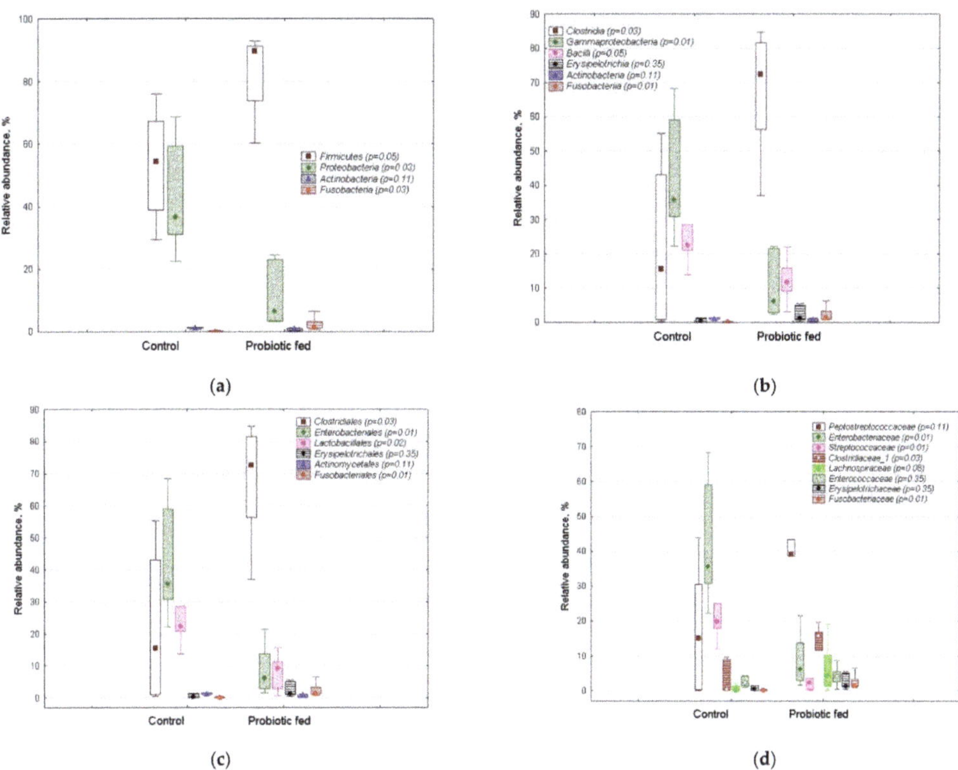

Figure 2. Relative abundance of taxon-specific sequences in fecal bacterial assemblages of ducks of the control and probiotic-fed groups: (**a**) phylum, (**b**) class, (**c**) order and (**d**) family levels. The markers show median, boxes show 25–75% percentiles, while the lines indicate fluctuation ranges. The *p*-values as estimated for each taxon by Mann–Whitney test are shown in brackets.

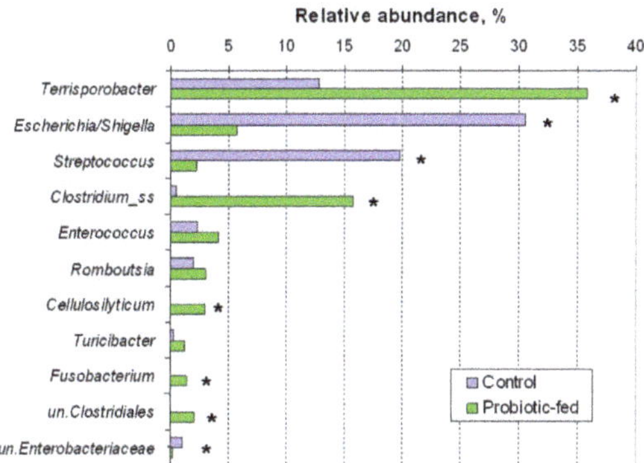

Figure 3. Relative abundance of genera in fecal bacterial assemblages of ducks of the control and probiotic-fed groups. Symbol * at the right of the columns denotes statistically significant difference between the groups (Mann–Whitney test, *p* ≤ 0.05). "un." stands for unclassified.

Of the total OTU number detected in the studied samples, only 13 OTUs, or 2%, were dominants, i.e., contributed $\geq$1% to the total number of sequences (Table 1). The number of dominant OTUs per sample varied from three to nine in the control group, with just two OTUs being common for all samples (*Escherichia/Shigella* sp. and *Streptococcus* sp.). In the probiotic-fed group the number of dominant OTUs varied from 5 to 17 per sample, with three OTUs being common for all samples (*Escherichia/Shigella* sp., *Terrisporobacter* sp. and *Romboutsia sedimentorum*). Thus only one OTU, namely *Escherichia/Shigella* sp., was common for all studied samples. Its relative abundance varied from 30 to 68% in samples of the control group, and from 6 to 20% in samples of the probiotic-fed group.

3.2. OTUs' Relative Abundance in Duck Fecal Bacterial Assemblages

The relative abundance of some OTUs found in the bacterial assemblages of ducks is shown in Table 2. As mentioned above, the ultimate dominant in the control group was *Escherichia/Shigella* sp. The second major dominant OTU in the control group was *Streptococcus* sp. In fecal assemblages of the probiotic-fed group the abundance of this bacterium was almost 10 times lower.

Table 2. Relative abundance (%) of bacterial OTUs, dominant in the fecal assemblages of ducks of the control and/or probiotic-fed groups.

	OTU	Control Group	Probiotic-Fed Group	*p*-Value
1	*Escherichia/ Shigella sp.* [1]	30.6	5.7	0.012
2	*Terrisporobacter sp.*	12.9	36.4	0.037
3	*Streptococcus sp.*	20.0	2.2	0.012
6	*Enterococcus cecorum*	1.2	3.8	0.210
7	*unc. Clostridiaceae_1* [2]	0.2	3.7	0.012
8	*Clostridium_ss^3 sp.*	0.3	1.5	0.295
10	*Cellulosilyticum sp.*	0.1	1.3	0.094
11	*Turicibacter sanguinis*	0.3	1.2	0.403
12	*unc. Clostridiales*	0.01	1.9	0.094
13	*Clostridium_ss^3 sp.*	0.04	1.04	0.210
14	*Fusobacterium sp.*	0.11	1.41	0.012
15	*Cellulosilyticum lentocellum*	0.04	1.48	0.037
19	*Romboutsia sedimentorum*	2.0	3.2	0.403

[1] The lines with statistically significant ($p \leq 0.05$) difference are highlighted in bold; [2] "unc." stands for unclassified; [3] "ss" stands for *sensu stricto*.

Overall, 70 OTUs were found to have differential relative abundance ($p \leq 0.05$) in fecal microbiota of the studied groups. Most of these OTUs were minor or rare members, contributing much less than 1% into the total number of sequence reads.

Four dominant OTU were found to have increased ($p \leq 0.05$) abundance in fecal bacteriobiomes of the probiotic-fed group. *Terrisporobacter* sp. sequences were almost three times more abundant in the probiotic-fed group, comprising one third of the entire bacteriobiome. A *Fusobacterium* sp. was also found increased in the probiotic-fed group (Table 2).

3.3. Biodiversity Indices of the Duck Fecal Bacterial Assemblages

Biodiversity indices serve to compact information about communities, assemblages, guilds, etc. of living organisms; thus, the indices are useful for comparing large arrays of metagenomic data. Therefore, for each studied sample, i.e., an array with the number of sequence reads for each OTU, we calculated α-biodiversity indices (Table 3).

Table 3. Alpha-biodiversity indices (median) of fecal bacterial assemblages of ducks of the control and probiotic-fed group.

Index	Control Group	Probiotic-Fed Group	*p*-Value
Total number of identified OTUs	208	114	0.095
Dominance (D)	0.31	0.17	0.222
Simpson (1-D)	0.69	0.83	0.222
Shannon	1.92	2.49	0.151
Evenness [1]	0.02	0.06	0.032
Brillouin	1.91	2.49	0.151
Menhinick [1]	0.85	0.51	0.032
Margalef	19	9	0.095
Equitability	0.33	0.49	0.095
Fisher-alpha	27	13	0.095
Berger-Parker	0.46	0.36	0.841
Chao-1	246	127	0.095

[1] The lines with statistically significant ($p \leq 0.05$) difference are highlighted in bold.

The probiotic-fed group showed a tendency for decreased OTU richness, as indicated by the number of OTUs, Chao-1, Fisher's alpha, Margalef and Menchinik indices, and increased evenness, with Shannon and Brilluin indices, on the contrary, tending to increase in the probiotic-fed group.

3.4. Production Performance of Ducks

The data on production characteristics for the entire groups, i.e., consisting of 16 birds each, were normally distributed: ANOVA showed that probiotic supplementation accounted for 43% of the bird body mass variance at day 60 and for 30% of the growth rate variance. Thus due to the beneficial effect of probiotic-enriched feed ducks' production characteristics improved, as the probiotic-fed ducks demonstrated (Table 4) higher both daily mass increase rate (by 4.0 g/bird) and total body mass at the end of the feeding (by 235 g/bird).

Table 4. Production characteristics of ducks fed with conventional (control) and probiotic-supplemented feed (probiotic-fed group).

Characteristic	Control Group (n = 16) [2]	Probiotic-Fed Group (n = 16)	*p*-Value [1]
Living mass of a 1-day-old duck, g/bird	57.8 ± 5.8 [2]	57.9 ± 6.3	0.931
Living mass of a 60-day-old duck, g/bird	2772 ± 222	3007 ± 141	0.001
Average daily gain, g/bird per day	45.2 ±3.7	49.2 ± 2.3	0.001
Feed intake, kg/kg bird mass	3.35	2.85	

[1] The lines with statistically significant ($p \leq 0.05$) difference are highlighted in bold. [2] Performance was assessed for a bigger sets of birds than the fecal microbiome.

Supplementing conventional duck feed with probiotic resulted in 0.5 kg less consumption per 1 kg of duck living mass. The ducks in the control group consumed on average 0.72 kg more feed, as compared to the probiotic-fed ducks.

4. Discussion

The finding that two bacterial phyla, namely *Firmicutes* and *Proteobacteria*, prevailed in the ducks' feces agrees with the results obtained in other studies: for instance, the representatives of *Firmicutes*, *Proteobacteria* and *Bacteroidetes* were reported to account for at least 90% of bacteriobiomes in duck's ileum and cecum [28–31]. As for the rectum (as in our study), recently *Firmicutes* and *Proteobacteria* were found to account for 75% of the Muscovy ducks' rectum bacteriobiome [32], with *Proteobacteria* abundance being twice lower as in the control group in our study (15% vs. 30%).The difference may be attributed to the difference in duck species and sex, as well as other factors; this is an area that is still poorly investigated.

The fact that we did not explicitly detect *B. subtilis* and *B. licheniforms*, the major components in the probiotic preparation, may be due to the relatively short V3–V4 amplicon sequences, not allowing discrimination between closely related species [33]. Our result that the probiotic bacteria consumed by the ducks with their feed did not reside in the gut complies with the fact that *Bacillus* representatives are not common for the gut microbiota of poultry [31,34].

The occurrence of an *Escherichia/Shigella* bacterium in the probiotic-fed group was more than five times lower. The genus represents important pathogens of humans and animals [35], therefore the change was most likely beneficial for birds' health and welfare. Among the *Streptococcus* genus some serious pathogens for humans and animals were reported before: for example, the ones capable of causing meningitis in ducks [36]. Thus it seems that *Streptococcus* sp. in our study was not a beneficial bacterium, so decrease in the probiotic-fed ducks might have contributed to their enhanced production performance. The drastically decreased abundance of these two harmful bacteria, i.e., *Escherichia/Shigella* sp. and *Streptococcus* sp., in the probiotic-fed group confirms the antagonistic and hence beneficial impact of the *Bacillus*-based probiotic on the major opportunistic pathogens of the fecal microbiota of ducks. It should be noted that the ducks in this study harboring abundant *Escherichia/Shigella* and *Streptococcus* were apparently healthy, which means that even high abundance of a potential pathogen's sequence reads in a bacteriobiome is not be immediately manifested as an actual instance of a disease.

Although some authors claim that *Riemerella anatipestifer* is one of the most common bacterial pathogens of ducks [32], in our study none of the assigned OTUs were classified into *Riemerella*.

The finding that most of the differentially abundant OTUs in fecal bacterial assemblages of the conventionally and probiotic-fed groups were minor or rare members suggests that low-abundant OTUs may be important for the host adjustments to shifts in environmental conditions. Such OTUs in the gut microbiome may have systemic interactions with potentially important consequences for the microbial performance within a host organism.

There is evidence about the pathogenicity of the *Terrisporobacter* genus for humans [37], but for animals and poultry we could not find such information. Some *Terrisporobacter* genus representatives are known as chemoorganotrophs, while others are chemolithoautotrophs, or acetogenic bacteria [38,39], capable of decomposing plant material in anaerobic conditions. Thus this bacterium is beneficial for host functioning, and its increased abundance in the probiotic-fed ducks' feces also confirms the positive effect of the *Bacillus*-based probiotic on the fecal microbiota of ducks.

Although recently a *Fusobacterium* sp. was reported to be associated with decreased production of hens, thus likely being an opportunistic pathogen [40], *Fusobacteria* phylum representatives are common and often dominant members of the gut microbiota of wild ducks and geese [41]. In view of the latter the increased relative abundance of the bacterium in the probiotic-fed group can also be considered promoting intestinal health of the ducks. As for *Cellulosilyticum lentocellum*, another bacterium with increased abundance in the feces of the probiotic-fed ducks, it is known as a slow cellulose-degrader and member of the healthy animal fecal bacteriobiome [42]. Increased presence of an unclassified *Clostridiaceae_1* OTU, important anaerobic degraders of plant polymers [43], in the fecal bacteriobiome of probiotic-fed ducks, can be considered beneficial and hence might have contributed to higher production performance of the ducks. Overall, increased abundance of *Clostridiales* representatives in probiotic-fed ducks corroborates the use of these bacteria for novel probiotic formulations: recently some of the latter were shown to exert beneficial influence on Peking duck performance [44].

It should be noted that the available information about the influence of probiotic-enriched feed on ducks' gut microbiota is inconclusive, as both beneficial [28] and neutral effects were reported earlier [45]. One of the reasons for such a discrepancy may be because two independent groups of birds, with and without probiotic supplementation, are usually compared: implementing the repeated measures design, i.e., sampling the same member of a group before and after the probiotic treatment, is more difficult in practice even in case of feces. Other reasons may be associated with regional/country differences between the studied duck groups such as lifespan, feed and its supplementation, medication, raising conditions, genetics [28,30], etc.

Compared with the α-biodiversity indices for the gut microbiota of ducks reported earlier [30], in our study Shannon and Chao-1 indices were lower. The dominance indices were found to show tendency to be lower in probiotic-fed group, while equitability tended to be higher. Therefore, overall α-diversity seemed to be increasing in the probiotic-fed group, which is generally regarded as positive.

Our finding that probiotic supplementation decreased feed consumption per unit of living mass of the ducks indirectly corroborates the results about increased abundance of beneficial, particularly plant material fermenting, bacteria, which most likely translated into more efficient transformation of nutrients in the gut and consequent more efficient utilization of nutrients by host organisms, i.e., ducks.

The found beneficial effect of probiotic-enriched feed on ducks' production characteristics agrees with the improved production performance of probiotic-fed Pitalah ducks [46]. The Pekin ducks' productivity performance in our study corroborates the beneficial influence of *Bacillus*-based supplementation on egg quality and biochemical properties of blood of Shaoxing ducks [47], and on gut microbiota established with lysine-yielding *Bacillus subtilis* on a locally domesticated Chinese duck breed [48]. Enhanced Pekin ducks' production, associated with beneficial changes in ducks' gut microbiota due to *Bacillus*-based probiotic supplementation, is also in line with improved growth performance shown by the Cherry Valley ducks [13]. Therefore, the studied *Bacillus*-based probiotic formulation a promising basis for further improvement [6] and use.

Finally, we want to stress that it is difficult to compare studies on duck intestinal micrtobiome diversity due to substantive differences in methodology, beginning from the studied groups (species, breed, age, raising conditions, site of sample collection in the gut, etc.) and all the way to amplification (primers), sequencing (platforms) and bioinformatic tools (software and databases). Therefore, there is an urgent need for a comprehensive meta-analysis of the duck gut microbiome data, hopefully resulting in recommendations for a more standardized research approach.

We also want to emphasize, albeit truistically, that case–control design, often used to infer the medication/supplementation-associated effects in humans and animals, prevents following directly, i.e., in one and the same individual, the dynamics of the properties of interest, i.e., bacteriobiome diversity as in our study. Therefore, repeated measures' design should be implemented if and when possible and feasible, despite the objective difficulties of doing so in studies with animals. Such a design helps to move closer to the cause–effect mechanisms of microbiome shifts, rather than be confined to association/correlation relations, as most of the microbiome studies do.

5. Conclusions

Our study aimed at comparing the structure and composition of fecal microbiota, as determined using 16S rRNA gene amplicon sequencing, in ducks receiving conventional and *Bacillus*-based probiotic supplemented feed. This is the first profile of gut bacteriobiome of domestic ducks in Russia and as such can be used as a regional reference in further research as well as a tiny contribution for constructing the global pattern. Duck fecal bacteriobiome was found to be drastically dominated by just two phyla (*Firmicutes* and *Proteobacteria*), represented by three classes (*Clostridia*, *Bacilli* and *Gammaproteobacteria*). *Escherichia/Shigella* sp. turned out to be the major and sole common dominant in all samples. Fecal bacteriobiome of probiotic-fed ducks differed from the conventionally fed control in the relative abundance of some dominant OTUs, mainly the pathogenic ones (*Escherichia/Shigella* sp., *Streptococcus* sp.). A number of minor and rare members of bacterial assemblages (12% of the total number of OTUs) also displayed differential abundance; however, it was difficult to infer their physiological and/or pathogenic significance. The *Bacillus* bacteria, contained in the probiotic preparation used in the study, could not survive in the gut and were eliminated. Supplementation of the conventional feed with *Bacillus*-based probiotic resulted in pronounced shifts towards the more beneficial gut microbiota of ducks. The increased bacteriobiome α-diversity in the probiotic-fed group enhance gut microbiota and hence ducks' resilience towards adverse environmental effects. The bacterial OTUs, found to be the significantly related to the probiotic supplementation, provide a framework for further research on bacteria functioning and interactions within gut microbiota in order to improve birds' health and, as a consequence, both industrial and small farm poultry production. The studied *Bacillus*-based probiotic is promising for the development of improved formulations for specifically targeted interventions to modify gut microbiota of ducks. Such formulations can be effective alternatives for growth-promoting antibiotics, but there is still a great need to understand the role of poultry gut microbiota in the prophylaxis, growth and health promoting mechanisms.

Author Contributions: Conceptualization, V.I.P. and N.S.Z.; methodology, V.I.P.; software, M.R.K.; validation, T.Y.A., N.B.N. and A.V.K.; formal analysis, N.B.N.; investigation, A.V.K.; resources, V.I.P.; data curation, N.B.N.; writing—original draft preparation, N.B.N.; writing—review and editing, N.A.L., M.R.K. and V.I.P.; visualization, T.Y.A.; supervision, M.R.K.; project administration, V.I.P.; funding acquisition, V.I.P. All authors have read and agreed to the published version of the manuscript.

Funding: This research was funded by the MINISTRY OF SCIENCE AND HIGHER EDUCATION OF THE RUSSIAN FEDERATION (project number 121031300042-1).

Institutional Review Board Statement: The study was conducted according to the guidelines of the Declaration of Helsinki, and approved by the Institutional Review Board of STOLYPIN OMSK STATE AGRARIAN UNIVERSITY (protocol code 23/11 23 August 2018).

Informed Consent Statement: All experimental procedures involving ducks met the guidelines approved by the in-stitutional animal care and use committee and were performed in accordance with the Russian National Law concerning the care of animals for research purposes, as well as in compliance with the European Commission Directive 2010/63/EU on the protection of animals used for scientific purposes.

Data Availability Statement: The read data reported in this study were submitted to the GenBank under the study accession PRJNA523560.

Acknowledgments: In this section, you can acknowledge any support given which is not covered by the author contribution or funding sections. This may include administrative and technical support, or donations in kind (e.g., materials used for experiments).

Conflicts of Interest: The authors declare no conflict of interest. The funders had no role in the design of the study; in the collection, analyses, or interpretation of data; in the writing of the manuscript, or in the decision to publish the results.

References

1. FAOSTAT. 2021. Available online: http://www.fao.org/faostat/en/#data/QA (accessed on 20 April 2021).
2. Oviedo-Rondón, E.O. Holistic view of intestinal health in poultry. *Anim. Feed Sci. Technol.* **2019**, *250*, 1–8. [CrossRef]
3. Patterson, J.A.; Burkholder, K.M. Application of prebiotics and probiotics in poultry production. *Poult. Sci.* **2003**, *82*, 627–631. [CrossRef] [PubMed]
4. Sorokulova, I.B. A comparative study of the biological properties of Biosporin and other commercial *Bacillus*-based preparations. *Mikrobiol. Zhournal* **1997**, *59*, 43–49. (In Russian)
5. Mingmongkolchai, S.; Panbangred, W. *Bacillus* probiotics: An alternative to antibiotics for livestock production. *J. Appl. Microbiol.* **2018**, *124*, 1334–1346. [CrossRef]
6. Kogut, M.H. The effect of microbiome modulation on the intestinal health of poultry. *Anim. Feed Sci. Technol.* **2019**, *250*, 32–40. [CrossRef]

7.   Franciosini, M.P.; Costarelli, S.; Cobellis, G.; Trabalza-Marinucci, M. Effects of dietary *Lactobacillus acidophilus* and *Bacillus subtilis* on laying performance, egg quality, blood biochemistry and immune response of organic laying hens. *J. Anim. Physiol. Anim. Nutr.* **2016**, *100*, 977–987. [CrossRef]

8.   Zamanizadeh, A.; Mirakzehi, M.T.; Agah, M.J.; Saleh, H.; Baranzehi, T. A comparison of two probiotics *Aspergillus oryzae* and *Saccharomyces cerevisiae* on productive performance, egg quality, small intestinal morphology, and gene expression in laying Japanese quail. *Ital. J. Anim. Sci.* **2021**, *20*, 232–242. [CrossRef]

9.   Wang, H.; Ha, B.D.; Kim, I.H. Effects of probiotics complex supplementation in low nutrient density diet on growth performance, nutrient digestibility, faecal microbial, and faecal noxious gas emission in growing pigs. *Ital. J. Anim. Sci.* **2020**, *20*, 163–170. [CrossRef]

10.   Liu, X.; Liu, W.; Deng, Y.; He, C.; Xiao, B.; Guo, S.; Zhou, X.; Tang, S.; Qu, X. Use of encapsulated Bacillus subtilis and essential oils to improve antioxidant and immune status of blood and production and hatching performance of laying hens. *Ital. J. Anim. Sci.* **2020**, *19*, 1583–1591. [CrossRef]

11.   Park, I.; Lee, Y.; Goo, D.; Zimmerman, N.P.; Smith, A.H.; Rehberger, T.; Lillehoj, H.S. The effects of dietary *Bacillus subtilis* supplementation, as an alternative to antibiotics, on growth performance, intestinal immunity, and epithelial barrier integrity in broiler chickens infected with Eimeria maxima. *Poult. Sci.* **2020**, *99*, 725–733. [CrossRef]

12.   Liu, X.; Peng, C.; Qu, X.; Guo, S.; Chen, J.F.; He, C.; Zhou, X.; Zhu, S. Effects of Bacillus subtilis C-3102 on production, hatching performance, egg quality, serum antioxidant capacity and immune response of laying breeders. *J. Anim. Physiol. Anim. Nutr.* **2019**, *103*, 182–190. [CrossRef] [PubMed]

13.   Guo, M.; Hao, G.; Wang, B.; Li, N.; Li, R.; Wei, L.; Chai, T. Dietary Administration of *Bacillus subtilis* Enhances Growth Performance, Immune Response and Disease Resistance in Cherry Valley Ducks. *Front. Microbiol.* **2016**, *7*, 1975. [CrossRef] [PubMed]

14.   Khattab, A.; El Basuini, M.; El-Ratel, I.T.; Fouda, S.F. Dietary probiotics as a strategy for improving growth performance, intestinal efficacy, immunity, and antioxidant capacity of white Pekin ducks fed with different levels of CP. *Poult. Sci.* **2021**, *100*, 100898. [CrossRef] [PubMed]

15.   Directive 2010/63/EU of the European Parliament and of the Council of 22 September 2010 on the Protection of Animals Used for Scientific Purposes. Available online: https://data.europa.eu/eli/dir/2010/63/oj (accessed on 27 April 2021).

16.   Probiotic Plus LLC. Available online: https://www.probiotic-plus.ru (accessed on 27 April 2021). (in Russian).

17.   Federal Service for Proprietary Rights. Patent RU 2,437,563C1, 27 December 2011.

18.   QIAGEN. Quick-Start Protocol. *DNeasy PowerSoil Kit*. 2016. Available online: https://www.qiagen.com/au/resources/resourcedetail?id=91cf8513-a8ec-4f45-921e-8938c3a5490c&lang=en (accessed on 27 April 2021).

19.   Fadrosh, D.W.; Ma, B.; Gajer, P.; Gajer, P.; Sengamalay, N.; Ott, S.; Brotman, R.M.; Ravel, J. An improved dual-indexing approach for multiplexed 16S rRNA gene sequencing on the Illumina MiSeq platform. *Microbiome* **2014**, *2*, 6. [CrossRef] [PubMed]

20.   Igolkina, A.A.; Grekhov, G.A.; Pershina, E.V.; Samosorova, G.G.; Leunova, V.M.; Semenov, A.N.; Baturina, O.A.; Kabilov, M.R.; Andronov, E.E. Identifying components of mixed and contaminated soil samples by detecting specific signatures of control 16S rRNA libraries. *Ecol. Ind.* **2018**, *94*, 446–453. [CrossRef]

21.   Edgar, R.C. UPARSE: Highly accurate OTU sequences from microbial amplicon reads. *Nat. Methods* **2013**, *10*, 996–998. [CrossRef]

22.   Edgar, R.C. UNOISE2: Improved error-correction for Illumina 16S and ITS amplicon reads. *bioRxiv* **2016**. [CrossRef]

23.   Wang, Q.; Garrity, G.M.; Tiedje, J.M.; Cole, J.R. Naïve Bayesian Classifier for Rapid Assignment of rRNA Sequences into the New Bacterial Taxonomy. *Appl. Environ. Microbiol.* **2007**, *73*, 5261–5267. [CrossRef]

24.   Fauth, E.; Bernardo, J.; Camara, M.; Resetarits, W.J.; Van Buskirk, J.A.; McCollum, S. Simplifying the Jargon of Community Ecology: A Conceptual Approach. *Am. Nat.* **1996**, *147*, 282–286. [CrossRef]

25.   Hsieh, T.C.; Ma, K.H.; Chao, A. iNEXT: An R package for rarefaction and extrapolation of species diversity (Hill numbers). *Methods Ecol. Evol.* **2016**, *7*, 1451–1456. [CrossRef]

26.   Hammer, Ø.; Harper, D.A.T.; Ryan, P.D. PAST: Paleontological Statistics Software Package for Education and Data Analysis. *Palaeontol. Electron.* **2001**, *4*, 1–9.

27.   Hughes, J.B.; Hellmann, J.J. The Application of Rarefaction Techniques to Molecular Inventories of Microbial Diversity. *Methods Enzymol.* **2005**, *397*, 292–308. [CrossRef]

28.   Vasaï, F.; Ricaud, K.B.; Cauquil, L.; Daniel, P.; Peillod, C.; Gontier, K.; Tizaoui, A.; Bouchez, O.; Combes, S.; Davail, S. *Lactobacillus sakei* modulates mule duck microbiota in ileum and ceca during overfeeding. *Poult. Sci.* **2014**, *93*, 916–925. [CrossRef] [PubMed]

29.   Best, A.A.; Porter, A.L.; Fraley, S.M.; Fraley, G.S. Characterization of Gut Microbiome Dynamics in Developing Pekin Ducks and Impact of Management System. *Front. Microbiol.* **2017**, *4*, 2125. [CrossRef] [PubMed]

30.   Wang, S.; Chen, L.; He, M.; Shen, J.; Li, G.; Tao, Z.; Wu, R.; Lu, L. Different rearing conditions alter gut microbiota composition and host physiology in Shaoxing ducks. *Sci. Rep.* **2018**, *8*, 7387. [CrossRef] [PubMed]

31.   Chen, X.; Zheng, M.; Lin, F.; Cheng, X.; Xiao, S.; Chen, S.; Chen, S. Impacts of novel duck reovirus infection on the composition of intestinal microbiota of Muscovy ducklings. *Microb. Pathog.* **2019**, *137*, 103764. [CrossRef] [PubMed]

32.   Yang, H.; Lyu, W.; Lu, L.; Shi, X.; Li, N.; Wang, W.; Xiao, Y. Biogeography of microbiome and short-chain fatty acids in the gastrointestinal tract of duck. *Poult. Sci.* **2020**, *99*, 4016–4027. [CrossRef] [PubMed]

33.   Jensen, E.A.; Berryman, D.E.; Murphy, E.R.; Carroll, R.K.; Busken, J.; List, E.O.; Broach, W.H. Heterogeneity spacers in 16S rDNA primers improve analysis of mouse gut microbiomes via greater nucleotide diversity. *Biotechniques* **2019**, *67*, 55–62. [CrossRef]

34. Pandit, R.J.; Hinsu, A.T.; Patel, N.V.; Koringa, P.G.; Jakhesara, S.J.; Thakkar, J.R.; Shah, T.M.; Limon, G.; Psifidi, A.; Guitian, J.; et al. Microbial diversity and community composition of caecal microbiota in commercial and indigenous Indian chickens determined using 16s rDNA amplicon sequencing. *Microbiome* **2018**, *6*, 115. [CrossRef]
35. Gaastra, W.; Kusters, J.G.; van Duijkeren, E.; Lipman, L.J.A. *Escherichia fergusonii*. *Vet. Microbiol.* **2014**, *172*, 7–12. [CrossRef]
36. Li, M.; Gu, C.; Zhang, W.; Li, S.; Liu, J.; Qin, C.; Su, J.; Cheng, G.; Hu, X. Isolation and characterization of *Streptococcus gallolyticus* subsp. *pasteurianus causing meningitis in ducks*. *Vet. Microbiol.* **2013**, *162*, 93036. [CrossRef]
37. Cheng, M.P.; Domingo, M.C.; Lévesque, S.; Yansouni, C.P. A case report of a deep surgical site infection with Terrisporobacter glycolicus/T. Mayombei and review of the literature. *BMC Infect. Dis.* **2016**, *29*, 529. [CrossRef]
38. Müller, V.; Frerichs, J. *Acetogenic Bacteria*; John Wiley & Sons Ltd.: Hoboken, NJ, USA, 2013. [CrossRef]
39. Groher, A.; Weuster-Botz, D. General medium for the autotrophic cultivation of acetogens. *Bioprocess Biosyst. Eng.* **2016**, *39*, 1645. [CrossRef] [PubMed]
40. Kollarcikova, M.; Kubasova, T.; Karasova, D.; Crhanova, M.; Cejkova, D.; Sisak, F.; Rychlik, I. Use of 16S rRNA gene sequencing for prediction of new opportunistic pathogens in chicken ileal and cecal microbiota. *Poult. Sci.* **2019**, *98*, 2347–2353. [CrossRef]
41. Wang, W.; Zheng, S.; Li, L.; Yang, Y.; Liu, Y.; Wang, A.; Sharshov, K.; Li, Y. Comparative metagenomics of the gut microbiota in wild greylag geese (*Anser anser*) and ruddy shelducks (*Tadorna ferruginea*). *MicrobiologyOpen* **2019**, *8*, e00725. [CrossRef]
42. Miller, D.A.; Suen, G.; Bruce, D.; Copeland, A.; Cheng, J.F.; Detter, C.; Goodwin, L.A.; Han, C.S.; Hauser, L.J.; Land, M.L.; et al. Complete genome sequence of the cellulose-degrading bacterium *Cellulosilyticum lentocellum*. *J. Bacteriol.* **2011**, *193*, 2357–2358. [CrossRef]
43. Wachemo, A.C.; Tong, H.Y.; Hairong, Z.X.; Korai, R.M.; Li, X. Continuous dynamics in anaerobic reactor during bioconversion of rice straw: Rate of substance utilization biomethane production and changes in microbial community structure. *Sci. Total Environ.* **2019**, *687*, 1274–1284. [CrossRef] [PubMed]
44. Liu, Y.; Jia, Y.; Liu, C.; Ding, L.; Xia, Z. RNA-Seq transcriptome analysis of breast muscle in Pekin ducks supplemented with the dietary probiotic Clostridium butyricum. *BMC Genom.* **2018**, *28*, 844. [CrossRef] [PubMed]
45. Even, M.; Davail, S.; Rey, M.; Tavernier, A.; Houssier, M.; Bernadet, M.D.; Gontier, K.; Pascal, G.; Ricaud, K. Probiotics Strains Modulate Gut Microbiota and Lipid Metabolism in Mule Ducks. *Open Microbiol. J.* **2018**, *12*, 71–93. [CrossRef] [PubMed]
46. Wizna, Z.; Abbas, M.H.; Mahata, M.E.; Fauzano, R. Effect of *Bacillus amyloliquefaciens* as a Probiotic on Growth Performance Parameters of Pitalah Ducks. *Int. J. Poult. Sci.* **2017**, *16*, 147–153. [CrossRef]
47. Li, W.F.; Rajput, I.R.; Xu, X.; Li, Y.L.; Lei, J.; Huang, Q.; Wang, M.Q. Effects of Probiotic (*Bacillus subtilis*) on Laying Performance, Blood Biochemical Properties and Intestinal Microflora of Shaoxing Ducks. *Int. J. Poult. Sci.* **2011**, *10*, 583–589. [CrossRef]
48. Xing, Y.; Wang, S.; Fan, J.; Oso, A.O.; Kim, S.W.; Xiao, D.; Yang, T.; Liu, G.; Jiang, G.; Li, Z.; et al. Effects of dietary supplementation with lysine-yielding *Bacillus subtilis* on gut morphology, cecal microflora, and intestinal immune response of Linwu ducks. *J. Anim. Sci.* **2015**, *93*, 3449–3457. [CrossRef] [PubMed]

Article

Effect of Inorganic Zinc on Selected Immune Parameters in Chicken Blood and Jejunum after *A. galli* Infection

Viera Karaffová [1,*], Viera Revajová [1], Emília Dvorožňáková [2], Ľubomíra Grešáková [3], Martin Levkut [1], Zuzana Ševčíková [1], Róbert Herich [1] and Mikulas Levkut [1,4]

[1] Department of Morphological Disciplines, University of Veterinary Medicine and Pharmacy, Komenského, 73041 81 Košice, Slovakia; viera.revajova@uvlf.sk (V.R.); martin.levkut@uvlf.sk (M.L.); zuzana.sevcikova@uvlf.sk (Z.Š); robert.herich@uvlf.sk (R.H.); mikulas.levkut@uvlf.sk (M.L.)
[2] Institute of Parasitology, Slovak Academy of Sciences, Hlinkova, 3040 01 Košice, Slovakia; dvoroz@saske.sk
[3] Institute of Animal Physiology Centre of Biosciences of the Slovak, Academy of Sciences, Šoltésovej, 4040 01 Košice, Slovakia; gresakl@saske.sk
[4] Institute of Neuroimmunology, Dúbravská cesta, 9845 10 Bratislava, Slovakia
* Correspondence: viera.karaffova@uvlf.sk; Tel.: +421-905-871-840

Abstract: Ascaridiosis in poultry results in a reduction in body weight gain, egg production, as well as microelement levels. Infected poultry have higher demands on feed with the addition of essential elements including zinc. The effects of the infection by *Ascaridia galli* and the supplementation of inorganic zinc on the immune status of broilers were monitored through evaluation of the relative expression of selected genes (interleukins, IFN-γ, and TNF-α) by real-time PCR, haematology parameters by microscopy, and quantitative changes of lamina propria lymphocytes by flow cytometry in day 7 and day 14 of the study. We observed that the enrichment of the diet with inorganic zinc has a positive effect on the relative percentage of CD4+ lamina propria lymphocytes in the jejunum and on heterophil counts in blood. In addition, it was concluded that inorganic zinc has an anti-inflammatory effect (downregulation of TNF-α and IL-17) and activates IgA-producing cells in the jejunum of chicks infected with *A. galli*.

Keywords: zinc; imunity; *Ascaridia galli*; chicken

Citation: Karaffová, V.; Revajová, V.; Dvorožňáková, E.; Grešáková, Ľ.; Levkut, M.; Ševčíková, Z.; Herich, R.; Levkut, M. Effect of Inorganic Zinc on Selected Immune Parameters in Chicken Blood and Jejunum after *A. galli* Infection. *Agriculture* **2021**, *11*, 551. https://doi.org/10.3390/agriculture11060551

Academic Editor: István Komlósi

Received: 7 May 2021
Accepted: 15 June 2021
Published: 16 June 2021

Publisher's Note: MDPI stays neutral with regard to jurisdictional claims in published maps and institutional affiliations.

1. Introduction

In the poultry industry, most chickens with outdoor access are often exposed to a wide range of parasites, e.g., *A. galli*. In this regard, *A. galli* poses a serious biological threat due to its direct life cycle and ability to survive extreme environmental conditions. Ascaridiosis in poultry results in a reduction in body weight gain, egg production, ruffled feathers, drooped wings, high mortality, and other secondary pathological symptoms [1].

Chickens become infected by the ingestion of infective eggs. In the gut's lumen, ingested eggs release larvae, where they molt and stay for approximately 10 days. Larvae gradually penetrate the inner lining of the gut, where they spend 1–7 weeks and molt again. Then, they return to the lumen of the intestine, where they develop into adult worms and the females begin to produce eggs. Larvae in the gut's lining destroy the tissues around them and cause enteritis, which is frequently associated with haemorrhagic exudate. Similarly, adult worms cause mechanical damage of the intestinal wall, thereby contributing to malnutrition. In addition, they compete for nutrients and cause bowel obstruction [2].

Nematodes, such as *A. galli*, activate both cellular and humoral immune responses in the host organism [3,4]. In general, *A. galli* infections have been observed to stimulate classical Th2 immune responses in laying hens [4,5]. Specifically, an increase in serum antibodies IgY and the influx of CD4+ and CD8+ T cells at the site of infection have been recorded during *A. galli* infections [6]. Upregulation of IL-4 and IL-13, but not

IFN-γ gene expression, was observed in chickens' intestine at 14 days post-infection [7]. This contributes to the hypothesis that Th2 polarization predominates during *A. galli* nematode infections. During the migration of parasites in the intestine, pro-inflammatory Th1 response is usually suppressed, which allows, among other things, wound healing of the host [8]. Lambrecht and Hammad [9] reported a positive association between type 2 immunity and IL-17 during eosinophilic inflammation in a mouse model. A similar correlation was found mainly in mice with parasites migrating through the lungs [10,11]. However, another investigation is needed for a deeper understanding of the mechanism regulating *A. galli*-induced Th1 and Th2 immune responses in broilers.

Studies of the relationship between poultry and *A. galli* revealed an adverse impact of *A. galli* on the mineral balance of the host with a reduction in microelement levels in the liver and muscles [12,13]. For this reason, infected poultry have higher demands on feed, with the requirement for the addition of essential elements including zinc.

Zinc (Zn) is an essential mineral involved in many biochemical processes and associated with a wide range of physiological disorders, including weight loss, growth retardation, and nervous and immune system disorders [14]. In addition, zinc is an essential factor in the gene expression of proteins required for growth and development, maintaining cell wall integrity, free radical sequestration, and protection against lipid peroxidation [15].

Zinc deficiency affects cells involved in both innate and acquired immunity at the level of proliferation and maturation. T cell function and the balance between different subsets of helper T-lymphocytes are notably sensitive to changes in zinc levels in the organism. Acute zinc deficiency results in the impairment of innate immunity through the reduction in chemotaxis and phagocytosis of mononuclear cells and has negative impacts on acquired immunity, which causes thymic atrophy with subsequent T cell lymphopenia. Specifically, chronic zinc deficiency results in a significant increase in the production of pro-inflammatory cytokines (for example IL-17), which may initiate the development of autoimmune diseases [16].

Accordingly, the effect of *A. galli* infection and the supplementation of inorganic zinc on the broilers' immune status were monitored through evaluation of the relative expression of selected genes (interleukins, IFN-γ, and TNF-α), white blood cell counts (peripheral blood), and the quantification of immunocompetent cells, such as jejunal lamina propria lymphocytes, on study days 7 and 14.

2. Material and methods

2.1. Chickens

The chickens were handled and killed according to state regulations. The specific experiment was approved by the Ethics Committee of the Veterinary Medicine and Pharmacy followed by the Committee for Animal Welfare of Ministry of Agriculture of the Slovak Republic (permit number 836/17-221).

A total of 48 chicken broilers of hybrid COBB500 35-day-old were included in the study that lasted 14-days. The birds were randomly divided into 4 groups: control (C), *A. galli* (Ag), zinc (Z), and a combination (*A. galli* + zinc) (Ag + Z). Chickens were fed a commercial diet BR1 (Table 1) ad libitum. The health status of birds was monitored twice a day by visual inspection. Values of room temperature, feed, and water consumption as well as any clinical signs of adverse conditions were recorded daily.

Table 1. Composition of commercial diet BR1.

Ingredients g/kg	BR1
Wheat	290
Maize	300
Soybean meal	320
Rapeseed oil	40
Fish meal	20
Limestone	12
Dicalcium phosphate	10
Sodium chloride	2
DL-methionine	1
Vitamin-mineral mix	5
Composition by analysis (g/kg dry matter)	
dry matter	899.9
crude protein	232.7
fat	64.5
dietary fiber	22.7
ash	53
Ca	90.4
P total	69.6

On the second day of the study, birds of the Ag and Ag+Z groups were perorally infected with 500 embryonated *A. galli* eggs in 0.5 mL of phosphate buffered saline (PBS) per bird (via a plastic Pasteur pipette). Zinc and Ag + Z groups were individually subjected to daily peroral administration of aqueous solution of inorganic $ZnSO_4$ at a concentration of 50 mg/0.5 mL PBS from day 1 to day 12 of the experiment. To simulate the same stress manipulation, an equal volume of saline was applied to the control group with a Pasteur pipette. Blood samples were withdrawn by vein puncture from vena subclavia, the chickens were euthanized by intra-abdominal injection of xylazine (Rometar 2%, SPOFA, Prague, Czech Republic) and ketamine (Narkamon 5%, SPOFA) at doses of 0.7 mL/kg body weight, and samples from jejunum were collected during necropsy. Two samplings were performed on days 7 and 14 of the study.

2.2. Infective Material and Inoculation

A. galli eggs were isolated from adult female worm uterus, obtained from the gut of naturally infected chickens, according to Permin et al. [17], by a gentle mechanical maceration in 0.5 N NaOH. The eggs were embryonated in 0.1 N NaOH in the dark for 4 weeks at 28 °C. During the incubation, the egg suspensions were oxygenated three times a week by stirring. Subsequently, eggs embryonation was evaluated microscopically on a weekly basis starting from day 14. Ultimately, the embryonated *A. galli* eggs were stored in 0.1 N NaOH at 4–6 °C and regularly oxygenated until the application to the experimental chickens.

2.3. Homogenization of Jejunum and Isolation of Total RNA of Interleukins (IL-4, IL-17), IFN-γ, and TNF-α Gene

Jejunal samples ($n = 6$) (20 mg weighted pieces) were immediately placed in RNA Later solution (Qiagen, Manchester, UK) and stored at −70 °C before RNA purification and transcription as described in Karaffová et al. [18].

2.4. Relative Expression of Interleukins, IFN-γ, and TNF-α Gene in Quantitative Real-Time PCR (qRT-PCR)

The mRNA levels of interleukins, IFN-γ, and TNF-α genes were determined. Moreover, the mRNA relative expression of reference gene, coding GAPDH (glyceraldehyde-3-phosphate dehydrogenase), was selected based on confirmed expression stability using the geNorm program. The primer sequences, annealing temperatures, and times for each primer used for qRT-PCR are listed in Table 2. All primer sets allowed cDNA amplification efficiencies between 94% and 100%.

Table 2. List of primers used for the chicken cytokine mRNA quantification.

Primer	Sequence 5′–3′	Annealing/Temperature Time	References
IL-4 Fw	AGCACTGCCACAAGAACCTG	60 °C /30 s	[19]
IL-4 Rev	CCTGCTGCCGTGGGACAT		
IL-17 Fw	TATCAGCAAACGCTCACTGG	59 °C /30 s	[20]
IL-17 Rev	AGTTCACGCACCTGGAATG		
IFN-γ Fw	GCCGCACATCAAACACATATCT	59 °C /30 s	[21]
IFN-γ Rev	TGAGACTGGCTCCTTTTCCTT		
TNF-α Fw	AATTTGCAGGCTGTTTCTGC	59 °C /30 s	[22]
TNF-α Rev	TATGAAGGTGGTGCAGATGG		
GAPDH Fw	CCTGCATCTGCCCATTT	59 °C /30 s	[23]
GAPDH Rev	GGCACGCCATCACTATC		

Amplification and detection of target products were performed using the CFX 96 RT system (Bio-Rad, Hercules, CA, USA) and Maxima SYBR Green qPCR Master Mix (Thermo Scientific, Waltham, MA, USA). Subsequent qRT-PCR to detect relative expression of mRNA in selected parameters was performed for 36 cycles under the following conditions: initial denaturation at 95 °C for 2 min, subsequent denaturation at 95 °C for 15 s, and annealing (Table 2) and extension step 2 min at 72 °C. A melting curve from 50 °C to 95 °C with readings at every 0.5 °C was produced for each individual qRT-PCR plate. Analysis was performed after every run to ensure a single amplified product for each reaction. All reactions for real-time PCR were conducted in duplicate. We also confirmed that the efficiency of amplification for each target gene (including GAPDH) was essentially 100% in the exponential phase of the reaction where the quantification cycle (Cq) was calculated. The Cq values of the studied genes were normalised to an average Cq value of the reference gene (ΔCq) and the relative expression of each gene was calculated as $2^{-\Delta Cq}$.

2.5. White Blood Cell Count (WBC)

One ml of peripheral blood was taken from vena subclavia into Heparin (20 IU.mL^{-1} PBS). A total number of leukocytes were counted using the Bűrker chamber and Fried-Lukáčová solution (475 µL solution plus 25 µL blood) [24]. A white blood cell differentiation was performed, expressed in relative percentages utilising the light microscopy at 1000× magnification on blood smears after staining with Hemacolor (Merck, Darmstadt, Germany), and a count of 100 cells per slide was used. The absolute number of the different types of white blood cell count (G.l^{-1}= $10^9 \times 1^{-1}$) was determined as follows:

absolute leukocyte count × relative % of a different type of WBC/100 counted cells.

2.6. Isolation of Lamina Propria Lymphocytes (LPL)

Isolation of lymphocytes from jejunal mucosa (*n* = 6) was performed by a modification of the method of Solano-Aquilar et al. [25]. At first, the mucin was removed and, subsequently, intraepithelial lymphocytes were isolated [26]. Then, the intestine (cut into 0.5 mm pieces) was washed with 30 mL RPMI-1640 (Sigma, Darmstadt, Germany) for

15 min at 37 °C to remove the previous medium. The supernatant was discarded and the gut fragments were incubated in RPMI-1640 with collagenase type I (15 mg/60 mL RPMI; Sigma-Aldrich, St. Louis, MO, USA) for 1 h at 37 °C. The solution was slightly shaken every 5 min. Collagenase released LPL into the medium. The supernatant fluid was harvested, filtered, and immediately centrifuged at $600\times g$ for 10 min and resuspended in PBS (Sigma, Germany). Cells were washed two times in PBS (centrifugation 5 min at $250\times g$) and sediment was resuspended in 1 mL of PBS. Lymphocytes were counted in the Bűrker chamber by Tűrk solution (1:20 ratio) for correct dilution.

2.7. Staining of Lymphocytes by Direct Immunofluorescence

After the isolation of bloody and jejunal lymphocytes, their concentration was adjusted to 10^6/50 μL for immunophenotyping. Labelled mouse anti-chicken monoclonal antibodies CD4, CD8, IgM, and IgA (SouthernBiotech, Birmingham, AL, USA) at protocol-specified concentrations were added to lymphocytes followed by incubation (15 min) in the dark at room temperature. After being stained, the cells were washed once in PBS (centrifugation 5 min at $110\times g$), resuspended in 0.2 mLof PBS with 0.1% paraformaldehyde, and stored at 4 °C until measurement by flow cytometer.

2.8. Flow Cytometry Analysis of Stained Cells (FC)

FACScan cytometer and Cell Quest Program (Becton Dickinson, Heidelberg, Germany) were used to measure and analyse labelled bloody and jejunal LPL subpopulations. Gates were drawn around lymphocytes and the fluorescence data collected on at least 10,000 lymphocytes were analysed by two-parameter dot-plot histogram. The results are therefore expressed as the relative percentage of the lymphocyte subpopulation, which was positive for the specific monoclonal antibodies.

2.9. Statistical Analysis

Statistical analysis of data was performed using one-way ANOVA with Tukey post hoc analysis using the statistical program GraphPad PRISM version 6.00. Differences between the mean values for different treated groups were considered statistically significant at $p < 0.05$, $p < 0.01$, and $p < 0.001$. Values in figures are given as means or median in the case of relative gene expression with standard deviations ($\pm$SD).

3. Results

3.1. Relative Expression for Target Genes

The relative expression for IL-4 gene was significantly upregulated in groups infected by *A. galli* (Ag, Ag + Z) compared to the zinc group and control ($p < 0.05$) on day 14 of study (Figure 1). On the other hand, the relative expression of IL-17 gene was markedly upregulated on study day 7 mainly in Ag group and Ag + Z group than compared to zinc and control groups ($p < 0.01$; $p < 0.001$), as well as on day 14 (Figure 2). Similarly, the relative expression for IFN-γ gene was upregulated in the Ag group compared to zinc ($p < 0.05$) and control ($p < 0.001$), as well as in the Ag + Z group compared to control ($p < 0.001$), and significantly in the zinc group compare to control ($p < 0.001$); however, this was only observed on study day 7 (Figure 3). In a similar manner, TNF-α gene expression was markedly upregulated in the Ag group compared to the other groups ($p < 0.01$; $p < 0.001$) as well as in the combined group compared to zinc and control group ($p < 0.001$) on study day 7. However, on day 14 TNF-α gene expression was downregulated in the jejunum of infected groups compared to the zinc group and control ($p < 0.05$) (Figure 4).

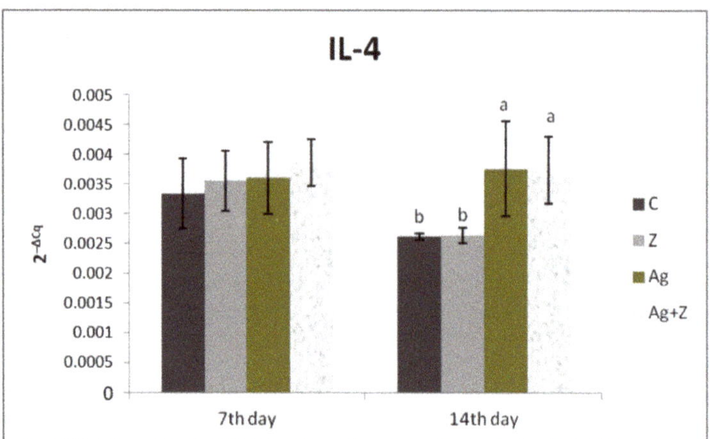

Figure 1. Relative expression of IL-4 gene in the jejunum of chickens treated with inorganic $ZnSO_4$ and infected by *A. galli*. Results at each time point are the median of $2^{-\Delta Cq}$. Means with different superscripts are significantly different. [ab] $p < 0.05$.

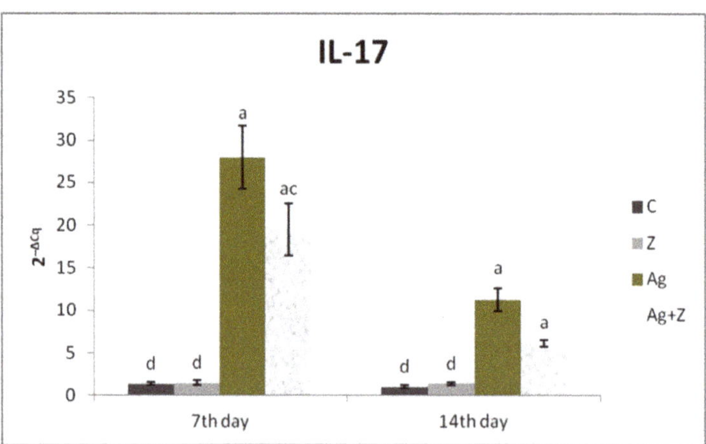

Figure 2. Relative expression of IL-17 gene in the jejunum of chickens treated with inorganic $ZnSO_4$ and infected by *A. galli*. Results at each time point are the median of $2^{-\Delta Cq}$. Means with different superscripts are significantly different. [ac] $p < 0.01$; [ad] $p < 0.001$.

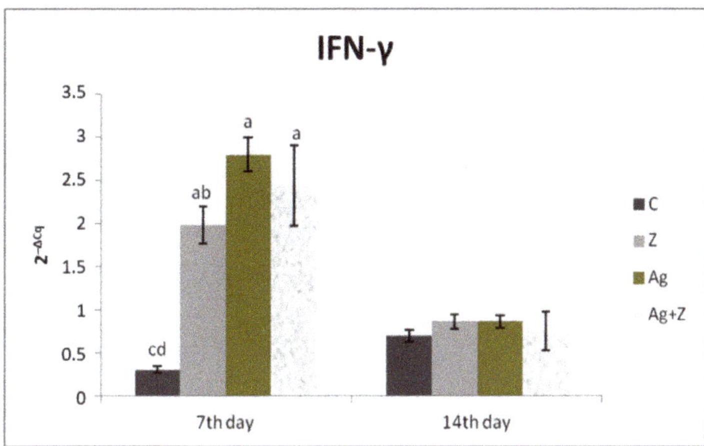

Figure 3. Relative expression of IFN-γ gene in the jejunum of chickens treated with inorganic $ZnSO_4$ and infected by *A. galli*. Results at each time point are the median of $2^{-\Delta Cq}$. Means with different superscripts are significantly different. [ab] p <0.05; [ad] $p < 0.001$.

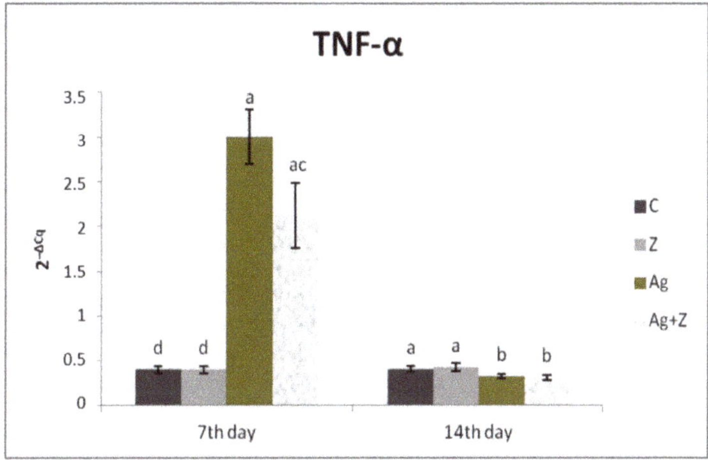

Figure 4. Relative expression of TNF-α gene in the jejunum of chickens treated with inorganic $ZnSO_4$ and infected by *A. galli*. Results at each time point are the median of $2^{-\Delta Cq}$. Means with different superscripts are significantly different. [ab] $p < 0.05$; [ac] $p < 0.01$; [ad] $p < 0.001$.

3.2. White Blood Cell Count (WBC)

Absolute number of eosinophils in blood of chickens was the highest in the infected group compared to the control group ($p < 0.01$), zinc group ($p < 0.001$), and AG + Z group ($p < 0.05$) on study day 7. Similar significant differences persist on day 14 in comparison with the control group and zinc groups ($p < 0.001$) and the same trend was noticed for the Ag + Z group compared to the zinc group and control ($p < 0.001$) (Figure 5a).

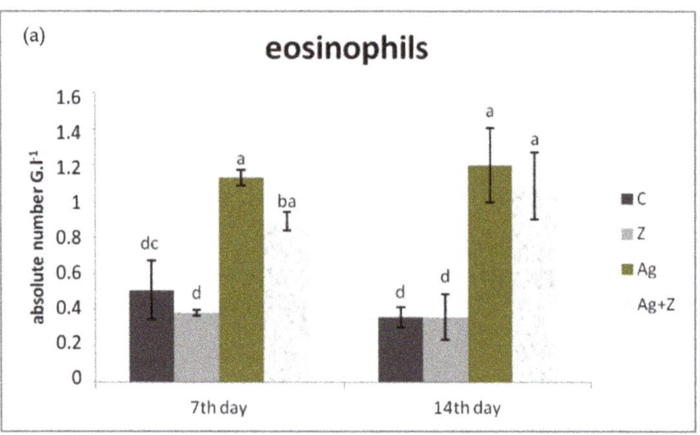

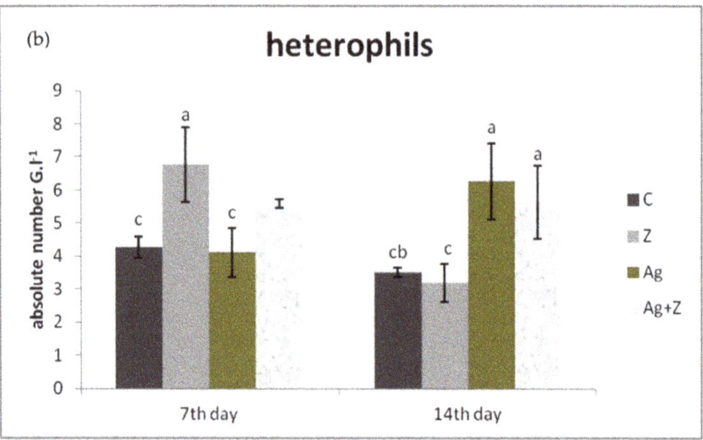

Figure 5. Effect of the administration of inorganic $ZnSO_4$ and infection by *A. galli* on the absolute number $(G.l^{-1})$ of (**a**) eosinophils and (**b**) heterophils in peripheral blood. Means with different superscripts are significantly different in Ag group. [ab] $p < 0.05$; [ac] $p < 0.01$; [ad] $p < 0.001$.

On the contrary, the number of heterophils was the highest in the zinc group compared to the control and Ag group ($p < 0.01$) on day 7. In the combined group the number of heterophils was higher compared to the Ag group and control, but not significantly ($p < 0.0085$). During second sampling, the highest number of heterophils was recorded in the Ag group compared to the control and zinc group ($p < 0.01$). Moreover, an increased the number of heterophils was recorded in the Ag + Z group compared to the control ($p < 0.01$) (Figure 5b).

3.3. Immunophenotyping of Lymphocytes

The effect of zinc on the relative percentage of CD4+ LPL was manifested in the zinc supplementation group compared to the other groups ($p < 0.001$) and control ($p < 0.01$) on study day 7. On the contrary, on day 14 the relative percentage of CD4+ was significantly improved in the infected group (Ag) compared to the zinc and control group ($p < 0.001$) as well as the combined group ($p < 0.01$) (Figure 6a). The proportion of CD8+ LPL was not significantly influenced by the administration of zinc or *A. galli* infection during both samplings (Figure 6b). The relative percentage of IgM+ cells was markedly stimulated by the combination of zinc and *A. galli* in comparison with the Ag ($p < 0.01$), the Z ($p < 0.001$)

group alone, and the control ($p < 0.01$) on study day 7. On the other hand, the opposite trend for IgM+ cells was noted during second sampling in the Ag group, where it was the highest compared to the control, Zn ($p < 0.001$), and Ag + Zn ($p < 0.01$) groups (Figure 6c). The same tendency was recorded for the relative percentage of IgA+ cells in jejunum in the combination group during the early stages of parasite infection compared to the control and Ag group ($p < 0.001$). On the second sampling, the proportion of IgA+ cells was almost equally highest in the zinc and Ag groups compared to the combination and control groups ($p < 0.01$; $p < 0.001$) (Figure 6d).

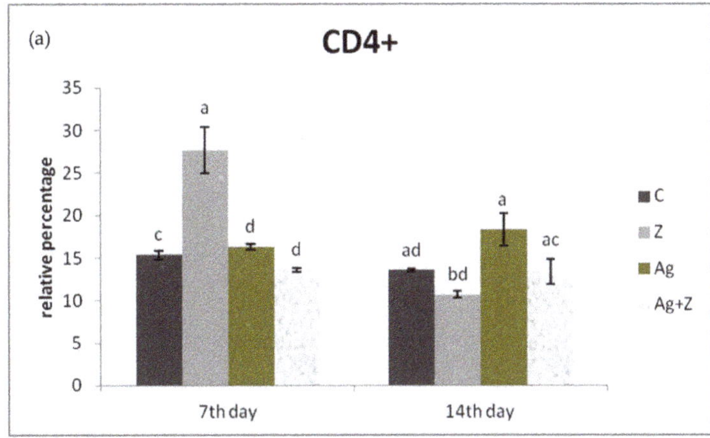

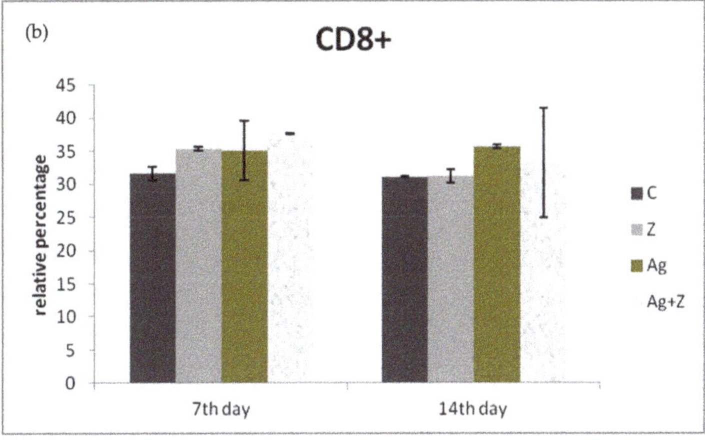

Figure 6. *Cont.*

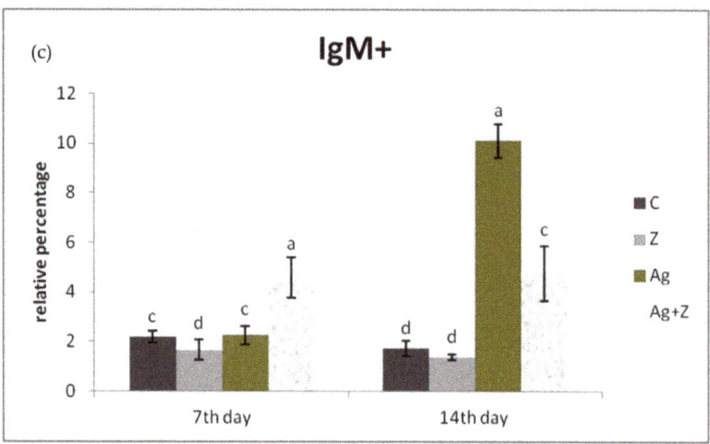

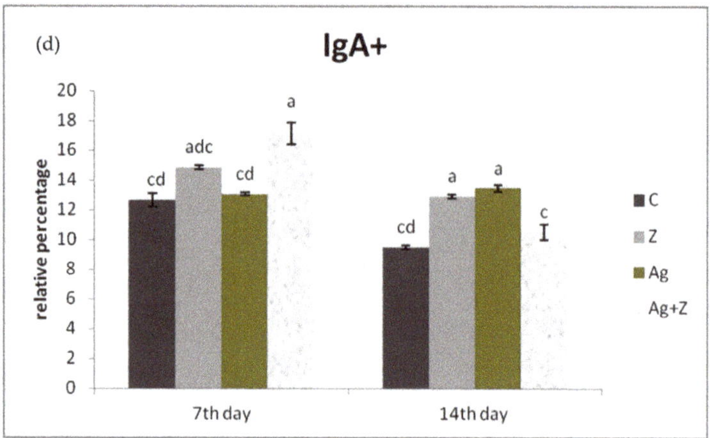

Figure 6. Effect of the administration of inorganic ZnSO$_4$ and infection by *A. galli* on the relative percentage of (**a**) CD4+, (**b**) CD8+ (**c**) IgM+, and (**d**) IgA+ LPL in the jejunum. Means with different superscripts are significantly different. [ab] $p < 0.05$; [ac] $p < 0.01$; [ad] $p < 0.001$.

4. Discussion

Generally, *A. galli* infections in chickens are accompanied by various clinical manifestations including retarded muscular and osteological development, anorexia, depression, altered hormone levels, and increased mortality [12]. Essentially, the infected chickens are more susceptible to secondary bacterial infections, which may be responsible for the fatal consequences associated with *A. galli* infections [4,27–29]. Therefore, it is necessary to support and modulate the immune response in infected chickens.

In our study, we observed increased IL-4 gene expression in both infected groups on study day 14. This indicates that Th2 cytokines indeed play crucial roles in the intestinal immune reactions during *A. galli* infection in broilers, especially in the later stages of infection [4]. In addition, Finkelman et al. [30] found that IL-4 deficient mice were more sensitive to nematode infections. The anti-inflammatory properties of IL-4 are well-documented along with their ability to suppress classical inflammation, which is necessary for wound repair functions [31].

During the supposed time of larval invasion into the intestinal mucosa (on study day 7), an increased expression of IL-17, IFN-γ, and TNF-α was observed in the infected groups. This may also indicate the involvement of Th1 component in immune responses during the early phase of infection. On the other hand, during the later stage of infection (day 14), gene expression of pro-inflammatory cytokines (IFN-γ and TNF-α), except for IL-17, was down-regulated in the infected groups and showed the onset of type 2 immunity. Remarkably, IL-17 gene expression was significantly upregulated in the infected groups in both samplings, mainly in the Ag group alone, which suggests close a relationship between the presence of nematode parasites and the activation of Th17 pathway. Furthermore, it could contribute to the pathology of this infection [32]. In the particular case of zinc, its supplementation suppressed IL-17 expression in both samplings. Similar results were presented by Cardenas et al. [33] when zinc supplementation diminished the production of IL-17 by Th17 cells in neonate mice, thereby contributing to the initiation of repair processes.

At the same time, zinc did not have a marked impact on the gene expression of other pro-inflammatory cytokine in the jejunum that may confirm its anti-inflammatory properties. Therefore, zinc supplementation in various forms may be one of the therapeutic options to restore normal mineral balance and simultaneously has a better effect due to its lower utilization and damage to the host's intestinal villi caused by *A. galli* infection [34]. Likewise, Sun et al. [35] observed that zinc deficiency caused shrinkage and flattening of jejunal villi in rats.

Eosinophilia is a characteristic feature of parasitic infections because the larval stages of nematodes can be killed by eosinophils that possess high phagocytic activity [36]. In fact, the onset of the immune response was also confirmed by elevated blood eosinophil levels in both infected groups and this indicates the presence of the parasite in the host. This result is consistent with the study by Tanwar and Mishra [37] and Kumar et al. [38]; they observed an increase in the number of eosinophils in the blood during intestinal helminthiasis, including *A. galli* in poultry. On the other hand, the number of eosinophils was decreased by zinc supplementation in both stages of infection, which confirmed that zinc inhibits the release of the preformed mediators from the basophils and eosinophils [39].

Mainly heterophils are involved in phagocytosis not only in microbial but also during parasitic infections. The phagocytic effect of heterophils, according to the study by Deka and Borah [40], may correlate with their increased amount in the host organism. In our study, infection by *A. galli* revealed the activation of heterophils in the later phase of infection, which could correlate with the destructive activity of the larval stage of *A. galli* in the mucosa of jejunum and the rupture of blood vessels. Similarly, in another study increased heterophils counts were observed on the third day after the infection with *A. galli* [41]. We assume that it also depends on the dose of infection or the immune status of the chickens. As we expected, the effect of zinc in peripheral blood was manifested by an elevation of heterophils during early stage of infection. In contrast, zinc deficiency impaired the oxidative burst of heterophils. It is likely that due to zinc having a central role in the activity of the enzyme superoxide dismutase during oxidative burst, it protects cells from radical oxygen molecules [42]. However, the stimulatory effect of zinc on the total number of heterophils was only temporary. We suppose that the decreased absolute number of heterophils in the Ag+Zn group during the later infection could be caused by a persistent infection, which is when the zinc reserve is depleted due to impaired intestinal absorption.

The most abundant immune cells localised in the subepithelial lamina propria of intestine are lamina propria lymphocytes, which represent the main executive component of the intestinal mucosal immune response. LPL are mostly helper Th-lymphocytes and plasma cells that produce most of the polymeric IgA [43]. In *A. galli*-infected birds we observed increased infiltrations of lamina propria with CD4+ lymphocytes on day 14 of the study. A recent study by Ruhnke et al. [36] reported an increase in intraepithelial CD4$^+$ but a decrease in the number of CD8+ cytotoxic T cell populations after experimental infections with *A. galli* in broiler chickens. In our study, on day 7 we noted only a moderate increase in CD8+ LPL population in the combined group (without statistical significance); however, it is not clear whether this was due to the influence of *A. galli* infection or inorganic zinc supplementation. In the later phase of infection (day 14), the highest percentage of CD8+ LPL was observed solely in the Ag group, which demonstrates the host's response to intestinal mucosal damage. This is in agreement with the study by Schwarz et al. [4] who showed that, in broiler chickens, a CD8 + T cell response can be expected about 14 days after *A. galli* infection.

On the flip side, an increase in CD4+ LPL recorded during first sampling in the zinc group alone indicates the potential of zinc to induce an immune response. Our results also confirmed the stimulation of the CD8+, IgM+, and IgA+ LPL in the combined group at the first sampling. It has been shown that zinc controls follicular B cell maintenance in the spleen as well as the regulation of the BCR signaling pathway [44–46]. Therefore, the effect of zinc on the humoral immune response is unequivocal. Collectively, the obtained results also point to the activation and cooperation of cellular and humoral immunity in the processes of control of *A. galli* parasites. In addition, zinc supplementation accelerates mucosal regeneration [35].

The highest relative percentage of IgM+ cells was noted in *A. galli* infected group (second sampling) together with the combined group (first sampling). It is correlated with the fact that, at the beginning of the disease process, the level of IgM usually increases, which is stimulated mainly during the first contact of the host with the parasitic antigen.

5. Conclusions

Based on our results, it can be stated that the enrichment of the diet with inorganic zinc has a positive effect on humoral and cell-mediated immunity in chickens infected with *A. galli*. In addition, supplementation of inorganic zinc successfully suppressed IL-17 gene expression and, thus, makes a significant contribution to the regulation of potential autoimmune reaction development. Despite the fact that the role of zinc as an important nutritional supplement is well-known, it is still necessary to investigate certain aspects of its use, especially in its application for therapeutic purposes.

Author Contributions: Conceptualization, V.K., V.R., Z.Š., M.L. (Mikulas Levkut), and R.H.; methodology, V.K., V.R., E.D., and Ľ.G.; formal analysis, V.R. and M.L. (Mikulas Levkut); data curation, V.K. and M.L. (Martin Levkut); writing—original draft preparation, V.K. and V.R.; writing—review and editing, V.R. and M.L. (Mikulas Levkut); supervision, Z.Š., R.H., and V.R.; funding acquisition, M.L. (Mikulas Levkut) All authors have read and agreed to the published version of the manuscript.

Funding: This work was supported by the Grant Agency for Science of Slovak Republic VEGA 1/0355/19.

Institutional Review Board Statement: The study was conducted according to the European directive 2010/63/EU on the protection of animals used for scientific purpose and approved by The Ethical Committee of the University of Veterinary Medicine and Pharmacy in Košice and the State Veterinary and Food Administration of the Slovak Republic approved the experimental protocol number 836/17-221 in May 2018.

Data Availability Statement: The data presented in this study are available within the article.

Conflicts of Interest: The authors have declared no conflict of interest.

References

1. Ferdushy, T.; Nejsum, P.; Roepstorff, A.; Thamsborg, S.M.; Kyvsgaard, N.C. *Ascaridia galli* in chickens: Intestinal localization and comparison of methods to isolate the larvae within the first week of infection. *Parasitol. Res.* **2012**, *111*, 2273–2279. [CrossRef] [PubMed]
2. Ramadan, H.H.; Znada, N.Y.A. Morphology and life history of *Ascaridia galli* in the domestic fowl that are raised in Jeddah. *Science* **1992**, *4*, 87–99. [CrossRef]
3. Marcos-Atxutegi, C.; Gandolfi, B.; Arangüena, T.; Sepúlveda, R.; Arévalo, M.; Simón, F. Antibody and inflammatory responses in laying hens with experimental primary infections of *Ascaridia galli*. *Vet. Parasitol.* **2009**, *161*, 69–75. [CrossRef] [PubMed]
4. Schwarz, A.; Gauly, M.; Abel, H.; Daş, G.; Humburg, J.; Rohn, K.; Breves, G.; Rautenschlein, S. Immunopathogenesis of *Ascaridia galli* infection in layer chicken. *Dev. Comp. Immunol.* **2011**, *35*, 774–784. [CrossRef] [PubMed]

5. Darmawi, U.; Balqis, M.; Hambal, R.; Tiuria, F.; Priosoeryanto, B.P. Mucosal mast cells response in the jejunum of *Ascaridia galli*-infected laying hens. *Media Peternak.* **2013**, *36*, 113–119. [CrossRef]
6. Norup, L.R.; Dalgaard, T.S.; Pleidrup, J.; Permin, A.; Schou, T.W.; Jungersen, G.; Fink, D.R.; Juul-Madsen, H.R. Comparison of parasite-specific immunoglobulin levels in two chicken lines during sustained infection with Ascaridia galli. *Vet. Parasitol.* **2013**, *191*, 187–190. [CrossRef] [PubMed]
7. Degen, W.G.J.; Daal, N.V.; Rothwell, L.; Kaiser, P.; Schijns, V.E.J.C. Th1/Th2 polarization by viral and helminth infection in birds. *Vet. Microbiol.* **2005**, *105*, 163–167. [CrossRef] [PubMed]
8. Allen, J.E.; Wynn, T.A. Evolution of Th2 immunity: A rapid repair response to tissue destructive pathogens. *PLoS Pathog.* **2011**, *7*, e1002003. [CrossRef]
9. Lambrecht, B.N.; Hammad, H. The immunology of asthma. *Nat. Immunol.* **2015**, *16*, 45–56. [CrossRef]
10. Sutherland, T.E.; Logan, N.; Rückerl, D.; Humbles, A.A.; Allan, S.M.; Papayannopoulos, V.; Stockinger, B.; Maizels, R.M.; Allen, J.E. Chitinase-like proteins promote IL-17-mediated neutrophilia in a tradeoff between nematode killing and host damage. *Nat. Immunol.* **2014**, *15*, 1116–1125. [CrossRef]
11. Ajendra, J.; Chenery, A.L.; Parkinson, J.E.; Chan, B.H.K.; Pearson, S.; Colombo, S.A.P.; Boon, L.; Grencis, R.K.; Sutherland, T.E.; Allen, J.E. IL-17A both initiates, via IFNγ suppression, and limits the pulmonary type-2 immune response to nematode infection. *Mucosal Immunol.* **2020**, *13*, 958–968. [CrossRef]
12. Dahl, C.; Permin, A.; Christensen, J.P.; Bisgaard, M.; Muhairwa, A.P.; Petersen, K.M.; Poulsen, J.S.; Jensen, A.L. The effect of concurrent infections with *Pasteurella multocida* and *Ascaridia galli* on free range chickens. *Vet. Microbiol.* **2002**, *86*, 313–324. [CrossRef]
13. Sharma, N.; Hunt, P.W.; Hine, B.C.; Ruhnke, I. The impacts of *Ascaridia galli* on performance, health, and immune responses of laying hens: New insights into an old problem. *Poult. Sci.* **2019**, *98*, 6517–6526. [CrossRef]
14. Mudroňová, D.; Nemcová, R.; Lauková, A.; Koščová, J.; Strompfová, V.; Györyová, K.; Szunyogová, E.; Lazar, G. Effect of *Lactobacillus fermentum* alone, and in combination with zinc(ii) propionate on *Salmonella Enterica* serovar. Dusseldorf in Japanese quails. *Biologia* **2006**, *6*, 797–801. [CrossRef]
15. Sloup, V.; Jankovská, I.; Nechybová, S.; Peřinková, P.; Langrová, I. Zinc in the animal organism: A review. *Sci. Agric. Bohem.* **2017**, *48*, 13–21. [CrossRef]
16. Bonaventura, P.; Benedetti, G.; Albarède, F.; Miossec, P. Zinc and its role in immunity and inflammation. *Autoimmun. Rev.* **2015**, *14*, 277–285. [CrossRef]
17. Permin, A.; Bojesen, M.; Nansen, P.; Bisgaard, M.; Frandsen, F.; Pearman, M. *Ascaridia galli* populations in chickens following single infections with different dose levels. *Parasitol. Res.* **1997**, *83*, 614–617. [CrossRef]
18. Karaffová, V.; Bobíková, K.; Levkut, M.; Revajová, V.; Ševčíková, Z.; Levkut, M. The influence of Farmatan® and Flimabend® on the mucosal immunity of broiler chicken. *Poult. Sci.* **2019**, *98*, 1161–1166. [CrossRef]
19. Truong, A.D.; Park, B.; Ban, J.; Hong, Y.H. The novel chicken interleukin 26 protein is overexpressed in T cells and induces proinflammatory cytokines. *Vet. Res.* **2016**, *47*, 65. [CrossRef]
20. Crhanova, M.; Hradecka, H.; Faldynova, M.; Matulova, M.; Havlickova, H.; Sisak, F.; Rychlik, I. Immune response of chicken gut to natural colonization by gut microflora and to *Salmonella enterica* serovar enteritidis infection. *Infect. Immun.* **2011**, *79*, 2755–2763. [CrossRef]
21. Berndt, A.; Wilhelm, A.; Jugert, C.; Pieper, J.; Sachse, K.; Methner, U. Chicken cecum immune response to *Salmonella enterica* serovars of different levels of invasiveness. *Infect. Immun.* **2007**, *75*, 5993–6007. [CrossRef]
22. Kolesarova, M.; Spisakova, V.; Matulova, M.; Crhanova, M.; Sisak, F.; Rychlik, I. Characterisation of basal expression of selected cytokines in the liver, spleen, and respiratory, reproductive and intestinal tract of hens. *Vet. Med.* **2011**, *56*, 325–332. [CrossRef]
23. De Boever, S.; Vangestel, C.; De Backer, P.; Croubels, S.; Sys, S.U. Identification and validation of housekeeping genes as internal control for gene expression in an intravenous LPS inflammation model in chickens. *Vet. Immunol. Immunopathol.* **2008**, *122*, 312–317. [CrossRef]
24. Fried, K.; Jantošovič, J. Blood sampling by cardiac punction. *Vet. Cas.* **1961**, *10*, 383–391. (In Slovak)
25. Solano-Aguilar, G.I.; Vengroski, K.G.; Beshah, E.; Lunney, J.K. Isolation and purification of lymphocyte subsets from gut-associated lymphoid tissue in neonatal swine. *J. Immunol. Methods* **2000**, *241*, 185–199. [CrossRef]
26. Bucková, B.; Revajová, V. Immunophenotyping of intraepithelial (IEL) and *lamina propria* lymphocytes (LPL) in the chicken intestine by flow cytometry. *Folia Vet.* **2014**, *58*, 75–77.
27. Eigaard, N.M.; Schou, T.W.; Permin, A.; Christensen, J.P.; Ekstrøm, C.T.; Ambrosini, F.; Cianci, D.; Bisgaard, M. Infection and excretion of *Salmonella* enteritidis in two different chicken lines with concurrent *Ascaridia galli* infection. *Avian Pathol.* **2006**, *35*, 487–493. [CrossRef]
28. Permin, A.; Christensen, J.P.; Bisgaard, M. Consequences of concurrent *Ascaridia galli* and *Escherichia coli* infections in chickens. *Acta Vet. Scand.* **2006**, *47*, 43–54. [CrossRef] [PubMed]
29. Pleidrup, J.; Dalgaard, T.S.; Norup, L.R.; Permin, A.; Schou, T.W.; Skovgaard, K.; Vadekær, D.F.; Jungersen, G.; Sørensen, P.; Juul-Madsen, H.R. *Ascaridia galli* infection influences the development of both humoral and cell-mediated immunity after Newcastle disease vaccination in chickens. *Vaccine* **2014**, *32*, 383–392. [CrossRef] [PubMed]

30. Finkelman, F.D.; Shea-Donohue, T.; Morris, S.C.; Gildea, L.; Strait, R.; Madden, K.B.; Schopf, L.; Urban, J.F. Interleukin-4- and interleukin-13-mediated host protection against intestinal nematode parasites. *Immunol. Rev.* **2004**, *201*, 139–155. [CrossRef] [PubMed]

31. Allen, J.E.; Sutherland, T.E. Host protective roles of type 2 immunity: Parasite killing and tissue repair, flip sides of the same coin. *Semin. Immunol.* **2014**, *26*, 329–340. [CrossRef]

32. Zhang, L.; Liu, R.; Song, M.; Hu, Y.; Pan, B.; Cai, J.; Wang, M. *Eimeria tenella*: Interleukin 17 contributes to host immunopathology in the gut during experimental infection. *Exp. Parasitol.* **2013**, *133*, 121–130. [CrossRef]

33. Cardenas, A.E.; Sánchez Tapia, M.T.; Muñoz Chacon, B.; Pastelin Palacios, R. Impact of zinc supplementation on the production of IL-17 in perinatal stages. In Proceedings of the 15th International Congress of Immunology (ICI), Milan, Italy, 22–27 August 2013.

34. Gabrashanska, M.; Teodorova, S.E.; Galvez-Morros, M.M.; Tsocheva-Gaytandzhieva, N.; Mitov, M.; Ermidou-Pollet, S.; Pollet, S. Administration of Zn-Co-Mn basic salt to chickens with ascaridiosis. *Parasitol. Res.* **2004**, *93*, 235–241. [CrossRef]

35. Sun, J.Y.; Wang, J.F.; Zi, N.T.; Jing, M.Y.; Weng, X.Y. Effects of zinc supplementation and deficiency on bone metabolism and related gene expression in rat. *Biol. Trace Elem. Res.* **2011**, *143*, 394–402. [CrossRef]

36. Ruhnke, I.; Andronicos, N.M.; Swick, R.A.; Hine, B.; Sharma, N.; Kheravii, S.K.; Wu, S.B.; Hunt, P. Immune responses following experimental infection with *Ascaridia galli* and necrotic enteritis in broiler chickens. *Avian Pathol.* **2017**, *46*, 602–609. [CrossRef]

37. Tanwar, R.K.; Mishra, S. Clinico-HaematoBiochemical studies on intestinal helminthiasis in poultry. *Vet. Practitioner* **2001**, *2*, 137–140.

38. Kumar, R.; Sinha, S.R.P.; Verma, S.B.; Sinha, S. Haematological changes in the Japanese quails (Coturnix coturnix japonica) naturally infected with nematode *Ascaridia galli*. *Ind. Vet. Med. J.* **2003**, *27*, 297–299.

39. Sudadi, A.M.; Hariyati, R.; Dewi, A.M.K. Influence of zinc supplementation in eosinophil nasal mucous count and quality of life in moderate to severe persistent allergic rhinitis patient in ENT clinic Dr. Kariadi hospital Semarang in November 2016 until January 2017. *Bali Med. J.* **2019**, *8*, 88–93. [CrossRef]

40. Deka, K.; Borah, J. Haematological and biochemical changes in japanese quails Coturnix coturnix japonica and chickens due to Ascaridia galli infection. *Int. J. Poult. Sci.* **2008**, *7*, 704–710. [CrossRef]

41. Luna-Olivares, L.A.; Ferdushy, T.; Kyvsgaard, N.C.; Nejsum, P.; Thamsborg, S.M.; Roepstorff, A.; Iburg, T.M. Localization of *Ascaridia galli* larvae in the jejunum of chickens 3 days post infection. *Vet. Parasitol.* **2012**, *185*, 186–193. [CrossRef]

42. Ibs, K.H.; Rink, L. Zinc-altered immune function. *J. Nutr.* **2003**, *133*, 1452S–1456S. [CrossRef]

43. Dahan, S.; Roda, G.; Pinn, D.; Roth-Walter, F.; Kamalu, O.; Martin, A.P.; Mayer, L. Epithelial: Lamina *propria* lymphocyte interactions promote epithelial cell differentiation. *Gastroenterology* **2008**, *134*, 192–203. [CrossRef]

44. Hojyo, S.; Fukada, T. Roles of zinc signaling in the immune system. *J. Immunol. Res.* **2016**, *2016*, 6762343. [CrossRef]

45. Miyai, T.; Hojyo, S.; Ikawa, T.; Kawamura, M.; Irié, T.; Ogura, H.; Hijikata, A.; Bin, B.H.; Yasuda, T.; Kitamura, H.; et al. Zinc transporter SLC39A10/ZIP10 facilitates antiapoptotic signaling during early B-cell development. *Proc. Natl. Acad. Sci. USA.* **2014**, *111*, 11780–11785. [CrossRef]

46. Maiguma, M.; Suzuki, Y.; Suzuki, H.; Okazaki, K.; Aizawa, M.; Muto, M.; Tomino, Y. Dietary zinc is a key environmental modifier in the progression of IgA nephropathy. *PLoS ONE* **2014**, *9*, e90558. [CrossRef]

Article

Microbiological Effectivity Evaluation of New Poultry Farming Organic Waste Recycling

Edit Gorliczay [1], Imre Boczonádi [1,*], Nikolett Éva Kiss [1], Florence Alexandra Tóth [1], Sándor Attila Pabar [2], Borbála Biró [2], László Renátó Kovács [3,4] and János Tamás [1]

[1] Institute of Water and Environmental Management, University of Debrecen, 4032 Debrecen, Hungary; edit.gorliczay@agr.unideb.hu (E.G.); kiss.nikolett@agr.unideb.hu (N.É.K.); toth.florence@agr.unideb.hu (F.A.T.); tamas@agr.unideb.hu (J.T.)

[2] Department of Agri-Environmental Sciences, Szent Istvan University, 1118 Budapest, Hungary; pabarattila@gmail.com (S.A.P.); biro.borbala@gmail.com (B.B.)

[3] Department of Medical Microbiology, Faculty of Medicine, University of Debrecen, 4032 Debrecen, Hungary; kovacs.renato@med.unideb.hu

[4] Faculty of Pharmacy, University of Debrecen, 4032 Debrecen, Hungary

* Correspondence: boczonadi.imre@agr.unideb.hu

Abstract: Due to the intensification of the poultry sector, poultry manure is being produced in increasing quantities, and its on-site management is becoming a critical problem. Animal health problems can be solved by stricter the veterinary and environmental standards. The off-site coupled industrial chicken manure recycling technology (Hosoya compost tea) fundamentally affects the agricultural value of new organic-based products. Due to the limited information available on manure recycling technology-related microbiological changes, this was examined in this study. A pot experiment with a pepper test plant was set up, using two different soils (Arenosol, slightly humous Arenosol) and two different doses (irrigation once a week with 40 mL of compost tea: dose 1, D1; irrigation twice a week with 40 mL of compost tea: dose 2, D2) of compost tea. Compost tea raw materials, compost tea, and compost tea treated soils were tested. The products (granulated manure, compost tea) and their effects were characterized by the following parameters: aerobic bacterial count (log CFU/g), fluorescein diacetate activity (3′,6′-diacetylfluorescein, FDA, μg Fl/g soil), glucosidase enzyme activity (GlA; PNP/μmol/g), and identification of microorganisms in compost tea with matrix-assisted laser desorption ionization time-of-flight mass spectrometry (MALDI-TOF MS). Furthermore, we aimed to investigate how the microbiological indicators tested, and the effect of compost tea on the tested plant, could be interpreted. Based on our results, the microbiological characteristics of the treated soils showed an increase in enzyme activity, in the case of FDA an increase +0.26 μg Fl/g soil at D1, while the GlA increased +1.28 PNP/μmol/g with slightly humous Arenosol soil and increased +2.44 PNP/μmol/g at D1; and the aerobic bacterial count increased +0.15 log CFU/g at D2, +0.35 log CFU/g with slightly humous Arenosol and +0.85 log CFU/g at W8. MALDI-TOF MS results showed that the dominant bacterial genera analyzed were *Bacillus* sp., *Lysinibacillus* sp., and *Pseudomonas* sp. Overall, the microbial inducers we investigated could be a good alternative for evaluating the effects of compost solutions in soil–plant systems. In both soil types, the total chlorophyll content of compost tea-treated pepper (*Capsicum annuum* L.) had increased as a result of compost tea. D1 is recommended for Arenosol and, D2 for slightly humous Arenosol soil.

Keywords: microbiology; poultry farming waste; organic waste

Citation: Gorliczay, E.; Boczonádi, I.; Kiss, N.É.; Tóth, F.A.; Pabar, S.A.; Biró, B.; Kovács, L.R.; Tamás, J. Microbiological Effectivity Evaluation of New Poultry Farming Organic Waste Recycling. *Agriculture* **2021**, *11*, 683. https://doi.org/10.3390/agriculture11070683

Academic Editor: Vito Armando Laudicina

Received: 30 June 2021
Accepted: 16 July 2021
Published: 19 July 2021

Publisher's Note: MDPI stays neutral with regard to jurisdictional claims in published maps and institutional affiliations.

1. Introduction

The quick spread of intensive agricultural systems, the use of fertilizers, and rapid human population growth have had negative effects on soil fertility, mainly decreasing the soil organic carbon and the total soil nitrogen, and changing the composition of carbon and nitrogen, owing to the loss of soil organic matter through erosion and leaching [1,2], and thus resulting in unsustainable soil degradation [3,4]. Soil is a finite natural resource that is under pressure of increasing consumption, rapid population growth, and agricultural intensification [5]. Increasing consumption and population growth are driving up yields per unit area in crop production, while poultry production is on the rise to provide an intensive source of protein [6]. Over the last few decades there have been rapid changes in livestock production, with 61% of pork, 81% of poultry, and 86% of eggs now produced on intensive, industrial farms [7]. Generally, there has been a rapid change in how animal products are produced, processed, consumed, and marketed. Growth in livestock production, in both developed and developing countries, has been led by poultry [8]; while eggs and poultry meat have become the main source of animal protein [9]. Recently, African swine fever led to an increase in demand for poultry meat. This market is further complicated by the increase in EU internal production, the UK's exit from the EU, and the increased quota for Ukrainian poultry meat. Broiler production is the dominant sector in poultry meat production, being the second most produced and consumed meat in the EU after pork [10]. In Europe, broiler farms with more than 100,000 places are very common. In 2013, 891.4 million broiler chickens were bred on more than two million farms in the EU. Farms with more than 100,000 birds account for 38% of the total poultry population. It is estimated that 90% of broilers in the EU are reared in intensive indoor systems [11,12]. These closed, intensive farming systems, with high animal density, indoor housing, and the use of fast-growing breeds obtained by genetic selection can facilitate the spread of epidemics and certain zoonotic diseases (H1N1, H5N1 influenza, brucellosis, salmonellosis, leptospirosis, avian influenza, etc.) [13]. Animal health problems can cause serious social, economic, and environmental damage, and in some cases can also pose a threat to human health [14]. The epidemics did not end with African swine fever, with the emergence of H5N1 avian influenza in early February 2020. The outbreak of avian influenza in Central Europe in December 2019 limited the production in 2020 in terms of volume and has posed a major challenge for the animal health system and the poultry sector. In the current situation (due to ever stricter animal health regulations) manure is an "environmental pollutant" that livestock farmers are trying to dispose of. At the same time, the concept of the circular economy highlights the inescapable role of animal manure in soil management; there has always been a fundamental link between livestock and crop production in agriculture. It is necessary to establish a system of biodegradable organic matter management that focuses on the cycling of organic matter to maintain soil fertility in the long term and that creates a shared interest between crop and livestock farmers in the use of animal manure [15].

There are various different types of poultry manure, such as deep litter manure, broiler manure, and hen manure. The ratio of litter to manure and the moisture content causes variation among manures from different poultry houses. The quality of manure can vary depending upon many factors, including the age and diet of the flock, the moisture content, and the age of the manure [16]. Secondary pollution is making landfills less and less suitable for organic waste [17], and tightening environmental regulations mean that the landfilling of organic waste and by-products is not an option [18]. The use of energy (biogas production) or material (composting) recovery methods to manage organic wastes and by-products is widespread around the world [19,20]. Composting is considered a favorable option in many developing countries due to the lower investment and operating costs, need for scientific expertise, and technical complexity [21,22]. One way to treat litter manure is composting, which increases the quality of raw manure and reduces the environmental risks [23,24]. During the composting process, the volume and weight of manure are reduced, pathogens and weed seeds are destroyed [25], unpleasant odors are reduced [26], and nutrients and organic matter are stabilized [27].

The use of compost in agriculture is very important because it contributes to the increase of soil fertility [28,29] and can also be crucial in the treatment of plant diseases [30]. There is a growing demand for compost to be further utilized as a raw material to produce compost tea. Compost tea is a concentrated microbial solution, which is made by extracting nutrients and microorganisms from the compost using an extractant. Compost tea is made by mixing the compost with a distillated water (as a weak agent) and by incubating for a specified period of time, with or without active aeration (aerated compost tea, (ACT), or non-aerated compost tea, (NCT)), and produced with or without additives [31]. Compost teas made from matured compost contain high levels of bacteria (10^8–10^9/mL). Aerated compost teas contain even higher amounts of bacteria (up to 10^{10}–10^{11}/mL). However, if oxygen-deficient conditions exist during the production of a compost tea, the aerobic bacteria are no longer able to multiply and grow and die or enter to a dormant state [32]. The use of compost tea is widespread worldwide due to its wide applicability as a biostimulator [33,34]. The advantage of using compost tea instead of compost is that compost cannot be applied to the foliage by spraying or irrigation, and the availability of nutrients is better [35]. Several researchers have pointed to the effectiveness of organic manure, compost, and composting in horticultural technologies, and the positive effects of compost teas have been demonstrated on cumin [36], fennel [37], basil [38], pepper [34], lettuce [39], and tomatoes [40] with test plants. Compost tea reduces plant diseases [34,35], protects plant roots, provides nutrients to the plants, and improves plant health [41]; thus compost tea is a possible alternative to synthetic agents [38,39]. Following the application of these organic-based substances, the improvement in yield and quality can be attributed to the enhancement of the beneficial microbial communities in the soil, the improvement of plant mineral absorption, and the stimulation of phytohormones [42]. Various liquid fertilizers or their extracts are known to serve primarily as a source of soluble plant nutrients, growth promoters, and disease suppressants [43,44].

Based on the above, in this study we investigated the specific products produced by a novel approach to organic waste utilization technology. A pot experiment with a pepper test plant was set up, using two different soils (Arenosol, slightly humous Arenosol) and two different doses (irrigation once a week with 40 mL of compost tea: dose 1, D1; irrigation twice a week with 40 mL of compost tea: dose 2, D2) of compost tea. Compost tea raw materials, compost tea, and compost tea treated soils were tested. The products (granulated manure, compost tea) and their effects were characterized by the following parameters: aerobic bacterial count (log CFU/g), fluorescein diacetate activity (FDA, µg Fl/g soil), glucosidase enzyme activity (GlA; PNP/µmol/g), and the identification of microorganisms in compost tea with MALDI-TOF MS. Furthermore, we aimed to investigate how the microbiological indicators tested and the effects of compost tea on the tested plants could be interpreted.

2. Materials and Methods

The examined product from which compost tea was produced, consisted of a mixture of broiler manure, hen manure, and straw pellets, which were treated by intensive composting with meat meal as an additive. The process was carried out under standard, controlled conditions from hatching, through feed production and mixing, to manure processing. A schematic of the process is shown in Figure 1.

The composting plant uses deep litter broiler manure and hen manure from a downstream hatchery as feedstock, and 60,000 tons of manure are produced annually at the company's various sites. The parent breeding is carried out with colored breeding pairs and the broiler is kept with a white meat hybrid Ross 308, and straw bedding is used for both breeds. The breeding is carried out with colored breeding pairs and the broiler is kept with a white meat Ross 308 hybrid. Both on the breeding farms and on the broiler farms, litter is produced with heat-treated straw pellets, whose high absorption capacity results not only in excellent litter quality but also in a low moisture, dry, and deep manure. In the manure processing plant, where the product under study originated, three continuous mode manure fermenters of the HOSOYA [45] (Hosoya Ltd., Kanagawa, Japan) type are operated, with controlled and regulated fermentation [46,47]. The fermenters with a stirring machine are used to store 30 tons of raw material every day of the year (1.5 tons of raw material (a mixture of broiler, hen manure, and straw pellets) is stored, which represents 953.84 kg of broiler manure, 476.93 kg of hen manure, and 69.23 kg of straw pellets (Table S1). The 2/3:1/3 mixing ratio of the raw materials (deep layer broiler manure and hen manure) is necessary because our preliminary studies showed that (at least) 1/3 of chicken manure is needed for composting of broiler manure, whose microbial composition helps to start the fermentation process.

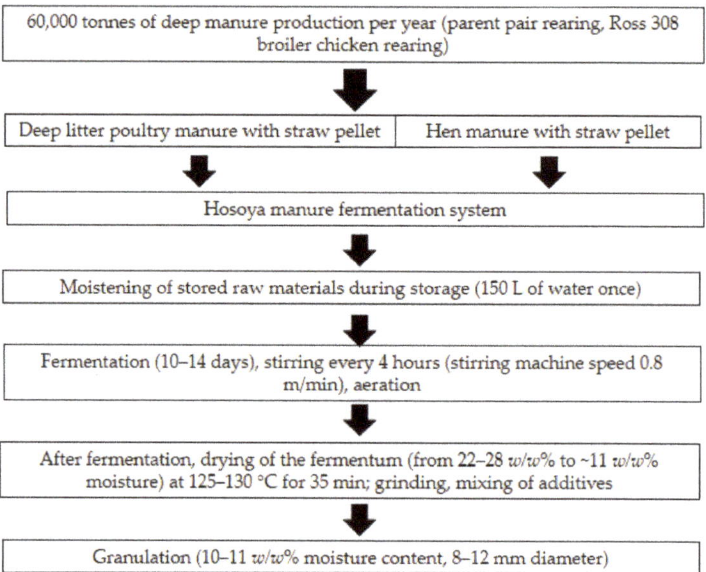

Figure 1. The schematic of the poultry manure utilization process.

Fermentation is carried out by air injection, mixing the raw material, and turning it over every 4 h, for 10 to 14 days. The incoming raw material has a moisture content of 20–40 $w/w\%$, depending on the type of raw material. In order for fermentation to start and be intensive, the raw material must be adjusted to a moisture content of 40–45 $w/w\%$ by adding water. The raw material is moistened once with 150 L of water when it is fed into the fermenters. At the end of fermentation (after 10 to 14 days) the moisture content of the fermented fermentate is reduced to 22–28 $w/w\%$. The next technological process is the intensive fluid bed drying (Tema Process, Wapenveld, The Netherlands) (125–130 °C, 25–30 min) of the fermentum from 22–28 $w/w\%$ to 10–11 $w/w\%$ moisture. The prolonged exposure to heat above 125 °C kills all pathogenic bacteria. The dried fermentum is ground into a powder fraction and becomes the raw material for further finished products. The final product is packaged in granulated form, but before the granulation process is carried out, the natural fermentum is supplemented with complex organic nutrients as a raw material.

2.1. Compost Tea Preparation

The characteristics of the product used to prepare compost tea are detailed in Table 1.

The equipment used in the production of compost tea was sterilized in an AE-75 DRY autoclave (Raypa Ltd., Barcelona, Spain) with wet steam sterilization. After sterilization, the composted granulated manure mixture was weighed in a 0.7 l volume glass container. For the preparation of compost tea, distillated water was used as a weak agent. Compost tea was prepared in a 1/10 (w/v) compost to water ratio (CWR) based on Islam et al. [48] and Zhang et al. [49], and using an Unimax 1010 shaker (Heidolph Instruments GmbH & Co, Schwabach, Germany) at 130 rpm for 48 h; the compost tea temperature was set at 35 °C (Table S2), after the extraction time the compost tea was filtered through a filter paper (12–15 μm, VWR International, Debrecen, Hungary), and the filtered tea was used for further processing.

Due to the concentration of the compost tea and the nutrient requirements of the test plant, the suspension was applied to the peppers at a five-fold dilution.

Table 1. Characteristics of evaluated product.

Parameters	Value
Dry matter content (*w/w*%)	87.60
Moisture content (*w/w*%)	12.40
pH	6.99
Organic matter content (*w/w*%)	73.03
Nitrogen-content (*w/w*%)	4.86
Phosphorus pentoxide content (*w/w*%)	6.88
Potassium oxide content (*w/w*%)	4.04

2.2. Compost Tea Treatments and Design of Pot Experiments

The selected test plant for testing the effect of compost tea was white sweet pepper (*Capsicum annuum* L.). During the experiments, 1 kg of slightly humous Arenosol (SHA) and Arenosol (A) soil was weighed into pots and the pepper seedlings were planted.

The selected soils are classified as "Arenosols" according to the World Base Reference of Soil Resources (WRB). Soils in this category are characterized by low water holding capacity, high water permeability, and low nutrient content, all of which lead to rapid water stress [50]. Soils that are poorly vegetated and susceptible to wind erosion belong to this major group [51]. The main characteristics of the sub-categorized soils are presented in Table 2.

Table 2. Characteristics of the soils used for the experiments.

Measured Parameters	Slightly Humous Arenosol	Arenosol
pH (KCl-extract)	5.76	6.13
Total water soluble salts (*w/w*)%	0.02	0.05
Carbonate content (*w/w*)%	<0.100	<0.100
Organic carbon content (humus content) (*w/w*)%	1.57	0.67
Phosphorus-pentoxide (mg/kg) (AL-extract)	176	131.2
Potassium-oxide (mg/kg) (AL-extract)	351	177.96
Nitrate (mg/kg) (KCl-extract)	12.3	7.42

The Arenosol soils are characterized by a partially developed topsoil layer with low humus content and no subsurface clay accumulation. In addition, the aggregate thickness of the finer textured layers is less than 15 cm, "the proportion of coarse debris within ≤100 cm of the mineral soil surface is <40 *v/v*% in all layers" [50]. Arenosol and slightly humous Arenosol soils are also characterized by rapid mineralization of organic matter, lack of organic colloid, poor water management, poor nutrient supply, and drought sensitivity. The conditions necessary for biological soil formation processes in the formation of these main soil types are only present for a short period of time and their impact is therefore limited. Of the soils selected, Arenosol soil had inferior properties compared to the humic sandy soil, as reflected in the parameters tested. There was a minimal difference in pH, as well as a higher total water-soluble salinity in the Arenosol soil.

Each treatment was set up in three replicates. In the pot experiments, 40-40 mL of a five-fold diluted compost tea was applied by seeding once (Dose 1, D1) and twice (Dose

2, D2) a week. On the other days of the week, the seedlings were irrigated with distilled water to a level of 70 *w/w%* of field water capacity. The experiment was terminated at the fourth (Week 4, W4) and eighth weeks (Week 8, W8).

2.3. Microbiological Analysis

The following microbiological parameters were used to examine the granulated product, compost tea produced, and compost tea treated soils (Table 3).

Table 3. Evaluated microbiological indicators.

Measured Microbiological Indicators	Tested Samples	References
Fluorescein diacetate hydrolysis activity (FDA)	Granulated product	[52,53]
	Compost tea	
	Soils treated with compost tea	
Most probable number of microorganisms (aerobic bacteria: nutrient medium)	Granulated product	[54]
	Compost tea	
	Soils treated with compost tea	
β-Glucosidase enzyme activity (GlA)	Granulated product	[55]
	Compost tea	
	Soils treated with compost tea	
MALDI-TOF MS	Compost tea	[56]

Monitoring of the total microbial activity is a suitable method for measuring organic matter cycling, as more than 90% of the energy passes through microbial degraders. The FDA enzyme assay is one that shows how many microorganisms in the soil are engaged in life activity, i.e., degradative (catabolic) activity. The FDA test can be used to show the livingness of the soil and its actual functionality [51]. The β-glucosidase enzyme activity assay was chosen because it is generally positively correlated with soil organic matter content.

2.4. Identification of Microorganisms in Compost Tea

To identify the microorganisms in compost tea, a sterilized pelleted sample of a mixture of deep litter broiler manure, deep litter hen manure, and straw pellets (953.84 kg broiler manure, 476.92 kg chicken manure and 69.23 kg straw pellets) was used, which had been subjected to a 14-day composting process; 1:10 CWR, 48 h extraction time, 35 °C extraction temperature; the compost tea was prepared and filtered on filter paper (12–15 μm, VWR International, Debrecen, Hungary) prior to the tests and refrigerated at +4 °C until the tests. The identification procedure was performed once.

Single colonies from freshly grown isolates from given solid medium were picked in duplicate on a ground-steel target plate. Afterwards, 1μL formic acid solution was added to each spot, then 1 μL of matrix solution was pipetted onto each spot, and the plate was air dried at room temperature. Mass spectra were generated with a Microflex Biotyper (Bruker Daltonics, Billerica, MA, USA) using the standard settings. In case of each sample, mass fingerprints were acquired using flexControl version 3.0 software (Brucker Daltonics, Billerica, MA, USA), analyzed over a mass range from 2000 to 20,000 Dalton, and compared with the Bruker Daltonics database. This software generates a result list with score values suggesting the reliability of identification. The received score values are interpreted as unreliable identification when a score is lower than 1.7, as a probable genus identification when a score is between 1.7 and 1.99, and as a secure genus identification when a score is >2.0. The results were probable and highly probable species identification for 2.0–2.29 and ≥2.3, respectively [56].

2.5. Examination on the Pepper Test Plant

The pot experiments were terminated after 4 and 8 weeks, and the height of the plants (cm) was measured according to Slezák [57], while the total chlorophyll content (µg/g) was determined and calculated according to Szabó et al. [58].

2.6. Statistical Analysis

Statistical analyzes were performed using R software in an R Studio user environment (version 3.6.2.) [59]. The Shapiro–Wilk normality test was used to examine the distribution of the data, and then the type of test to be used for further analyzes was selected as a function of the distribution. To verify statistical differences between the different treatments, one-way analysis (Duncan-test) of variance was used at a $p < 0.05$ level of significance.

3. Results

3.1. Microbiological and Chemical Characteristics of Broiler and Hen Manure

The dry matter content (*w/w*%) of broiler manure (65–70 *w/w*%) and hen manure (63–67 *w/w*%) was high due to the breeding technology, as the dry matter content of the manure decreases at the end of the 6-week rotation for broiler manure. The pH of the manures was slightly alkaline to neutral (in the case of broiler manure the pH was 6.91–7.40, and in the case of hen manure the pH was 6.59–6.82), while the specific conductivity (11.10–12.78 mS/cm) and total nitrogen content (2.14–2.75 *w/w*%) were almost the same. The organic matter content, was higher in hen manure (66.18 *w/w*%) compared to broiler manure (58.81 *w/w*%). The difference in breeding technology also showed a higher biological activity for chicken manure.

In addition to the physical and chemical characteristics of the broiler and chicken manure, the microbiological characteristics were also investigated. The number of microorganisms in poultry manure is very high, up to 10^{10} CFU/g (Colony Forming Unit/g), and Gram-positive bacteria (*Actinomycetes* sp., *Bacillus* sp.) account for 90%. The presence of *Actinomycetes* sp. and *Lactobacillus* sp. is beneficial because they prevent the development of pathogens in the manure [60]. The aerobic bacteria in poultry manure were of the order of 10^9 CFU/g [61]. In the poultry manure, aerobic bacteria of the genus *Enterococcus* were predominant. The bacterial plate count of digestible aerobic bacteria in chicken manure was of the order of 10^8 CFU/g, while in broiler manure the aerobic bacteria count was three orders of magnitude, or a thousand times, less (10^5 CFU/g). In contrast, the most probable numbers of microorganisms in both the broiler manure and hen manure were of the same order of magnitude (10^3 CFU/g).

3.2. Microbiological Characteristics of the Evaluated Product and Compost Tea

Compost tea produced from granulated manure had a slightly acidic-neutral pH (pH 6.59 ± 0.06), which was similar to that of a mixture of broiler manure and hen manure. The electrical conductivity of the compost tea was high (14.82 ± 0.05 mS/cm), which can be explained by the high electrical conductivity of the two types of manure and the high content of water soluble salt in the manure. The nitrate concentration (1002.22 ± 40.55 mg/L) and the ammonium concentration (1426.67 ± 46.90 mg/L) in compost tea were high, indicating that the non-aerated system was not anaerobic, but oxidative. The high potassium concentration (1000.00 ± 223.61 mg/L) can be explained by the fact that potassium is highly soluble in aqueous media, yet the phosphate concentration is the lowest of the ions tested, which can be explained by the low phosphorus concentration in the initial compost and the low water-soluble phosphorus concentration (Table S1).

In general, the biological activity of the starting products in a solid form was higher than that of the extracted compost tea made from them (Table 4).

Table 4. Microbiological characteristics of evaluated product and compost tea.

Indicators	Evaluated Product	Compost Tea
FDA (µg Fl/g soil) *	16.74	12.41
GlA (PNP/µmol/g)	174.98	6.67
Culturable aerobic bacteria ($\log_{10}$CFU/g)	5.97	6.36

* µg Fl/g soil: µg Fluorescein/g soil.

The detectable values of the aerobic bacteria count were very high in the evaluated product, within which only one order of magnitude difference could be detected. Due to the magnitude of the variance, an order of magnitude (10-fold) difference also occurred within the treatment. However, the germ count values of aerobic bacteria were not proportional to the activities of any of the enzymes tested. The added bacteria did not correlate with β-glucosidase activity (G1A) or FDA, which indicates sugar metabolism, so they did not indicate active metabolism. However, their presence was confirmed by this method, which may suggest that they may be activated during subsequent use.

The biological activity of the starting products in solid form was higher than that of the extracted compost tea made from this. Non-aerated compost teas usually have no added nutrients other than the starting material. In these compost teas, there is a high probability that the solution will be low in oxygen or anoxic, anaerobic.

However, it was observed, that after 10-fold dilution, the difference in the FDA measurement did not become ten times smaller, as would be expected from the degree of dilution, but much higher values were obtained. FDA activity, which indicates degradable metabolism, is a good indication of the role of microbes in the breakdown of organic matter. However, the low values of GlA indicated that the sugar utilization capacity was not high or may have already been incorporated into the body weight of microorganisms engaged in catabolic activity. The numbers are expressed in logarithms based on 10, so the differences between the individual numbers in the germ count show $10–100\times$ differences between the samples according to the logarithm.

The preparation of compost tea was thus primarily conducive to the growth of bacteria, although the growth did not reach an order of magnitude of 0.5 compared to the original starting products. The aerobic bacterial count of the products and the solutions did not follow the previous trends, as in the present case compost tea soluble and relatively easy-to-absorb nutrients were revealed and ideal conditions created primarily for bacterial growth However, high aerobic bacterial germ counts do not correlation with enzyme activities.

On the other hand, the enzymes are specific and selectively show the activity of the participating microbes, so only a certain part of the microorganisms can be functional, depending on the enzyme measured. In addition, soils and compost are not only inhabited by bacteria, but also by so-called whole soil food web organisms. The soil–food web includes not only bacteria and fungi, but also nematodes, plant roots, algae, and some macroscopic animals. Microorganisms in the compost tea were identified using MALDI-TOF MS. This method provides efficient, rapid, and accurate results for the determination of protein mass spectra, and is thus useful for the identification of microorganism strains and the evaluation of their relatedness [56]. Nevertheless, this technology has not yet been used for the identification of microorganisms in compost tea. Based on this, our aim was to investigate how the agglomeration process affects the microbial composition and thus the microbiological properties of compost tea. The microbial communities of the compost tea were predominantly bacteria. The dominant bacterial genera were analyzed as *Bacillus* sp., *Lysinibacillus sp.*, and *Pseudomonas* sp. These identified bacterial genera were found in the range of 10^3–10^4 CFU/mL in the compost solution. The results show that the heat treatment in the production process significantly reduced the number of bacteria, which in result reduced the environmental risk of the manure.

3.3. Microbiological Characteristics of Soils Treated with Compost Tea

Among the enzymes, the activity of the fluorescein diacetate (FDA) enzyme was examined. In general, the slightly humous Arenosol showed (Figure 2.) higher values, as did the compost tea with which the treatments were performed. Outliers may also have been influenced by the water content of the soil. This is because the values of FDA activity can also be strongly influenced by abiotic environmental factors, such as the water and organic matter content of a given soil and other soil physical properties. Higher values were observed in the 8-week-old, older samples, indicating an increase in FDA enzyme activity with the age of the plants; in accordance with previous experience and literature data [59].

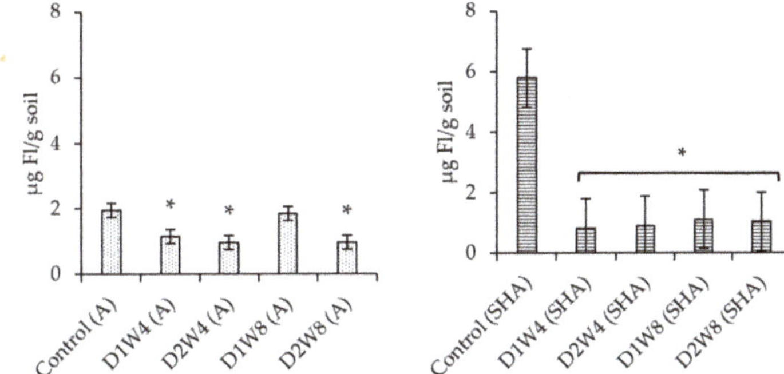

Figure 2. FDA enzyme activity of Arenosol and slightly humous Arenosol. The codes means the following treatments: Control (A): Control Arenosol. D1W4 (A): Dose 1 week 4 (Arenosol). D2W4 (A): Dose 2 week 4 (Arenosol). D1W8 (A): Dose 1 week 8 (Arenosol). D2W8 (A): Dose 2 week 8 (Arenosol). Control (SHA): Control slightly humous Arenosol. D1W4 (SHA): Dose 1 week 4 (slightly humous Arenosol). D2W4 (SHA): Dose 2 week 4 (slightly humous Arenosol). D1W8 (SHA): Dose 1 week 8 (slightly humous Arenosol). D2W8 (SHA): Dose 2 week 8 (slightly humous Arenosol). * Indicates significant difference at $p < 0.05$ (calculated by Duncan-test) between control and compost tea treated soils.

Lower enzyme activity values could be detected in the treated soil without plants, as plants increase enzyme activities. When mixing compost tea into the soil, it was also observed that dilution of compost products with high enzyme activity occurred. A significantly lower microbial activity could be detected in the Arenosol than in the slightly humous Arenosol soil. However, when the values of the two soils were averaged from a plant point of view, the peppers increased the FDA enzyme activity values. There was less activity in compost tea treated soils.

The level of FDA activity was found to be uniform in the soils at weeks 4 and 8 of the growing season, although after week 8 the activity tended to be lower. In the control, the value tended to be higher, but due to the large standard deviation, we could not detect an appreciable difference between the control and treated soils. However, the activity of the original granulated product was significantly higher than these.

It can be stated that the activity of the two soils was similar, although significantly larger standard deviations were obtained in the Arenosol compared to the slightly humous Arenosol soil (Figure 3). With the treatment of the product, it was necessary to increase the intake of organic matter in the Arenosol, and thus increase the activity. However, increasing the product dose did not result in a proportionally higher FDA activity. The values of the samples taken after 8 weeks tended to be higher, which is proportional to the age and growth of the plant, but this could also be detected only as a tendency, it did not prove to be significant, due to the large standard deviations. FDA activity is a good indicator of the microbiological and degradative activity of soils. Studies have demonstrated the high activity value found in original compost. This became lower immediately after mixing into the soil, as a result of the obvious dilution. This was not significantly increased by doubling the dose. However, it can also be stated that the activity could be increased to sand-like values in Arenosol, which were significantly poorer in organic matter than before. The pepper test plants also increased the FDA values of the soils at the same level, which tended to improve with the physiological condition of the plants with age. A similar conclusion can be drawn for the enzyme glucosidase as for the examination of FDA enzyme activity.

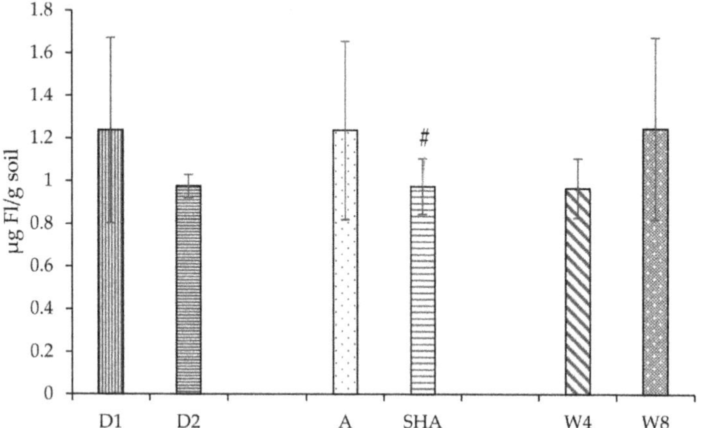

Figure 3. Summary of FDA enzyme activity results by treatments (doses), soil types, sampling weeks. The codes are as follows: D1: Dose 1. D2: Dose 2. A: Arenosol. SHA: slightly humous Arenosol. W4: Week 4. W8: Week 8. # indicates significant difference at $p < 0.05$ (calculated by Duncan-test) between Arenosol and slightly humous Arenosol soils.

In the study of glucosidase enzyme, the sand soil values improved the most between the two soils, as organic matter supplementation had a greater effect on increasing the initial activity in Arenosol (Figures 4 and 5).

The untreated activity was markedly increased by the pepper, but the measured enzyme activity was balanced. β-glucosidase activity values also increased with the age of the plants, this dose-effect of the product was only proven in D2.

The number of aerobic bacteria also supports the above mentioned observations (Figure 6). Even in the case of the control soil, it can be seen that the values of the Arenosol are 0.5–1 orders of magnitude lower compared to the slightly humous Arenosol, which is not surprising in the case of Arenosol. Here it is more noticeable than the number of aerobic bacteria increased after the application of compost tea, and this was also confirmed by previous enzyme studies.

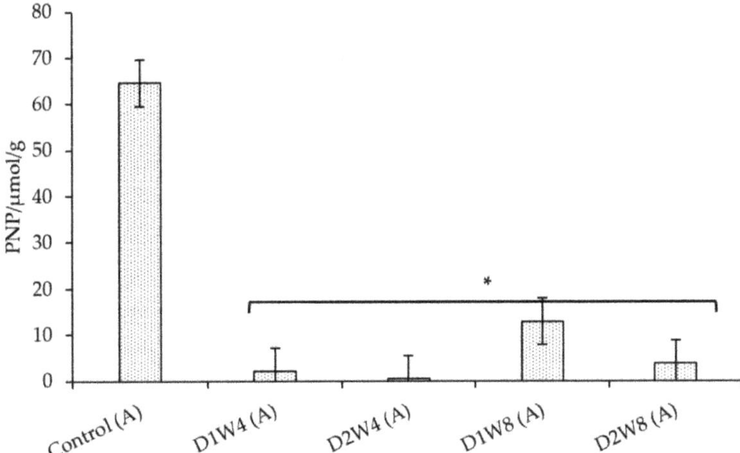

Figure 4. Glucosidase enzyme activity (GlA) of Arenosol. The codes mean the following treatments: Control (A): Control Arenosol. D1W4 (A): Dose 1 week 4 (Arenosol). D2W4 (A): Dose 2 week 4 (Arenosol). D1W8 (A): Dose 1 week 8 (Arenosol). D2W8 (A): Dose 2 week 8 (Arenosol). * indicates significant difference at $p < 0.05$ (calculated by Duncan-test) between control and compost tea treated soils.

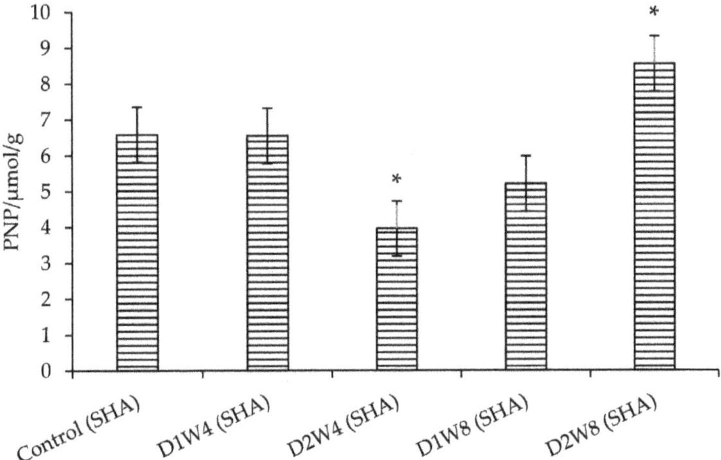

Figure 5. Glucosidase enzyme activity (GlA) of slightly humous Arenosol control (SHA): Control slightly humous Arenosol. D1W4 (SHA): Dose 1 week 4 (slightly humous Arenosol). D2W4 (SHA): Dose 2 week 4 (slightly humous Arenosol). D1W8 (SHA): Dose 1 week 8 (slightly humous Arenosol). D2W8 (SHA): Dose 2 week 8 (slightly humous Arenosol). * indicates significant difference at $p < 0.05$ (calculated by Duncan-test) between control and compost tea treated soils.

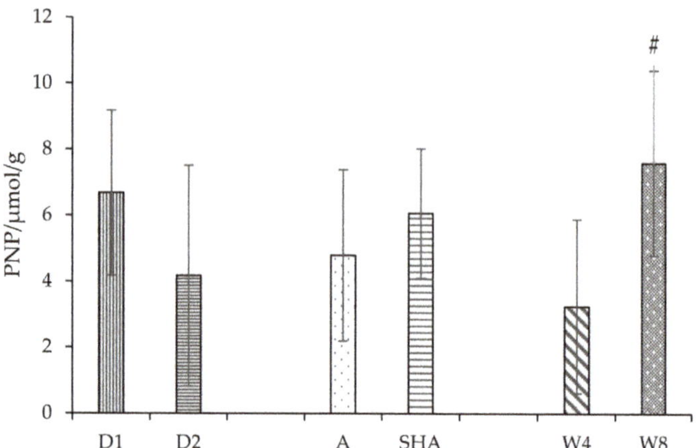

Figure 6. Summary of GlA enzyme activity results by treatments (doses), soil types, and sampling weeks. The codes are as follows: D1: Dose 1. D2: Dose 2. A: Arenosol. SHA: slightly humous Arenosol. W4: Week 4. W8: Week 8. # indicates significant difference at $p < 0.05$ (calculated by Duncan-test) between W4 and W8.

It can be stated that the values of the aerobic bacterial count became higher as a result of the treatment (Figure 7). In the treated soil, the effect of the treatment also showed improvements (Figure 8). However, peppers were also able to raise germ count values. Similarly, no single or double dose of compost tea caused an increase in germ count, although the detectable values could rise to the same level as the compost tea.

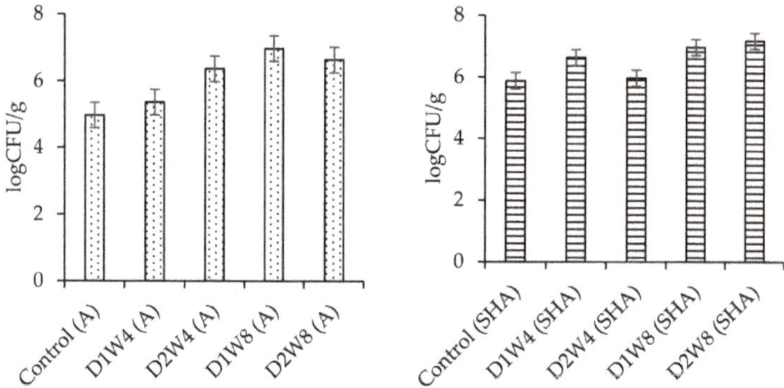

Figure 7. Number of aerobic bacteria, number (log CFU/g) of Arenosol and slightly humous Arenosol. The codes means the following treatments: Control (A): Control Arenosol. D1W4 (A): Dose 1 week 4 (Arenosol). D2W4 (A): Dose 2 week 4 (Arenosol). D1W8 (A): Dose 1 week 8 (Arenosol). D2W8 (A): Dose 2 week 8 (Arenosol). Control (SHA): Control slightly humous Arenosol. D1W4 (SHA): Dose 1 week 4 (slightly humous Arenosol). D2W4 (SHA): Dose 2 week 4 (slightly humous Arenosol). D1W8 (SHA): Dose 1 week 8 (slightly humous Arenosol). D2W8 (SHA): Dose 2 week 8 (slightly humous Arenosol).

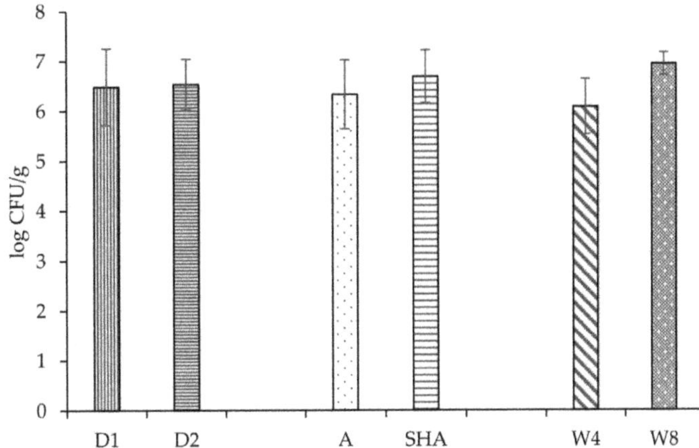

Figure 8. Summary of aerobic bacteria number (log CFU/g) results by treatments (Doses), soil types, and sampling weeks. The codes are as follows: D1: Dose 1. D2: Dose 2. A: Arenosol. SHA: slightly humous Arenosol. W4: Week 4. W8: Week 8.

The germ counts that could be cultured can be compared in a manner proportionate to the results of the enzyme assay, especially with FDA activity. Examination of germ counts showed that the number of aerobic microorganisms increased with compost tea treatment. The peppers were able to utilize the product and the values also improved with the age of the plant. At the application dose, even the first dose caused an increase, the double dose did not give a better result.

3.4. Effect of Compost Tea on the Pepper Test Plant

In addition to the microbiological measurements, the effect of compost tea on the pepper was determined by measuring plant shoot length (cm) and total chlorophyll content (µg/g).

At the fourth week for D2 of Arenosol soil, the longest average plant shoot length (41.00 ± 2.65 cm) was measured (Figure 9). This treatment was significantly different from the control when the fourth week of treatment is considered. At W8, the pepper shoot length varied between 37.67 ± 2.89 cm and 40.33 ± 2.08 cm. D2W4, D1W8, and D2W8 were statistically in the same group, indicating that the eighth week of treatment did not result in significantly longer shoot length. There was no significant difference in plant shoot length between treatments (Table S2).

In Arenosol soil (Figure 10), the total chlorophyll content of plants increased as a result of the treatments compared to the control.

For the fourth week of treatments, the D2W4 had the highest total chlorophyll content (4135.49 ± 344.17 µg/g). This was the only treatment that was significantly different from the control and the other treatments in both the fourth and eighth week. The lowest total chlorophyll content was measured for D1W8 (3028.96 ± 356.25 µg/g). All treatments except D2W4 were in the same statistic group as the control (Table S3). In W8 of measurements, the total chlorophyll content of the treated plants was lower than in the W4.

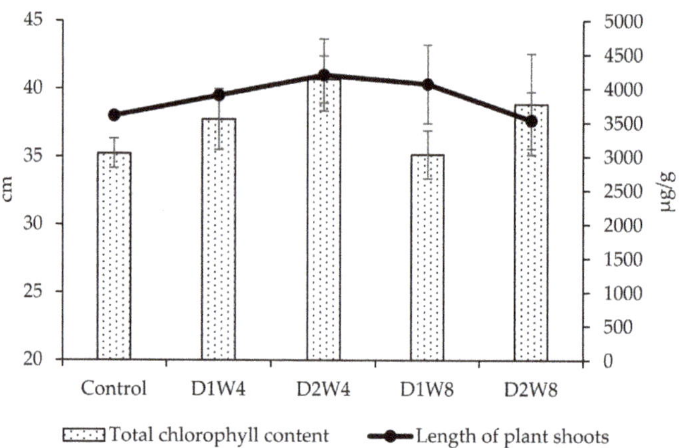

Figure 9. Changes in total chlorophyll content and plant shoot length of Arenosol soil. The codes are as follows: D1W4: Dose 1 Week 4. D2W4: Dose 2 Week 4. D1W8: Dose 1 Week 8. D2W8: Dose 2 Week 8.

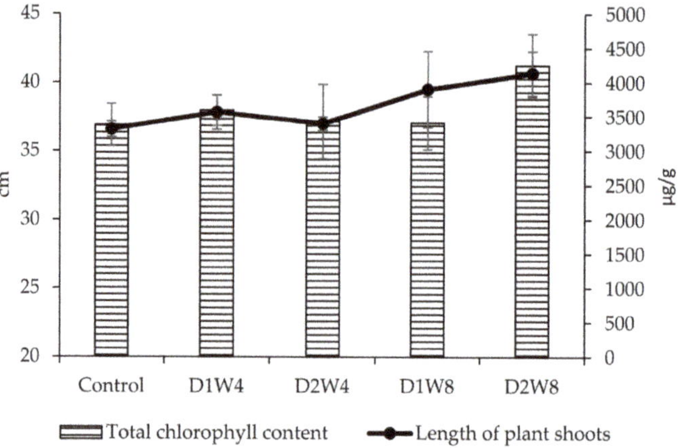

Figure 10. Changes in total chlorophyll content and plant shoot length of slightly humous Arenosol soil. The codes are as follows: D1W4: Dose 1 Week 4. D2W4: Dose 2 Week 4. D1W8: Dose 1 Week 8. D2W8: Dose 2 Week 8.

On in the slightly humous Arenosol did the length of the pepper shoots also increase with increasing dose in the fourth week (Figure 10). The fourth week treatments were significantly different from the control, but the treatments were statistically in the same group, with no significant difference between D1 and D2 (Table S3). Similar results were obtained for the eighth week treatment, as the treatments were statistically in the same group, but there was a significant difference between the fourth and eighth week treatments. The longest shoot length was measured for D2W8 (40.66 ± 1.61 cm), and the lowest shoot length was measured for the control (36.58 ± 0.58 cm). These results demonstrate the positive effect of compost tea on the pepper.

In the slightly humous Arenosol (Figure 10), there was no increase in total chlorophyll content compared to the control as a result of the treatments (Table S3). The fourth week treatments formed a statistical group with the control, no significant increase in total chlorophyll content was detected between these treatments. The eighth week treatments were significantly different from each other and from the fourth week treatments. The highest total chlorophyll content was measured for D2W8 (4249.62 ± 458.76 µg/g), and the lowest total chlorophyll content was measured for the control (3379.17 ± 310.64 µg/g). These results demonstrate that the compost solution applied on slightly humous Arenosol soil had a positive effect on the total chlorophyll content of the pepper.

In the Arenosol soil, there was no significant difference compared to the control (Figure S1), but in the slightly humous Arenosol, the effect of the treatments was statistically proven by W8 (Figure S2).

4. Discussion

In the case of industrial organic matter management technologies, less attention is paid to microbiology, and the physical, chemical, and engineering approach predominates. However, when using organic materials and fertilizers, the whole process must be considered, so it is important to examine the microbiology of the raw materials and the products made from them. The starting point for our investigations was that compost tea made from materials from conventional technology had been investigated before [62–65], but compost tea made from industrial compost had previoulsy not been prepared, investigated, and tested

The bacterial plate counts of cultivable aerobic bacteria in hen manure were of the order of 10^8 g, while in broiler manure the aerobic bacteria count was three orders of magnitude, or a thousand times, less (10^5 CFU/g). In contrast, the most probable number of microorganisms in both broiler manure and hen manure were of the same order of magnitude (10^3 CFU/g). Chen-Jiang [58], investigated the microbiological composition of organic nutrient supplements in poultry manure and poultry manure-based organic nutrient supplements. Their results showed that the concentration of microorganisms was 10^{10} CFU/g, and 90% of the microbial community was composed of Gram-positive bacteria, e.g., *Clostridia, Bacilli, Lactobacilli*.

Based on our results, the biological activity of the starting products was found to be higher (FDA: +4.33 µg/g, GlA: +168.31 PNP/µmol/g) than that of the compost tea made from them. The aerobic bacterial counts of the products and compost tea did not follow these trends, since in this case the solution had a higher value compared to the product. The most probable reason for this is that with compost tea, relatively easily absorbed nutrients were revealed and ideal conditions were created for bacterial growth in particular. The low oxygen levels in non-aerated compost tea, which results in anaerobic conditions, can lead to aerobic microorganisms becoming inactive and anaerobic microbes multiplying. Short anaerobic periods can increase the diversity if aerobic organisms do not die or become inactive [36]. Long anaerobic conditions mean that many organisms become inactive or die, and nutrients are lost. However, microbial nutrients added to solutions should be used with caution, as studies [43] have shown that the addition of molasses or other simple sugars to compost tea can lead to the growth of *Escherichia coli*, *Salmonella*, and *Listeria*. Compost solutions are dominated by bacteria: aerated compost solutions are dominated by aerobic bacteria, while non-aerated compost tea are dominated by facultative anaerobic bacteria. The compost used should be stable and pathogen-free [31,32]. A previous study [66] showed that nitrogen-rich feedstocks, such as composts containing manure, result in compost tea with a higher bacterial content. Kim et al. [39] also reported that the microbial communities in different compost teas were predominantly bacteria [67]. Aerated compost tea was characterized by a predominance of aerobic bacteria, yet non-aerated compost teas were dominated by facultative anaerobic organisms [31]. The dominant bacterial phylums detected in the non-aerated compost tea were *Actinobacteria, Bacteroidetes, Firmicutes, Proteobacteria, Verrucomicrobia, Chloroflexi, Planctomycetes,* and *Acidobacteria* [68]. In studies by González-Hernández et al. [68], the most abundant group of microorganisms in compost tea was total aerobic bacteria (2.0×10^7 CFU/mL), followed by *N*-fixing bacteria (1.4×10^5 CFU/mL) and *Actinobacteria* (7.4×10^4 CFU/mL).

For products containing living microorganisms, it is important to preserve the survival of microbes for as long as possible, at least until the actual use. It is necessary to start from the highest possible cell counts in order to maintain an adequate and efficient cell count in the compost tea despite the destruction caused by environmental stress factors. The product under test was formulated in a dry state or with a very low water content, and therefore requires micro-organisms that can survive in a 'dormant', inactive state and that will revive after application when the environmental conditions (temperature, water content) are ideal for their functionality.

Several studies have been conducted on the microbial characterization of compost tea [48,69], using 16S rDNA sequencing [70], plate count, or MPN [71] methods to determine the density of cultivable bacteria found in compost tea. The MALDI-TOF method used to identify microorganisms in compost tea is considered a new technology. Based on our results, the MALDI-TOF method may be a good alternative for species-level identification of microorganisms in compost teas. The results obtained were in agreement with the literature, as previous studies showed that the following groups of microorganisms were found in compost tea: *Bacillus, Pseudomonas, Micrococcus, Staphylococcus, Clavibacter, Lactobacillus*, and other bacterial species [71], as well as *Actinomycetes*, yeasts, and *Trichoderma* sp., *Aspergillus* sp., *Penicillium* sp., and other filamentous fungi species [71]. With the MALDI-TOF MS method we can identify not only beneficial, but also pathogenic, microorganisms at the species level; but in our case this was not a problem, because the heat treatment (125–130 °C) applied during composting and granulation significantly reduced the potential risk.

FDA activity is a good indicator of the microbiological and degradative activity of soils. The FDA activity values can also be strongly influenced by abiotic environmental factors, such as the water and organic matter content of the soil and other soil physical properties. Measurements demonstrated a high activity value (16.74 µg/g) in the granulated product. This was lowered immediately after mixing into the soil as a result of apparent dilution. This value was not significantly increased by doubling the dose. Although it was possible to increase the activity in Arenosol, which was originally much poorer in organic matter, to values similar to those in slightly humous Arenosol. Furthermore, the pepper test plant increased the FDA values of the soils to the same level, which also tended to improve with the age of the plant physiologically. Similar results were obtained by Elbl et al. [72], who found that compost applied at higher doses had a positive effect on FDA growth (+95% increase compared to the control). Furthermore, according to Komilis et al. [73], the hydrolysis of FDA can be used as an indicator of microbial activity in relation to the state of the soil environment. Based on their results, they concluded that the application of organic fertilizers has positive effects on soil enzyme activity. Tian et al. [74] reported that FDA is widely accepted as an accurate and simple method for determining total soil microbial activity, and observed a direct effect of organic matter application in the form of compost on the increase of soil microbial activity.

The decreasing trend of the FDA is probably explained by the fact that only a single snapshot was available during the tests, and it is possible that the enzyme peak in the compost tea had already occurred long before and was already low at the time of measurement, while in the untreated soil it occurred later. This faster peak may have occurred because the compost tea has a rapid mineralization process and the microbes multiply rapidly and then undergo a rapid death; hence the low FDA enzyme activity at that time. Alidadi et al. [75] obtained similar results in their study of dehydrogenase enzyme activity, as the enzyme activity decreased from day 75 of composting (maturation phase). Lazcano et al. [76] found that a high activity level of microorganisms is due to the high amount of water-soluble carbon in their starting substrates. The stabilization of dehydrogenase activity is attributed to the complete degradation of available organic matter [77]. This point therefore represents the maturation time of the compost. Based on these results, in our case, the readily available nutrients in the compost tea were taken up by the plant, leaving no available nutrients for the microorganisms in the soil.

For the β-glucosidase enzyme, sandy soil showed the greatest improvement of the two soil types, as organic matter supplementation had a greater effect on increasing the initial activity in sand. The glucosidase activity without treatment was greatly increased by pepper (+4.35 PNP/μmol/g), but the enzyme activity was balanced by treatment. The glucosidase activity values also increased with plant age, but the dose effect remains unproven. Results from Vinhal-Freitas et al. [78] showed that the decrease in β-glucosidase enzyme activity was significantly influenced by mixing compost into the soil and that the enzyme activity was higher (+10–15 PNP/μmol/g) with higher (20 g/kg) compost dosages. An increase in β-glucosidase activity after application of compost consisting of municipal solid residues was reported by several authors [79,80]. The detection of the enzyme β-glucosidase is related to the breakdown of cellulose synthesized by fungi, bacteria, and other soil organisms. Compost, however, is a stable organic waste (compared to, for example, uncomposted residues) that provides more resistant C compounds [81] and which is hydrolyzed more slowly by the enzymes.

Aerobic bacterial counts are comparable to enzyme assay results, especially FDA activity. Our results showed that the number of aerobic microorganisms increased with the effect of compost tea treatment. The peppers were able to utilize the applied compost tea in a sonic manner and the values improved with the age of the plant. At the application dose, the first dose already caused an increase (+0.15 log CFU/g), the double dose did not give better results, as was expected. It can be concluded that the activity of the two soils was similar, however, the results showed that it was the sandy soil that needed to be treated with the product to increase the organic matter input, and thus the activity. Bacteria such as *Enterobacteria* sp., *Nitrobacter* sp., *Pseudomonads* sp., *Bacillus* sp., *Staphylococcus* sp. and various *Actinomycetes* sp., as well as fungi such as *Trichoderma* sp. have been isolated from properly matured composts [82]. Subgroups of these species, known as "facultative anaerobes", live in low oxygen environments but can also grow under aerobic conditions.

The results of Sifatullah et al. [83] showed that anaerobic tea had a higher bacterial count (4×10^{10}, 4.2×10^{10}, 4.3×10^{10} logCFU/g) than aerated compost tea. The observed high microbial counts were due to the closed container anaerobic compost tea, which may be useful for disease control.

Considering the fertilizing effect of compost tea, Wang et al. [84] reported there was no effect on the yield of zucchini using a chicken manure based vermicompost tea (1:10, *w/v*, continuously aerated for 12 h) as fertilizer. Moreover, Hewidy et al. [85] showed no effect on broccoli yield of a compost tea obtained from the organic fraction of municipal solid wastes and pruning residues (1:5, *w/v*, continuously aerated for 24 h).

The electrical conductivity of the compost tea applied to the peppers was high (14.82 mS/cm), which resulted in a concentrated soil solution, and which may also cause yellowing of leaf edges. This may be the reason for the increase in total chlorophyll content at the four-week eradication period, whereas the eight-week eradication period was characterized by a decrease in chlorophyll content. Nitrogen deficiencies may also have developed during treatments, which may have resulted in chlorotic yellowing [86]. In the Arenosol soil, there was no significant differences compared to the control, but in the slightly humous Arenosol, the effect of the treatments was statistically proved by W8. Similar results for total chlorophyll content were reported by Pérez-López et al. [87], who observed that the use of composted manure increased the total chlorophyll content of sweet peppers (*Capsicum annuum* L.).

In this case study we evaluated a new technological approach for poultry farming organic waste. We investigated the effect of a compost tea made from a mixture of poultry manure and hen manure on a test pepper (*Capsicum annuum* L.) plant grown in a pot experiment. We investigated the potential microbiological indicators of the new utilization process and how the microbiological indicators and the mechanism of action of compost tea can be interpreted in the case of the indicator plant.

Overall, compost solutions can be used to keep organic fertilizers in circulation, and thus products suitable for organic nutrient replenishment can be used instead of fertilizers. One of the main objectives of the European Union's Green Deal [88] is to promote the efficient use of resources through the transition to a clean, circular economy, and one of the ways of achieving this in agricultural practice is to develop products based on organic fertilizers.

Supplementary Materials: The following are available online at https://www.mdpi.com/article/10.3390/agriculture11070683/s1, Table S1: Characteristics of evaluated product base materials (Broiler and hen manure), Table S2: Changes of plant shoot length (cm) and total chlorophyll content (μg/g) in Arenosol soil, week 4 and week 8 of the treatments ($p < 0.05$)*, Table S3: Changes of plant shoot length (cm) and total chlorophyll content (μg/g) in slightly humous Arenosol soil, week 4 and week 8 of the treatments ($p < 0.05$)*, Figure S1: Summary of total shoot length (cm) results by treatments (Doses), soil types, sampling weeks. The codes are as follows: D1: Dose 1. D2: Dose 2. A: Arenosol. SHA: slightly humous Arenosol. W4: Week 4. W8: Week 8, Figure S2. Summary of total chlorophyll content (μg/g) results by treatments (Doses), soil types, sampling weeks. The codes are as follows: D1: Dose 1. D2: Dose 2. A: Arenosol. SHA: Slightly humous Arenosol. W4: Week 4. W8: Week 8.

Author Contributions: Conceptualization, J.T. and B.B.; methodology, E.G., L.R.K. and S.A.P.; investigation, E.G., N.É.K. and F.A.T.; writing—original draft preparation, E.G., S.A.P. and I.B. All authors have read and agreed to the published version of the manuscript.

Funding: The APC was funded by EFOP-3.6.3-VEKOP-16-2017-00008 project. The project is cofinanced by the European Union and the European Social Fund.

Institutional Review Board Statement: Not applicable.

Data Availability Statement: The datasets generated and/or analyzed during the current study are available from the corresponding author on reasonable request.

Acknowledgments: This research was supported by EU grant to Hungary GINOP 2.2.1.-15-2017-00043. The publication is supported by the EFOP-3.6.3-VEKOP-16-2017-00008 project. The project is co-financed by the European Union and the European Social Fund.

Conflicts of Interest: The authors declare that they have no conflict of interest.

References

1. Purwanto, B.H.; Alam, S. Impact of intensive agricultural management of carbon and nitrogen dynamics in the humid tropics. *Soil Sci. Plant Nutr.* **2019**, *66*, 50–59. [CrossRef]
2. Juhos, K.; Madarász, B.; Kotroczó, Z.; Béni, Á.; Makádi, M.; Fekete, I. Carbon sequestration of forest soils is reflected by changes in physicochemical soil indicators–A comprehensive discussion of a long-term experiment on a detritus manipulation. *Geoderma* **2021**, *385*, 114918. [CrossRef]
3. Kopittke, P.M.; Menzies, N.W.; Wang, P.; McKenna, B.A.; Lombi, E. Soil and the intensification of agriculture for global food security. *Environ. Int.* **2019**, *132*, 105078. [CrossRef] [PubMed]
4. Madarász, B.; Jakab, G.; Szalai, Z.; Juhos, K.; Kotroczó, Z.; Tóth, A.; Ladányi, M. Long-term effects of conservation tillage on soil erosion in Central Europe: A random forest-based approach. *Soil Tillage Res.* **2021**, *209*, 104959. [CrossRef]
5. Livi Bacci, M. *A Concise History of World Population*, 6th ed.; Wiley/Blackwell: Hoboken, NJ, USA, 2017.
6. Garg, M.R. Balanced feeding for improving livestock productivity–Increase in milk production and nutrient use efficiency and decrease in methane emission. In *FAO Animal Production and Health Paper*; Makkar, H.P.S., Ed.; FAO: Rome, Italy, 2012; pp. 1–30. Available online: http://www.fao.org/docrep/016/i3014e/i3014e00.pdf (accessed on 26 June 2021).
7. MacLeod, M.; Gerber, P.; Mottet, A.; Tempio, G.; Falcucci, A.; Opio, C.; Vellinga, T.; Henderson, B.; Steinfeld, H. *Greenhouse Gas Emissions from Pig and Chicken Supply Chains–A Global Life Cycle Assessment*; Food and Agriculture Organization of the United Nations (FAO): Rome, Italy, 2012.
8. Narrod, C.; Tiongco, M.; Costales, A. Global poultry sector trends and external drivers of structural change. In *FAO Animal Production and Health Proceedings, Proceedings of the International Poultry Conference on Poultry in the 21st century: Avian influenza and beyond, Bangkok, Thailand, 5–7 November 2007*; Thieme, O., Pilling, O., Eds.; FAO: Rome, Italy, 2008.
9. Kumal, A.; Patyal, A. Impacts of intensive poultry farming on 'one health' in developing countries: Challenges and remedies. *Explor. Anim. Med Res.* **2020**, *10*, 100–111.
10. Augère-Granier, M.L. The EU Poultry Meat and Egg Sector. Main Features, Challenges and Prospects. European Parliamentary Research Service. European Union. 2019. Available online: https://www.europarl.europa.eu/RegData/etudes/IDAN/2019/644195/EPRS_IDA(2019)644195_EN.pdf (accessed on 26 June 2021).

11. Agribusiness Consulting. Intensive Broiler Farming in the EU: Impact on the Environment, Human Health and Animal Welfare; Final report; 2018. Available online: https://www.afisapr.org.br/attachments/article/1377/Envi-impact-of-Intensive-broiler-farming-FR.pdf (accessed on 26 June 2021).

12. CIWF. The Life of: Broiler Chickens. 2013. Available online: https://www.ciwf.org.uk/media/5235306/The-lifeof-Broiler-chickens.pdf (accessed on 26 June 2021).

13. Sobsey, M.D.; Khatib, L.A.; Hill, V.R.; Alocilja, E.; Pillai, S. Pathogens in animal wastes and the impacts of waste management practices on their survival, transport and fate. In *Animal Agriculture and the Environment: National Center for Manure and Animal Waste Management White Papers*, 1st ed.; ASABE: St. Joseph, MI, USA, 2006; pp. 609–666. Available online: https://fyi.extension.wisc.edu/manureirrigation/files/2014/03/ASABE_2006_Pathogens-in-Animal-Wastes-and-Impacts-of-Waste-Management-Practices.pdf (accessed on 26 June 2021).

14. Peyraud, J.L.; MacLeod, M. Future of EU Livestock. How to Contribute to a Sustainable Agricultural Sector? Final report. 2020. Available online: https://op.europa.eu/en/publication-detail/-/publication/b10852e8-0c33-11eb-bc07-01aa75ed71a1/language-en# (accessed on 26 June 2021).

15. Herman Ottó Institute. *Guidance for Defining Best Available Techniques for the Certification of Intensive Poultry Production*, 1st ed.; Ministry of Rural Development: Budapest, Hungary, 2020; p. 155.

16. Amanullah, M.M.; Sekar, S.; Muthukrishnan, P. Prospects and Potential of Poultry Manure. *Asian J. Plant Sci.* **2010**, *9*, 172–182.

17. Abdel-Shafy, H.I.; Mansour, M.S.M. Solid waste issue: Sources, composition, disposal, recycling and valorization. *Egypt. J. Pet.* **2018**, *27*, 1275–1290. [CrossRef]

18. Directive (EU) 2018/850 of the European Parliament and of the Council of 30 May 2018 Amending Directive 1999/31/EC on the landfill of Waste (Text with EEA Relevance). Available online: https://eur-lex.europa.eu/legal-content/EN/TXT/PDF/?uri=CELEX:32018L0850&from=EN (accessed on 26 June 2021).

19. Wang, Q.; Awasthi, M.K.; Zhang, Z.; Wong, J.W.C. Sustainable Composting and Its Environmental Implications. *Sustain. Resour. Recovery Zero Waste Approaches* **2019**, 115–132.

20. Pabar, S.A.; Mónok, D.; Kotroczó, Z.; Biró, B. Soil microbial parameters and synergies between bean growth and microbial inoculums as a dependence of five soils with different characteristics. *Hung. Agric. Eng.* **2020**, *37*, 27–33.

21. Onwosi, C.O.; Igbokwe, V.C.; Odimba, J.N.; Eke, I.E.; Nwankwoala, M.O.; Iroh, I.N.; Ezeogu, L.I. Composting technology in waste stabilization: On the methods, challenges and future prospects. *J. Environ. Manag.* **2017**, *190*, 140–157. [CrossRef]

22. Bernal, M.P.; Alburquerque, J.A.; Moral, R. Composting of animal manures and chemical criteria for compost maturity assessment. A review. *Bioresour. Technol.* **2019**, *100*, 5444–5453. [CrossRef]

23. Brake, J.D. A Practical Guide for Composting Poultry Litter. *MAFES Bull.* **1992**, *981*, 265.

24. Haga, K. Development of composting technology in animal waste treatment–Review. *Asian Aust. J. Anim. Sci.* **1999**, *12*, 604–606. [CrossRef]

25. Gilroyed, B.; Hao, X.; Larney, F.J.; McAllister, T.A. Greenhouse gas emissions from cattle feedlot manure composting and anaerobic digestion as a potential mitigation strategy. In *Understanding Greenhouse Gas Emissions from Agricultural Management 1072*; Guo, L., Gunasekara, A.S., McConnell, L.L., Eds.; American Chemical Society: Washington, DC, USA, 2012; pp. 419–441.

26. Tiquia, S.M.; Tam, N.F.Y. Elimination of phytotoxicity during co-composting of spent pig-manure sawdust litter and pig sludge. *Bioresour. Technol.* **1998**, *65*, 43–49. [CrossRef]

27. Michel, F.C.; Forney, L.J.; Huang, A.J.F.; Drew, S.; Czu, P.M.; Lindeberg, J.D.; Reddy, C.A. Effects of tuning frequency, leaves to grass mix ratio and windrow vs pile configuration ont he composting of yard trimmings. *Compos. Sci. Util.* **1996**, *4*, 26–43. [CrossRef]

28. Scotti, R.; Pane, C.; Spanicci, R.; Palese, A.M.; Piccolo, A.; Celano, G.; Zaccardelli, M. On-farm compost: A useful tool to improve soil quality under intensive farming systems. *Appl. Soil Ecol.* **2016**, *107*, 13–23. [CrossRef]

29. Simon, L.; Tamás, J.; Kovács, E.; Kovács, B.; Biró, B. Stabilisation of metals in mine spoil with amendments and growth of red fescue in symbiosis with mycorrhizal fungi. *Plant Soil Environ.* **2006**, *52*, 385–391. [CrossRef]

30. Pant, A.P.; Radovich, T.J.K.; Hue, N.V.; Talcott, S.T.; Krenek, K.A. Vermicompost extracts influence growth, mineral nutrients, phytonutrients and antioxidant activity in pak choi (*Brassica rapa* cv. *Bonsai*, Chinensis group) grown under vermicompost and chemical fertilizer. *J. Sci. Food Agric.* **2009**, *89*, 2383–2392. [CrossRef]

31. Ingham, E.R. *The Compost Tea Brewing Manual*; Soil Foodweb, Incorporated: Corvallis, OR, USA, 2005.

32. Ingham, E.R. *The Compost Tea Brewing Manual*; Soil Foodweb, Incorporated: Corvallis, OR, USA, 2000.

33. Hargraves, J.C.; Adl, M.S.; Warman, P.R. Are compost teas an effective nutrient amendment in the cultivation of strawberries? Soil and plant tissue effects. *J. Sci. Food Agric.* **2009**, *89*, 390–397. [CrossRef]

34. Zaccardelli, M.; Pane, C.; Villecco, D.; Palese, A.M.; Celano, G. Compost tea spraying increases yield performance of pepper (*Capsicum annuum* L.) grown in greenhouse under organic farming system. *Ital. J. Agron.* **2018**, 229–234. [CrossRef]

35. Shaban, H.; Fazeli-Nasab, B.; Alahyari, H.; Alizadeh, G.; Shahpesandi, S. An owerview of the benefits of compost tea on plant and soil structure. *Adv. Bioresearch* **2015**, *6*, 154–158.

36. Safwat, M.S.; Badran, F.S. Efficiency of organic and bio-fertilizers in comparison with chemical fertilization on growth yield and essential oil of cumin plants. In Proceedings of the 9th Conference of Medicinal and Aromatic Plants, Cairo, Egypt, 3–5 October 2002.

37. Azzaz, N.A.; Hassan, E.A.; Hamad, E.H. The chemical constituent and vegetative and yielding characteristics of Fennel plants treated with organic and bio-fertilizer instead of mineral fertilizer. *Aust. J. Basic Appl. Sci.* **2009**, *3*, 579–587.

38. Khalid, K.; Hendawy, S.F.; El-Gezawy, E. Ocimum basilicum L. production under organic farming. *Res. J. Agric. Biol. Sci.* **2006**, *2*, 25–32.

39. Kim, M.J.; Shim, C.K.; Kim, Y.K.; Hong, S.J.; Park, J.H.; Han, E.J.; Kim, J.H.; Kim, S.C. Effect of aerated compost tea on the growth promotion of lettuce, soybean, and sweet corn in organic cultivation. *Plant Pathol. J.* **2015**, *31*, 259–268. [CrossRef] [PubMed]

40. Morales-Corts, M.R.; Pérez-Sánchez, R.; Gómez-Sánchez, M.Á. Efficiency of garden waste compost teas on tomato growth and its suppressiveness against soilborne pathogens. *Sci. Agric.* **2018**, *75*, 400–409. [CrossRef]

41. Scheuerell, S.J.; Mahaffee, W.F. Compost tea: Principles and prospects for plant disease control. *Compos. Sci. Util.* **2002**, *10*, 313–338. [CrossRef]

42. Scheuerell, S.J.; Mahaffee, W.F. Compost tea as a container medium drench for suppressing seedling damping-off caused by Pythiumultimum. *Phytopathology* **2004**, *94*, 1156–1163. [CrossRef] [PubMed]

43. Ingham, D.T.; Millner, P.D. Factors Affecting Compost Tea as a Potential Source of Escherichia coli and Salmonella on Fresh Produce. *J. Food Prot.* **2007**, *70*, 828–834. [CrossRef]

44. Scheuerell, S.J.; Mahaffee, W.F. Variability associated with suppression of gray mold (*Botrytis cinerea*) on geranium by foliar applications of nonaerated compost teas. *Plant Dis.* **2006**, *90*, 1201–1208. [CrossRef] [PubMed]

45. Georgakakis, D.; Krintas, T. Optimal use of the Hosoya system in composting poultry manure. *Bioresour. Technol.* **2000**, *3*, 227–233. [CrossRef]

46. Magyar, T.; Mayes, R.; Nagy, P.T. Assessment of composting processes in an automated aerobic fermentation system based on key parameters. *Recycl. Sustain. Dev.* **2020**, *13*, 1–8. [CrossRef]

47. Nagy, P.T.; Karanja, M.; Magyar, T. Study of the mineralisation of pelletized chicken manure at different soil moisture content of a sandy soil. *Nat. Resour. Sustain. Dev.* **2020**, *10*, 101–114. [CrossRef]

48. Islam, M.K.; Yaseen, T.; Traversa, A.; Ben Kheder, M.; Brunetti, G.; Cocozza, C. Effects of the main extraction parameters on chemical and microbial characteristics of compost tea. *Waste Manag.* **2015**, *52*, 62–68. [CrossRef]

49. Zhang, W.; Han, D.Y.; Dick, W.A.; Davis, K.R.; Hoitink, H.A.J. Compost and compost water extract-induced systemic acquired resistance in cucumber and Arabidopsis. *Phytopathology* **1998**, *88*, 450–455. [CrossRef]

50. Food and agriculture organization of the united nations (FAO). *World Reference Base for Soil Resources 2014. International Soil Classification System for Naming Soils and Creating Legends for Soil Maps*; World Soil Resources Reports No. 106; FAO: Rome, Italy, 2015; Available online: http://www.fao.org/3/i3794en/I3794en.pdf (accessed on 20 June 2021).

51. Spaargaren, O. *Mineral Soils Conditioned by Parent Material: Andosols, Arenosols*; 2nd European summer school on soil survey; Office for Official Publications of the European Communities: Luxembourg, 2006; pp. 93–97.

52. Schnürer, J.; Rosswall, T. Fluorescein Diacetate Hydrolysis as a Measure of Total Microbial Activity in Soil and Litter. *Appl. Environ. Microbiol.* **1982**, *43*, 1256–1261. [CrossRef]

53. Villányi, I.; Füzy, A.; Angerer, I.; Biró, B. Total catabolic enzyme activity of microbial communities. Fluorescein diacetate analysis (FDA). Understanding and Modelling Plant-Soil Interactions in the Rhizosphere Environment. In *Handbook of Methods Used in Rhizosphere Research*; Jones, D.L., Ed.; Chapter 4.2. Biochemistry; Swiss Federal Research Institute WSL: Birmensdorf, Switzerland, 2006; pp. 441–442. ISBN 3-905621-35-5.

54. Szegi, J. *Soil Microbiological Test Methods*, 1st ed.; Mezőgazda Kiadó: Budapest, Hungary, 1979; p. 311.

55. Tabatabai, M.A. Soil Enzymes. In *Methods of Soil Analysis*; John Wiley & Sons, Ltd.: Hoboken, NJ, USA, 2018; pp. 775–833.

56. Imre, A.; Rácz, H.V.; Antunovics, Z.; Rádai, Z.; Kovács, R.; Lopandic, K.; István, P.; Pfliegler, W.P. A new, rapid multiplex PCR method identifies frequent probiotic origin among clinical Saccharomyces isolates. *Microbiol. Res.* **2019**, *227*, 126298. [CrossRef]

57. Slezák, K.A. Salt Tolerance of White Peppers. Doctoral Thesis, Department of Vegetable and Mushroom Production, Szent István University, Gödöllő, Hungary, 2001.

58. Szabó, A.; Tamás, J.; Nagy, A. The influence of hail net on the water balance and leaf pigment content of apple orchards. *Sci. Hortic.* **2021**, *283*, 110112. [CrossRef]

59. R Core Team. *R: A Language and Environment for Statistical Computing*; R Foundation for Statistical Computing: Vienna, Austria, 2017; Available online: https://www.R-project.org/ (accessed on 30 April 2021).

60. Chen, Z.; Jiang, X. Microbiological Safety of Chicken Litter or Chicken Litter-based Organic Fertilizers: A Review. *Agriculture* **2014**, *4*, 1–29. [CrossRef]

61. Lu, J.; Sanchez, S.; Hofacre, C.; Maurer, J.J.; Harmon, B.G.; Lee, M.D. Evaluation of broiler litter with reference to the microbial composition as assessed by using 16S rRNA and functional gene markers. *Appl. Environ. Microbiol.* **2003**, *69*, 901–908. [CrossRef]

62. Green, V.S.; Stott, D.E.; Diack, M. Assay for fluorescein diacetate hydrolytic activity: Optimization for soil samples. *Soil Biol. Biochem.* **2006**, *38*, 693–701. [CrossRef]

63. Al-Dahmani, J.H.; Abbasi, P.A.; Miller, S.A.; Hoitink, H.A.J. Suppression of bacterial spot of tomato with foliar sprays of compost extracts under greenhouse and field conditions. *Plant Dis. J.* **2003**, *87*, 913–919. [CrossRef]

64. Riggle, D. Compost teas in agriculture. *BioCycle* **1996**, *37*, 65–67.

65. Noble, R.; Coventry, E. Suppression of soil-borne plant diseases with composts: A review. *Biocontrol Sci. Technol.* **2005**, *15*, 3–20. [CrossRef]

66. Weltzien, H.C. Biocontrol of foliar fungal diseases with compost extracts. In *Microbial Ecology of Leaves*; Andrews, J.H., Hirano, S.S., Eds.; Springer: New York, NY, USA, 1992; pp. 430–450.
67. Mengesha, W.K.; Powell, S.M.; Evans, K.J.; Barry, K.M. Diverse microbial communities in non-aerated compost teas suppress bacterial wilt. *World J. Microbiol. Biotechnol.* **2017**, *33*, 49. [CrossRef] [PubMed]
68. González-Hernández, A.I.; Suárez-Fernández, M.B.; Pérez-Sánchez, R.; Gómez-Sánchez, M.Á.; Morales-Corts, M.R. Compost Tea Induces Growth and Resistance against *Rhizoctonia solani* and *Phytophthora capsici* in Pepper. *Agronomy* **2021**, *11*, 781. [CrossRef]
69. Hegazy, M.I.; Hussein, E.; Salama, A.S.A. Improving physico-chemical and microbiological quality of compost tea using difference treatments during extraction. *Afr. J. Microbiol. Res.* **2015**, *11*, 763–770.
70. Naidu, Y.; Sariah, M.; Jugah, K.; Siddiqui, Y. Microbial starter for the enhancement of biological activity of compost tea. *Int. J. Agric. Biol.* **2010**, *12*, 51–56.
71. Naidu, Y.; Meon, S.; Siddiqui, Y. In vitro and in vivo evaluation of microbial-enriched compost tea on the development of powdery mildew on melon. *BioControl* **2012**, *57*, 827–836. [CrossRef]
72. Elbl, J.; Maková, J.; Javoreková, S.; Medo, J.; Kintl, A.; Lošák, T.; Lukas, V. Response of Microbial Activities in Soil to Various Organic and Mineral Amendments as an Indicator of Soil Quality. *Agronomy* **2019**, *9*, 485. [CrossRef]
73. Komilis, D.; Kontou, I.; Ntougias, S. A modified static respiration assay and its relationship with an enzymatic test to assess compost stability and maturity. *Bioresour. Technol.* **2011**, *102*, 5863–5872. [CrossRef]
74. Tian, W.; Wang, L.; Li, Y.; Zhuang, K.; Li, G.; Zhang, J.; Xiao, X.; Yungan, X. Responses of microbial activity, abundance, and community in wheat soil after three years of heavy fertilization with manure-based compost and inorganic nitrogen. *Agric. Ecosyst. Environ.* **2015**, *213*, 219–227. [CrossRef]
75. Alidadi, H.; Hosseinzadeh, A.; Najafpoor, A.A.; Esmaili, H.; Zanganeh, J.; Takabi, M.D.; Piranloo, F.G. Waste recycling by vermicomposting: Maturity and quality assessment via dehydrogenase enzyme activity, lignin, water soluble carbon, nitrogen, phosphorous and other indicators. *J. Environ. Manag.* **2016**, *182*, 134–140. [CrossRef]
76. Lazcano, C.; Gomez-Brandon, M.; Dominguez, J. Comparison of the effectiveness of composting and vermicomposting for the biological stabilization of cattle manure. *Chemosphere* **2008**, 1013–1019. [CrossRef] [PubMed]
77. Benitez, E.; Nogales, R.; Elvira, C.; Masciandaro, G.; Ceccanti, B. Enzyme activities as indicators of the stabilization of sewage sludges composting with Eisenia foetida. *Bioresour. Technol.* **1999**, *67*, 297–303. [CrossRef]
78. Vihnal-Freitas, I.C.; Wangen, D.R.B.; Ferreira, A.S.; Corrêa, G.F.; Wendling, B. Microbial and enzymatic activity in soli after organic composting. *Rev. Bras. Ciênc. Solo* **2010**, *34*, 757–764. [CrossRef]
79. Marcite, I.; Hernandez, T.; Garcia, C.; Polo, A. Influence of one or two successive annual applications of organic fertilisers on the enzyme activity of a soil under barley cultivation. *Bioresour. Technol.* **2001**, *79*, 147–154. [CrossRef]
80. Ros, M.; Pascual, J.A.; García, C.; Hernández, M.T.; Insam, H. Hydrolase activities, microbial biomass and bacterial community in a soil alter long-term amendment with different composts. *Soil Biol. Biochem.* **2006**, *38*, 3443–3452. [CrossRef]
81. Pascual, J.A.; Garcia, C.; Hernández, T.; Ayuso, M. Changes in the microbial activity of an arid soil amended with urban organic wastes. *Biol. Fertil. Soils* **1997**, *24*, 429–434. [CrossRef]
82. Droffner, M.L.; Brinton, W.F.; Evans, E. Evidence for the Prominence of Well Characterized Mesophyllic Bacteria in Thermophilic (50–70 °C) Composting Environments. *Biomass Bioenergy* **1995**, *8*, 191–195. [CrossRef]
83. Sifatullah, S.; Khan, H.; Ali, Z.; Rowaidullah, R.; Ali, A. Comparative Study of the Effect of Compost Tea (Aerated & Non-Aerated) of Agro-Sieved-Waste on Germination and Biomass Yield of Maize, Mungbean and Cauliflower. *Environ. Sci. Technol.* **2011**, *35*, 31–40.
84. Wang, K.H.; Radovich, T.J.K.; Pant, A.; Cheng, Z. Integration of cover crops and vermicompost tea for soil and plant health management in a short-term vegetable cropping system. *Appl. Soil Ecol.* **2014**, *82*, 26–37. [CrossRef]
85. Hewidy, M.; Traversa, A.; Kheder, M.B.; Ceglie, F.G.; Cocozza, C. Short-term effects of different organic amendments on soil properties and organic broccoli growth and yield. *Compos. Sci. Util.* **2015**, *23*, 207–215. [CrossRef]
86. Hodossi, S.; Kovács, A.; Terbe, I. (Eds.) *Growing Vegetables in Field*; Mezőgazda Publishing: Budapest, Hungary, 2010; 355p.
87. Pérez-López, A.J.; López-Nicolas, J.M.; Núñez-Delicado, E.; Del Amor, F.M.; Carbonell-Barrachina, A.A. Effects of agricultural practices on color, carotenoids composition, and minerals contents of sweet peppers, cv. Almuden. *J. Agric. Food Chem.* **2007**, *3*, 8158–8164. [CrossRef] [PubMed]
88. European Commission. Communication from the Commission to the European Parliament, the European Council, the Council, the Economic and Social Committee and the Committee of the Regions. The European Green Deal. 2019. Available online: https://eur-lex.europa.eu/resource.html?uri=cellar:b828d165-1c22-11ea-8c1f-01aa75ed71a1.0002.02/DOC_1&format=PDF (accessed on 26 June 2021).

Article

Effect of Feeding Low Protein Diets on the Production Traits and the Nitrogen Composition of Excreta of Broiler Chickens

Nikoletta Such [1], László Pál [1], Patrik Strifler [1], Boglárka Horváth [2], Ilona Anna Koltay [1], Mohamed Ali Rawash [1], Valéria Farkas [1], Ákos Mezőlaki [3], László Wágner [1] and Károly Dublecz [1,*]

[1] Institute of Physiology and Nutrition, Georgikon Campus, Hungarian University of Agriculture and Life Sciences, Deák Ferenc Street 16, 8360 Keszthely, Hungary; such.nikoletta.amanda@phd.uni-mate.hu (N.S.); pal.laszlo@uni-mate.hu (L.P.); strifler.patrik@phd.uni-mate.hu (P.S.); koltay.ilona.anna@phd.uni-mate.hu (I.A.K.); mohamedali@post.agr.cu.edu.eg (M.A.R.); farkas.valeria@uni-mate.hu (V.F.); wagner.laszlo@uni-mate.hu (L.W.)

[2] UBM Feed Zrt., Kisvasút Street 1, 2085 Pilisvörösvár, Hungary; boglarka.horvath@ubm.hu

[3] Agrofeed Ltd., Duna Kapu Square 10, 9022 Győr, Hungary; akos.mezolaki@agrofeed.hu

* Correspondence: dublecz.karoly@uni-mate.hu

Abstract: The main goal of the current study was to investigate the effects of feeding low protein (LP) diets on the performance parameters and excreta composition of broiler chickens. In total, 288 male Ross 308 day-old chickens were divided into two dietary treatment groups using six replicate pens with 24 chickens each. No LP diet was fed in the starter phase. The protein reduction in the grower and finisher phases were 1.8% and 2% respectively. Beside the measurements of production traits, on day 24 and 40 representative fresh excreta samples were collected, their dry matter, total N, NH_4^+-N and uric acid-N contents determined, and the ratio of urinary and fecal N calculated. Dietary treatments failed to cause significant differences in the feed intake, growth rate, and feed conversion ratio of animals. LP diets decreased the total nitrogen and uric acid contents of excreta significantly. The age of birds had also significant effect, resulting more reduction in the grower phase compared with the finisher. The ratio of urinary N was higher at day 40 compared with the age of day 24. The urinary N content of broiler chicken's excreta is lower than can be found in the literature, which should be considered in the ammonia inventory calculations.

Keywords: low protein diet; broiler chickens; N excretion

Citation: Such, N.; Pál, L.; Strifler, P.; Horváth, B.; Koltay, I.A.; Rawash, M.A.; Farkas, V.; Mezőlaki, Á.; Wágner, L.; Dublecz, K. Effect of Feeding Low Protein Diets on the Production Traits and the Nitrogen Composition of Excreta of Broiler Chickens. *Agriculture* **2021**, *11*, 781. https://doi.org/10.3390/agriculture11080781

Academic Editor: István Komlósi

Received: 19 July 2021
Accepted: 12 August 2021
Published: 16 August 2021

Publisher's Note: MDPI stays neutral with regard to jurisdictional claims in published maps and institutional affiliations.

1. Introduction

Air quality is receiving more attention worldwide. Among air pollutants ammonia is one of the most dangerous gases, contributing to biodiversity loss, general acidification of the environment and new type of deforestation [1]. Ammonia released into the atmosphere is linked to the formation of small particles with a diameter less than 10 micrometers (particular matter, PM10) [2], which can cause serious human respiratory damage (WHO, 2013). Agriculture is the main source of ammonia emission. According to the latest Hungarian air pollutant inventory report, agriculture is responsible to 92% for the total national emissions, of which about 66% takes the animal production and the ammonia emitting from the manure [3]. According to the National Emission Ceiling Directive 2016/2284, all member countries of the European Union must reduce their national emissions of air pollutants [4]. Hungary must decrease the ammonia emission by 32% until 2030, compared with 2005-year baseline. This obligation force urgent actions both on national and farm level.

Intensive poultry production using genotypes with high growth rate and egg production, resulting improved N-retention and less excretion. Dietary protein quality is a critical regulator of poultry growth, reproductive performance, and plays important role in the development of the gastrointestinal tract [5]. Protein-rich components are the most expensive ingredients in broiler diets and the member states of the European Union are dependent on the overseas soybean import [6]. A major concern for the modern poultry industry is to reduce feed cost and to optimize the protein supply of animals. Feeding low protein (LP) diets with increased crystalline amino acids could be a solution. Numerous experiments have been performed on the effect of feeding reduced protein diets on poultry growth, but the results are often contradictory [7]. Experiments in which the control and reduced protein diets had the same digestible lysine content and the diets complied with the recommendations of the "ideal protein principle" were able to achieve a 2% protein reduction without compromising the production traits [8]. Feeding LP diets has also been reported to decrease the N in excreta, thus reducing the N loss to the environment [9–11]. This is an important result, because intensive animal production systems, in particular pig and poultry farming are responsible for the emission of several air pollutants [12]. The main emitted gases are carbon monoxide, carbon dioxide and ammonia [13]. During the metabolism, the animal is utilizing only the 30–50% of the consumed nitrogen. The excreted part, mainly the urinary N is the potential source of NH_3 emission.

Several factors can influence this process, for example the species, age, live weight of animals or the protein content and amino acid composition of the diet, the housing conditions or manure management [14]. The largest proportion of nitrogen content of the chicken urine consists of uric acid and ammonia [15]. These two compounds take the total ammoniacal nitrogen (TAN), which is used in the so-called TAN-based flow models in the ammonia emission calculations [16].

The decomposition process of uric acid is carried out by urease enzyme which catalyzes the hydrolysis of urea into ammonia and carbon dioxide in aqueous medium, allowing ammonia volatilization to the atmosphere [17]. Several previous studies have shown a reduction in nitrogen excretion by about 10% with a percentage point decrease in the dietary crude protein content of broiler diets [18].

There are lots of scientific data on the ratio of excreted urinary and fecal nitrogen in mammals. However, since birds excrete feces and urine as a mixture, only few data exist on the urinary N content of poultry excreta and how this ratio change with the intensity of production, the age of birds, the dietary protein content and protein quality. The aim of this study was therefore to evaluate the effect of feeding LP diets on the performance parameters and proportion of the different nitrogenous compounds of chickens' excreta at two different age categories. The novelty of the trial is that the protein reduction of LP diets was in a range that can be applied in the practice. Furthermore, low protein diet effect on the excreta composition was studied not only at the end of the production, but also at day 24.

2. Materials and Methods

2.1. Birds and Experimental Design

The animal experiment was approved by the Institutional Ethics Committee (Animal Welfare Committee, Georgikon Campus, Hungarian University of Agriculture and Life Sciences, Deák Ferenc Street 16, 8360 Keszthely, Hungary). A total of 288 one-day-old male broilers (Ross 308) were allocated randomly to one of the 12 pens at a stocking rate of 24 birds per pen (10 bird/m^2). Each treatment was replicated 6 times. Chopped wheat straw was used as litter material.

2.2. Feed

Three phases fattening was used. The starter diet (0–10 days) was fed in mash, the grower (11–24 days) and finisher diets (25–40 days) in pelleted form. In the starter phase no LP diets were fed. The protein content of the control and LP diets in the grower phase were 21.31% and 19.49%; in the finisher phase 20.43% and 18.38%. Ad libitum feed and water were offered throughout the experiment. The composition and the measured nutrient contents of the experimental diets are shown in Tables 1 and 2. LP diets in the grower and finisher phase contained less soybean meal, but more sunflower meal and DDGS, compared with the control diets. Beside methionine, lysine, threonine, and valin, LP diets contained also crystalline isoleucine and arginine.

Table 1. Composition of experimental diets (g/kg as fed).

Ingredient	Starter	Grower		Finisher	
	Control	LP	Control	LP	Control
Corn	370	436	409	473	436
Wheat	100	100	100	100	100
Extracted soybean meal	356	113.5	243	77.2	217
Sunflower meal	50	150	100	150	100
DDGS corn	30	100	50	100	50
Sunflower oil	49.0	56.1	58.8	56.7	61.4
MCP	9.95	5.52	6.80	4.48	5.70
Limestone	18.6	17.3	16.4	16.8	15.8
L-lysine	3.71	8.48	4.06	8.27	3.42
DL-methionine	2.94	2.36	2.12	2.13	1.82
L-threonine	0.87	1.62	0.72	1.54	0.53
L-valine	0.47	1.07	0.12	0.90	-
L-isoleucine	-	1.17	-	1.32	-
L-arginine	-	0.42	-	0.59	-
Premix	5.00	4.00	4.00	5.00	5.00
Salt	2.85	2.36	2.67	2.35	2.67
Sodium bicarbonate	0.71	0.18	0.64	0.21	0.66
Phytase (Quantum blue)	0.10	0.10	0.10	0.10	0.10
Xylanase (Econase XT25)	0.10	0.10	0.10	0.10	0.10
Total	1000	1000	1000	1000	1000

Premix was supplied by UBM Ltd. (Pilisvörösvár, Pest megye, Hungary). The active ingredients contained in the premix were as follows (per kg of diet): Composition of the starter—grower premix—retinyl acetate—5.0 mg, cholecalciferol—130 µg, dl-alpha-tocopherol-acetate—91 mg, menadione—2.2 mg, thiamine—4.5 mg, riboflavin—10.5 mg, pyridoxin HCL—7.5 mg, cyanocobalamin—80 µg, niacin—41.5 mg, pantothenic acid—15 mg, folic acid—1.3 mg, biotin—150 µg, betaine—670 mg, Ronozyme® NP—150 mg, monensin-Na—110 mg (only grower), narasin—50 mg (only starter), nicarbazin—50 mg (only starter), antioxidant—25 mg, Zn (as $ZnSO_4 \cdot H_2O$)—125 mg, Cu (as $CuSO_4 \cdot 5H_2O$)—20 mg, Fe (as $FeSO_4 \cdot H_2O$)—75 mg, Mn (as MnO)—125 mg, I (as KI)—1.35 mg, Se (as Na_2SeO_3)—270 µg. Composition of the finisher premix—retinyl acetate—3.4 mg, cholecalciferol—97 µg, dl-alpha- tocopherol-acetate—45.5 mg, menadione—2.7 mg, thiamine—1.9 mg, riboflavin—5.0 mg, pyridoxin HCl—3.2 mg, cyanocobalamin—19 µg, niacin—28.5 mg, pantothenic acid—10 mg, folic acid—1.3 mg, biotin—140 µg, l-ascorbic acid—40 mg, betaine—193 mg, antioxidant—25 mg, Zn (as $ZnSO_4 \cdot H_2O$)—96 mg, Cu—9.6 mg, Fe (as $FeSO_4 \cdot H_2O$)—29 mg, Mn (as MnO)—29 mg, I (as KI)—1.2 mg, Se (as Na_2SeO_3)—350 µg.

Table 2. Measured nutrient contents of the experimental diets (%).

Ingredient	Starter	Grower		Finisher	
	Control	Control	LP	Control	LP
AMEn (MJ/kg)	12.65	13.00	13.00	13.20	13.20
Crude protein	23.95	21.31	19.49	20.43	18.38
Crude fat	7.03	8.68	8.81	8.01	8.05
Crude fiber	4.7	5.84	6.79	5.99	6.44
Ca	1.05	0.94	0,97	0.9	0.87
P	0.75	0.77	0,74	0.71	0.67
Lys	1.47	1.19	1.22	1.11	1.17
Met	0.60	0.53	0.56	0.51	0.51
Cys	0.36	0.33	0.31	0.32	0.30

Control—commercial maize-based diet; LP—Low protein diet.

2.3. Measurements

During the 40 day long fattening period, the weight gain, feed intake, and feed conversion were measured at pen basis at the end of each period (10th, 24th, and 40th day) and for the whole fattening.

On day 24 and 40 about 200 g fresh excreta samples were collected from each pen on nylon foils. Samples were mixed thoroughly, frozen, and stored at −20 °C until further processing. From these samples their dry matter, total N, NH_4^+-N, and uric acid-N contents were measured. The dry matter content of excreta samples was measured in exicator (100 °C for 24 h). The urinary N was calculated as the sum of uric acid-N and NH_4^+-N as described by O'Dell et al. (1960). Total N was determined according to the Kjeldahl method with a Foss-Kjeltec 8400 Analyzer Unit (Nils Foss Allé 1, DK-3400 Hilleroed, Denmark). The determination of NH_4^+-N content of the excreta was carried out according to the method of Peters [19]. The uric acid measurement based on the method of Marquardt et al. [20]. All N parameters were adjusted to dry matter basis.

2.4. Statistical Analyses

All data were analyzed using the Statistical Package for Social Science (SPSS) software version 22.0 (IBM Corp., Armonk, NY, USA). The performance parameters were analyzed by *t*-test, while the N contents of excreta by multivariate test of the general linear model, using the main factors of age and dietary treatments. Differences were considered significant at a level of $p \leq 0.05$.

3. Results

The effects of dietary treatments on growth performance of broiler chickens are presented in Table 3. Evaluating the dietary treatments in the grower and finisher phase, feeding LP diets slightly decreased the daily gain, feed intake, and increased the feed conversion ratio of chickens, but the differences were not significant.

Table 3. Effects of dietary treatments on the performance parameters of broiler chickens.

		Control	LP	Pooled SEM	*p*-Values
Daily gain	Starter	237.6	227.3	3.696	0.112
	Grower	901.8	842.3	10.386	0.500
	Finisher	1593.9	1538.3	17.206	0.983
	Cumulative	2495.8	2380.7	24.168	0.756
Feed intake	Starter	314.5	301.4	3.455	0.346
	Grower	1370.2	1332.0	17.079	0.753
	Finisher	2770.0	2786.0	28.394	0.815
	Cumulative	4140.3	4118.9	42.244	0.963
Feed conversion ratio	Starter	1.32	1.33	0.030	0.757
	Grower	1.52	1.58	0.023	0.405
	Finisher	1.74	1.81	0.022	1.240
	Cumulative	1.66	1.73	0.020	1.264

Control—commercial maize-based diet; LP—low protein diet; SEM—standard error of the mean.

Similarly, to the production parameters, dry matter content of the excreta was not influenced by the treatments either (Table 4). As expected, feeding LP diets decreased the total N, NH_4^+-N and uric acid N contents of excreta. However, the decrease was significant only at day 24. Comparing with the control, the main effects of LP diets were significant decrease in all investigated N compounds. The decrease in total, uric acid, NH_4^+, fecal and urinary N were 32%, 28%, 31%, 35%, and 30% respectively. The excreta of the 40-day old birds contained significantly lower amounts of total, uric acid, fecal, and urinary N, compared with the 24-day old chickens. The decreases were in this case 23%, 18%, 29%, and 15%. The age effect on the NH_4^+-N was not significant (Table 4). In the case of the total, uric acid, fecal and urinary N significant age x dietary treatment interactions were found. Its reason was the significant age effect on these parameters in the control group, which was not the case in the LP diet treatment.

Table 4. The amount of the total-N, NH_4^+-N, uric acid-N, and dry matter content of the excreta.

Treatments		Dry Matter	Total-N	NH_4^+-N	Uric Acid N	Fecal-N	Urinary-N
		%			mg/g DM		
day 24	C	16.18	46.70 [a]	3.86 [a]	15.34 [a]	27.50 [a]	19.20 [a]
	LP	17.00	28.24 [b]	2.59 [b]	9.53 [b]	16.13 [b]	12.11 [b]
day 40	C	16.81	32.25 [b]	3.47 [ab]	11.44 [b]	17.35 [b]	14.90 [b]
	LP	18.10	25.50 [b]	2.68 [b]	8.99 [b]	13.83 [b]	11.67 [b]
			Diet effect				
Diet	C	16.49	39.48 [a]	3.65 [a]	13.39 [a]	22.43 [a]	17.05 [a]
	LP	17.54	26.87 [b]	2.63 [b]	9.26 [b]	14.49 [b]	11.89 [b]
			Age effect				
Age	day 24	16.58	37.47 [a]	3.23	12.43 [a]	21.81 [a]	15.66 [a]
	day 40	17.45	28.87 [b]	3.07	10.21 [b]	15.58 [b]	13.29 [b]
Pooled SEM		0.378	2.129	0.164	0.639	6.940	3.779
Diet		0.179	* 0.000	* 0.001	* 0.000	* 0.001	* 0.000
Age		0.262	* 0.005	0.561	* 0.012	* 0.004	* 0.000
Age x Diet		0.760	* 0.045	0.359	* 0.048	* 0.056	* 0.061

Control—commercial maize-based diet; LP—Low protein diet; SEM—standard error of the mean; [a,b] means with different superscripts are significantly different ($p < 0.05$). *: significant at the $p < 0.05$ level

As it can be seen in Figure 1, the protein content of the diets did not modify the ratio of urinary and fecal N in the excreta. On the other hand, the excreta of the 24-day old animals contained less urinary N. This difference was significant ($p = 0.027$).

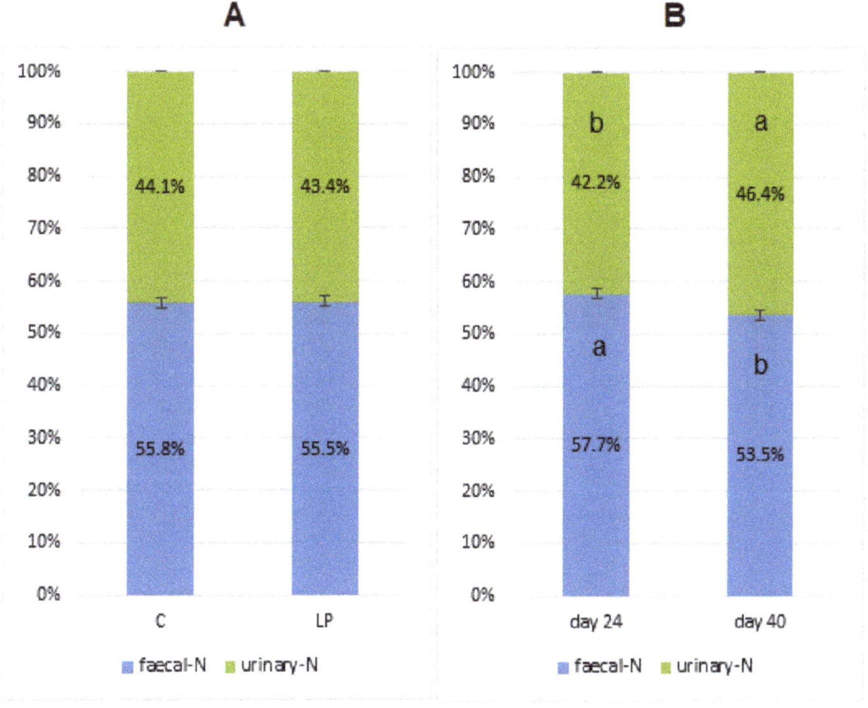

Figure 1. Effects of dietary protein content (**A**) and age of chickens (**B**) on the ratio of fecal and urinary nitrogen content of excreta. Control—commercial maize-based diet; LP—Low protein diet; SEM—standard error of the mean; a, b means with different superscripts are significantly different ($p < 0.05$).

One percent dietary protein reduction decreased the N excretion of chickens more at day 24, compared with the values at day 40 (Figure 2). The reason for this is the better feed conversion, higher N-retention of the younger animals. This means no constant factors can be used when the effects of LP diets are used for correcting the N-excretion and TAN-excretion of the animals. It should be also considered that the applied protein decrease means different relative protein reduction in the different phases—in the grower phase the relative protein reduction was 8.5%, while in the finisher period 10.0%.

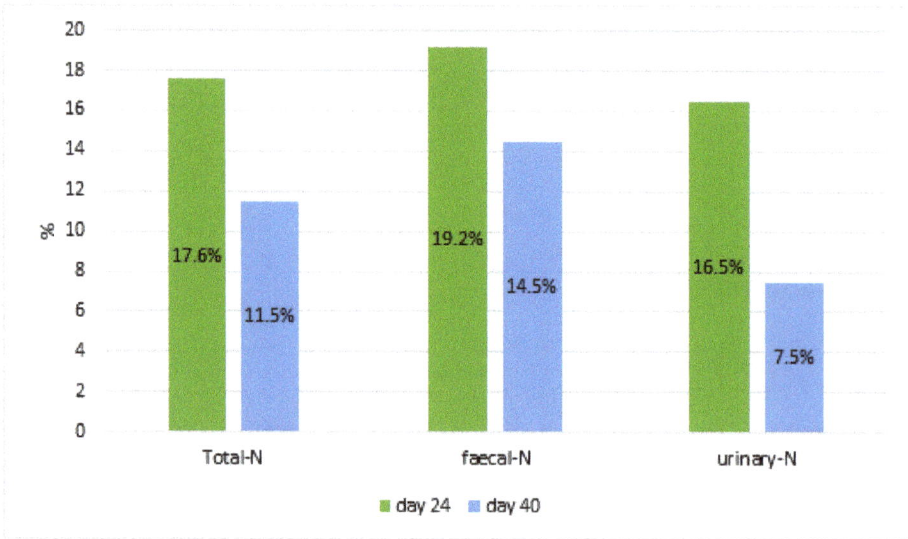

Figure 2. Effect of 1% dietary protein reduction on the decrease of excreted N of broiler chickens.

4. Discussion

Ammonia emission of poultry production is a real concern for all EU member countries and introduction of the best available practices are crucial its reduction. Nutrition has a big potential and feeding LP diets is one of the most effective options to reduce the emission. The main goal in this aspect is to find out the rate of protein reduction, which does not negatively affect the production traits and can even decrease the feeding costs [8]. Research on this often ends in unexpected results. Some experiments showed that the too high reduction cause depressive effect on performance parameters even with amino acid supplementation [21,22]. The discrepancies in responses often observed in the literature in chicks fed with low-protein amino acid-supplemented diets are related to the degree of CP reduction, the number of crystalline amino acid supplemented, the modified amino acid, the energy: protein and energy: amino acid ratios, the age of birds and the feedstuffs used [23]. Due to the changing amino acid ratios in reduced protein diets, certain amino acid deficiencies are to be expected [24]. In the case of avian species usually sulfur-containing amino acids are the first limiting, due to feather formation, but in LP diets lysine, threonine, valine, isoleucine, and arginine should also be supplemented [7]. In this experiment, the daily gain, feed intake, and feed conversion ratio of animals were the same with reduced CP diets. It means that the amino acid content of LP diets covered the requirements of the animals. These findings agree with other previous results [24], that in the case of broilers, 2% protein reduction can be achieved without impacts on the production results.

There are many factors that affect water intake of broilers, such as ambient temperature, feed intake, dietary composition, form of the feed, crude protein content, mineral metabolism, genetic factors, and the cation electrolyte balance of the feed [25–28]. According to the literature data LP diets can increase the dry matter content of the excreta and manure because of the lower water intake and urine excretion [28]. In this experiment dry matter content of excreta was not affected by the treatments, probably because of the limited, only 2% protein decrease. Similarly, in the study of van Emous et al. [29], when the protein content of the diet was 1.5% less, LP diet did not affect water intake and dry matter content of the manure, but the litter and manure samples of broilers had 8% lower total-N and 13% lower ammonia-N content.

If LP diets are used, 1% protein decrease can result about 10% decrease in the N-excretion [9]. In the study of Kerr [30], using amino acid supplemented LP diets with poultry and swine, 1% protein decrease of the diet reduced the N excretion of both species by 8.5% regardless of breed and body weight. Similar results have been published with laying hens and broilers [11,31–33]. In this study the age of chicken had significant impact on the N-excretion and on the ratio of the urinary and fecal N. In the case of 24-day old birds 1% protein decrease resulted 17.6% decrease in the total N excretion with 19.2% decrease in fecal and 16.5% in urinary N excretion. In the 40-day old chickens the decreases of the previously mentioned parameters were 11.5%, 14.5%, and 7.5% respectively. This means no constant values can be used in the N excretion calculations of LP diets. According to the knowledge of the authors, it is the first result that proves this age effect. It is also novel finding, that if LP diets are used, the decrease of fecal N is higher than the urinary N decrease. In the calculation of potential ammonia emission, it should also be considered.

Nitrogen-containing compounds from animal production are converted to gaseous ammonia by microbial activity [34]. Much of the ammonia released from manure comes from the hydrolysis of urea [35], or in the case of birds from the breakdown of uric acid [36].

Based on the available research results, uric acid represents 50–60% of the total N content of poultry excreta [37]. O'Dell et al. [15] found that the sum of uric acid and NH_4^+-N of the excreta gives approximately the total amount of urinary nitrogen in birds. In poultry, feces and urine are excreted together in a mixture, resulting fecal and urinary nitrogen separation difficulties. Our results prove, that the urinary N ratio of poultry excreta is less than that can be found in the literature. Using the default value of 70% TAN [16] of broiler chicken's excreta overestimates the real TAN excretion and thus the ammonia emission of poultry species. Interestingly dietary protein does not modify the fecal and urinary N ratio (55–56% vs. 43–44%). The reason for this can be, that in the case of LP diets the protein digestibility and protein utilization improves in the same manner.

Our results prove that not only the dietary factors, but also the age of birds influence the composition of the broiler's excreta. The urinary N ratio of excreta reflecting the differences in the N utilization of broiler chickens between the different age categories. The higher urinary-N content in the finisher phase means lower N-retention in this phase.

5. Conclusions

According to the results of this study 2% protein decrease of grower and finisher broiler diets did not affect the production traits but can decrease the N excretion of birds significantly. Feeding LP diets resulted into higher N-excretion reduction in younger birds. One percent protein reduction can decrease the N-excretion of 24-day old birds even by 17.6% and in the case of 40-day old animals by 11.5%. The protein content of diet does not modify the fecal: urinary N ratio of excreta, but the urinary N content of the 40-day old birds exceeds that of the 24-day old group. Due to the intensive growth rate and N-retention of broiler chickens, the urinary N ratio of excreta is lower, compared to the literature.

Author Contributions: Conceptualization, N.S. and K.D.; methodology, N.S. and L.W.; formal analysis, N.S. and K.D.; investigation, N.S., L.P., P.S., B.H., I.A.K., V.F., M.A.R. and Á.M.; data curation, L.W.; writing—original draft preparation, N.S. and K.D.; writing—review and editing, N.S., V.F. and K.D.; visualization, N.S.; supervision, K.D.; project administration, V.F., K.D. and N.S.; funding acquisition, K.D. All authors have read and agreed to the published version of the manuscript.

Funding: This research was supported by the Hungarian Government and the European Union, with the co-funding of the European Regional Development Fund in the frame of Széchenyi 2020 Program GINOP-2.3.2-15-2016-00054 project.

Institutional Review Board Statement: The animal experiment was approved by the Institutional Ethics Committee (Animal Welfare Committee, Georgikon Campus, Hungarian University of Agriculture and Life Sciences) under the license number MÁB-11/2019.

Informed Consent Statement: Not applicable.

Data Availability Statement: All data generated or analyzed during this study are included in this published article.

Acknowledgments: This research was supported by the ÚNKP-20-3-II new national excellence program of the Ministry for Innovation and Technology from the source of the national research, development, and innovation fund and was prepared with the professional support of the Doctoral Student Scholarship Program of the Cooperative Doctoral Program of the Ministry for Innovation and Technology financed from the National Research, Development, and Innovation Fund.

Conflicts of Interest: All authors declare no conflict of interest.

References

1. Task Force on Reactive Nitrogen. Available online: http://www.clrtap-tfrn.org/ (accessed on 18 July 2021).
2. Giannakis, E.; Kushta, J.; Bruggeman, A.; Lelieveld, J. Costs and Benefits of Agricultural Ammonia Emission Abatement Options for Compliance with European Air Quality Regulations. *Environ. Sci. Eur.* **2019**, *31*, 1–13. [CrossRef]
3. IIR. Hungary Informative Inventory Report 1990–2019. Available online: https://www.ceip.at/status-of-reporting-and-review-results/2021-submission (accessed on 18 July 2021).
4. European Parliament. Directive (EU) 2016/2284 of the European Parliament and of the Council of 14 December 2016 on the Reduction of National Emissions of Certain Atmospheric Pollutants, Amending Directive 2003/35/EC and Repealing Directive 2001/81/EC. Brussels: European Council. *Off. J. Eur. Union* **2016**, *344*, 1–34.
5. Kamran, Z.; Aslam Mirza, M.; Mahmood, S. Effect of Decreasing Dietary Protein Levels with Optimum Amino Acids Profile on the Performance of Broilers. *Pak. Vet. J.* **2004**, *24*, 165–168.
6. Dublecz, K.; Koltay, I.A.; Such, N.; Dublecz, F.; Husvéth, F.; Wágner, L.; Péterné Farkas, E.; Márton, A.; Farkas, V.; Pál, L. Lehetőségek a Takarmányok Nyersfehérje Tartalmának Csökkentésére Monogasztrikus Állatokban. *Állatteny. Takarm.* **2018**, *67*, 273–286.
7. Belloir, P.; Lessire, M.; Berri, C.; Lambert, W.; Corrent, E.; Tesseraud, S. Revisiting Amino Acid Nutrition. In Proceedings of the 20th ESPN, Prague, Czech Republic, 24–27 August 2015; pp. 27–34.
8. Belloir, P.; Lessire, M.; Van Milgen, J.; Schmidley, P.; Corrent, E.; Tesseraud, S. Reducing Dietary Crude Protein of Broiler: A Meta-Analysis Approach. In Proceedings of the 11émes Jourmées de la Researche Avicoles et Palmipédes a Foie Gras, Tours, France, 25–26 March 2015; pp. 1–6.
9. Giner Santonja, G.; Georgitzikis, K.; Scalet, B.; Montobbio, P.; Roudier, S.; Delgado Sancho, L. *Best Available Techniques (BAT) Reference Document for the Intensive Rearing of Poultry or Pigs: Industrial Emissions Directive 2010/75/EU (Integrated Pollution Prevention and Control)*; EUR 28674 EN; European Comission: Brussels, Belgium, 2017.
10. Sterling, K.G.; Vedenov, D.V.; Pesti, G.M.; Bakalli, R.I. Economically Optimal Dietary Crude Protein and Lysine Levels for Starting Broiler Chicks. *Poult. Sci.* **2005**, *84*, 29–36. [CrossRef] [PubMed]
11. Kamran, Z.; Sarwar, M.; Nisa, M.; Mahmood, M. Effect of Low Levels of Dietary Crude Protein with Constant Metabolizable Energy on Nitrogen Excretion, Litter Composition and Blood Parameters of Broilers. *Int. J. Agric. Biol.* **2010**, *12*, 401–405.
12. De Sousa, F.C.; de Tinôco, I.F.F.; da Silva, J.N.; de Jesus Fôlgoa Baptista, F.; Souza, C.F.; Lopes Silva, A. Gas Emission in the Poultry Production. *J. Anim. Behav. Biometeorol.* **2017**, *5*, 49–55. [CrossRef]
13. Angus, A.J.; Hodge, I.D.; Sutton, M.A. Ammonia Abatement Strategies in Livestock Production: A Case Study of a Poultry Installation. *Agric. Syst.* **2006**, *89*, 204–222. [CrossRef]
14. Battye, R.; Battye, W.; Overcash, C.; Fudge, S. *Development and Selection of Ammonia Emission Factors*; Final Report; U.S. Environmental Protection Agency: Washington, DC, USA, 1994.
15. O'dell, B.L.; Woods, W.D.; Laerdal, O.A.; Jeffay, A.M.; Savage, J.E. Distribution of the Major Nitrogenous Compounds and Amino Acids in Chicken Urine. *Poult. Sci.* **1960**, *39*, 426–432. [CrossRef]
16. EMEP/EEA. Air Pollutant Emission Inventory Guidebook 2019—Publications Office of the EU. Available online: https://op.europa.eu/en/publication-detail/-/publication/ce310211-4bc5-11ea-8aa5-01aa75ed71a1/language-en (accessed on 19 July 2021).
17. De Oliveira, M.C.; Almeida, C.V.; Andrade, D.O.; Rodrigues, S.M.M. Teor de Matéria Seca, PH e Amônia Volatilizada Da Cama de Frango Tratada Ou Não Com Diferentes Aditivos. *Rev. Bras. Zootec.* **2003**, *32*, 951–954. [CrossRef]

18. Belloir, P.; Méda, B.; Lambert, W.; Corrent, E.; Juin, H.; Lessire, M.; Tesseraud, S. Reducing the CP Content in Broiler Feeds: Impact on Animal Performance, Meat Quality and Nitrogen Utilization. *Animal* **2017**, *11*, 1881–1889. [CrossRef]
19. Peters, J.; Combs, S.; Hoskins, B.; Jarman, J.; Kovar, J.L.; Watson, M.; Wolf, A.; Wolf, N. Recommended Methods for Manure Analysis. In Proceedings of the ASA-CSSA-SSSA Annual Meeting Abstracts, Denver, CO, USA, 2–6 November 2003; ASA-CSSA-SSSA: Madison, WI, USA, 2003.
20. Marquardt, R.R.; Ward, A.T.; Campbell, L.D. A Rapid High-Performance Liquid Chromatographic Method for the Quantitation of Uric Acid in Excreta and Tissue Samples. *Poult. Sci.* **1983**, *62*, 2099–2105. [CrossRef]
21. Bregendahl, K.; Sell, J.; Zimmerman, D. Effect of Low-Protein Diets on Growth Performance and Body Composition of Broiler Chicks. *Poult. Sci.* **2002**, *81*, 1156–1167. [CrossRef] [PubMed]
22. Namroud, N.; Shivazad, M.; Zaghari, M.; Shahneh, A.Z. Effects of Glycine and Glutamic Acid Supplementation to Low Protein Diets on Performance, Thyroid Function and Fat Deposition in Chickens. *S. Afr. J. Anim. Sci.* **2010**, *40*, 238–244. [CrossRef]
23. Aletor, V.A.; Hamid, I.I.; Nieß, E.; Pfeffer, E. Low-Protein Amino Acid-Supplemented Diets in Broiler Chickens: Effects on Performance, Carcass Characteristics, Whole-Body Composition and Efficiencies of Nutrient Utilisation. *J. Sci. Food Agric.* **2000**, *80*, 547–554. [CrossRef]
24. Dublecz, K.; Husvéth, F.; Wágner, L.; Márton, A.; Koltay, I.; Such, N.; Rawash, M.A.; Mezőlaki, Á.; Pál, L.; Molnár, A. Feeding Low Protein Diets Poultry and Pig Diets—Physiological, Economic and Environmental Aspects. In Proceedings of the International Symposium on Animal Science, Herceg Novi, Montenegro, 3–8 June 2019; pp. 20–29.
25. Bailey, M. The water requirements of poultry. In *Recent Developments in Poultry Nutrition 2*; Nottingham University Press: Nottingham, UK, 1999; pp. 321–335.
26. Borges, S.A.; Da Silva, A.V.F.; Maiorka, A. Acid-Base Balance in Broilers. *World's Poult. Sci. J.* **2007**, *63*, 73–81. [CrossRef]
27. Koreleski, J.; Swiatkiewicz, S.; Arczewska-Włosek, A. The Effect of Different Dietary Potassium and Chloride Levels on Performance and Excreta Dry Matter in Broiler Chickens. *Czech J. Anim. Sci.* **2011**, *56*, 53–60. [CrossRef]
28. Ziaei, N.; Guy, J.H.; Edwards, S.A.; Blanchard, P.J.; Ward, J.; Feuerstein, D. Effect of Gender on Factors Affecting Excreta Dry Matter Content of Broiler Chickens. *J. Appl. Poult. Res.* **2007**, *16*, 226–233. [CrossRef]
29. Van Emous, R.A.; Winkel, A.; Aarnink, A.J.A. Effects of Dietary Crude Protein Levels on Ammonia Emission, Litter and Manure Composition, N Losses, and Water Intake in Broiler Breeders. *Poult. Sci.* **2019**, *98*, 6618–6625. [CrossRef] [PubMed]
30. Kerr, B.J. Nutritional strategies for waste reduction management. In *Nitrogen: New Horizons in Animal Nutrition and Health*; The Institution of Nutrition of the University of North Carolina: Chapel Hil, NC, USA, 1995; pp. 47–68.
31. Lopez, G.; Leeson, S. Response of Broiler Breeders to Low-Protein Diets. *Poult. Sci.* **1995**, *74*, 685–695. [CrossRef]
32. Lopez, G.; Leeson, S. Nitrogen of Manure from Older Broiler Breeders Fed Varying Quantities of Crude Protein. *J. Appl. Poult. Res.* **1995**, *4*, 390–394. [CrossRef]
33. Hernandez, F.; Megias, M.; Orengo, J.; Martinez, S.; López, M.J.; Madrid, J. Effect of Dietary Protein Level on Retention of Nutrients, Growth Performance, Litter Composition and NH_3 Emission Using a Multi-Phase Feeding Programme in Broilers. *Span. J. Agric. Res.* **2013**, *11*, 736–746. [CrossRef]
34. Zhang, R.; Ishibashi, K.; Day, D.L. Experimental Study of Microbial Decomposition in Liquid Swine Manure, and Generation Rates of Ammonia. In Proceedings of the Livestock Waste Management Conference, American Society of Agricultural Engineers, St. Joseph, MI, USA, 16–21 June 1991.
35. Varel, V.H.; Wells, J.E.; Miller, D.N. Combination of a Urease Inhibitor and a Plant Essential Oil to Control Coliform Bacteria, Odour Production and Ammonia Loss from Cattle Waste. *J. Appl. Microbiol.* **2007**, *102*, 472–477. [CrossRef] [PubMed]
36. Santoso, U.; Ohtani, S.; Tanaka, K.; Sakaida, M. Dried Bacillus Subtilis Culture Reduced Ammonia Gas Release in Poultry House. *Asian-Australas. J. Anim. Sci.* **1999**, *12*, 806–809. [CrossRef]
37. Nahm, K.H. Evaluation of the Nitrogen Content in Poultry Manure. *World's Poult. Sci. J.* **2003**, *59*, 77–88. [CrossRef]

agriculture

MDPI

Article

Inclusion of *Citrullus colocynthis* Seed Extract into Diets Induced a Hypolipidemic Effect and Improved Layer Performance

Mohamed I. Alzarah [1], Abdulaziz A. Alaqil [2], Ahmed O. Abbas [2,3], Farid S. Nassar [3], Gamal M. K. Mehaisen [3,*], Gouda F. Gouda [4], Hanaa K. Abd El-Atty [5] and Eman S. Moustafa [3]

[1] Department of Environmental and Natural Resources, College of Agricultural and Food Sciences, King Faisal University, P.O. Box 420, Al-Ahsa 31982, Saudi Arabia; malzarah@kfu.edu.sa

[2] Department of Animal and Fish Production, College of Agricultural and Food Sciences, King Faisal University, P.O. Box 420, Al-Ahsa 31982, Saudi Arabia; aalaqil@kfu.edu.sa (A.A.A.); aabbas@kfu.edu.sa (A.O.A.)

[3] Department of Animal Production, Faculty of Agriculture, Cairo University, Gamma Street, Giza 12613, Egypt; Fidsaber_nassar@agr.cu.edu.eg (F.S.N.); emansm23@agr.cu.edu.eg (E.S.M.)

[4] Animal Breeding Section, Department of Animal Production, Faculty of Agriculture, Ain Shams University, P.O. Box 68, Haddaik Shoubra 11241, Cairo, Egypt; gouda_fathi@yahoo.com

[5] Department of Poultry Breeding, Animal Production Research Institute, Agricultural Research Center, Dokki, Giza 12611, Egypt; Hanaa.amin@arc.sci.eg

* Correspondence: gamoka@cu.edu.eg; Tel.: +20-122-679-7270

Citation: Alzarah, M.I.; Alaqil, A.A.; Abbas, A.O.; Nassar, F.S.; Mehaisen, G.M.K.; Gouda, G.F.; Abd El-Atty, H.K.; Moustafa, E.S. Inclusion of *Citrullus colocynthis* Seed Extract into Diets Induced a Hypolipidemic Effect and Improved Layer Performance. *Agriculture* **2021**, *11*, 808. https://doi.org/10.3390/agriculture11090808

Academic Editor: István Komlósi

Received: 13 July 2021
Accepted: 24 August 2021
Published: 26 August 2021

Publisher's Note: MDPI stays neutral with regard to jurisdictional claims in published maps and institutional affiliations.

Abstract: *Citrullus colocynthis* (CC) has been known as a natural medicinal plant with wide biological activities, including antioxidant, anti-inflammatory, and antilipidemic effects. The aim of this study was to investigate the effect of inclusion of the ethanolic extract of CC seeds (ECCs) into layer diets on the lipid profile, stress indicators, and physiological and productive performance of laying hens. A total of 216 forty-week-old commercial Hy-Line brown laying hens were randomly assigned into four equal groups (3 birds × 18 replicates per group) that received a basal diet supplemented with 0, 0.5, 1.0, and 2.0 g/kg of ECCs for 12 consecutive weeks. The first group served as a control. The results showed that ECCs at 1.0 and 2.0 g/kg significantly ($p < 0.05$) improved the productive and physiological performance compared to the other groups. In addition, stress indicators examined in the laying hens, including lipid peroxidation (malondialdehyde (MDA)), corticosterone hormone (CORT), tumor necrosis factor alpha (TNFα), and heat shock protein 70 (HSP70), were significantly alleviated after inclusion of ECCs into layer diets at the three levels compared to the control group. Furthermore, all ECC levels induced a significant reduction in plasma triglyceride (TG) and cholesterol (CH) levels in the plasma, liver, and egg yolk, whereas the highest levels were obtained with 2.0 g/kg of ECCs. Particularly important, a high linear correlation ($R^2 = 0.60$–0.79) was observed between increasing doses of ECCs and MDA, liver CH, and egg yolk CH concentrations and egg weight, feed intake, and feed conversion ratio; moreover, the correlation was extremely high ($R^2 = 0.80$–0.100) with the level of TG, CH, low-density lipoprotein CH, high-density lipoprotein CH, and CORT. These results indicated that dietary supplementation with 2.0 g/kg of ECCs could be considered a successful nutritional approach to producing healthier, lower-cholesterol eggs for consumers, in addition to enhancing the physiological and productive performance of laying hens by alleviating the stress of intensive commercial production.

Keywords: *Citrullus colocynthis*; lipid profile; stress indicators; physiological status; productive performance; laying hens

1. Introduction

Commercial egg production has become a substantial trend in the poultry industry because chicken eggs provide a valuable source of high-quality nutrients for human food and health, including proteins, fatty acids, vitamins, and minerals [1]. However, the consumption of foods rich in dietary cholesterol may increase the cholesterol (CH) levels in the individual blood and low-density lipoprotein particles [2]. These events can be associated with a high incidence of atherosclerotic cardiovascular disease (CVD), especially in old-aged individuals [3,4]. Chicken egg yolks are a major source of dietary CH in the human diet (200 mg of CH per one large egg yolk) and contribute approximately 25% of the daily CH consumption in both children and adults in Western countries [5,6]. Egg consumption in old-aged or diabetic individuals shows a significant increase in serum CH in low-density lipoprotein cholesterol (LDL-CH) than in high-density lipoprotein cholesterol (HDL-CH), resulting in an increase in the LDL-CH/HDL-CH ratio, which is an event highly associated with the risk of coronary heart disease in some populations [7]. In this context, some dietary guidelines report that dietary cholesterol intake of more than 300 mg/day or consumption of more than four egg yolks per week could cause harm to human health [8]. In addition, intensive farming practices for commercial egg production lead to increasing stress and sensitivity of laying hens to any unfavorable conditions [9]. The physiological status, inflammation, antioxidant defense, and adrenal hormones of laying hens are factors that are influenced by stress [10]. In commercial poultry, stressors such as heat, fasting, dehydration, transportation, and caging cause an increase in corticosterone secretion, which is known in birds as a stress hormone [11,12]. Lipid peroxidation is another indicator for the oxidative stress status, and many studies have evidenced its correlation with atherosclerotic CVD [13,14]. Such stress markers and other markers such as plasma protein and T_3 hormone could be used, at the same time, as indicators reflecting the physiological performance of laying hens [15–17].

In recent years, a variety of natural products and medicinal plants have been widely accepted as alternative nutritional strategies for lowering egg cholesterol, maintaining the performance, and improving the productivity of laying hens [18–20]. *Citrullus colocynthis* (CC) belongs to the cucurbitaceous plant family and commonly grows in desert areas, including Egypt and Saudi Arabia [21]. This plant is well known in many countries as a traditional medicine for the treatment of some chronic diseases [22]. Previous research reported that CC has a wide range of biological activities that protect tissues and cells against oxidation [23–25], diabetes [23,26], hyperlipidemia [27,28], microbial infection [29,30], and inflammation [31,32]. It was also shown that CC possesses a high nutritional value due to its high contents of proteins, essential minerals, and beneficial fatty acids [33].

Most pharmacological and nutritional properties of CC have been studied in human or mammalian animals, such as rabbits and rats. For example, a significant decrease in total cholesterol, triglycerides, phospholipids, and lipid peroxidation was found in the serum and liver of diabetic rats treated with the ethanol extract of CC fruits at a dosage of 300 mg/kg [23]. Additionally, administration of CC seeds to hyperlipidemic patients [27] or rabbits [28] induced a significant reduction in serum cholesterol and triglycerides. An anti-inflammatory effect was also observed in rats intraperitoneally injected with 4 mg/kg of a water extract of fresh CC [32]. In contrast, there is no updated research available on the possible effect of CC administration to poultry species. Therefore, the current study was carried out to evaluate, for the first time, the impact of inclusion of extracted CC seeds into layer diets on the lipid profile, stress indicators, physiological status, and productive performance of laying hens.

2. Materials and Methods

2.1. Ethical Statement

All experimental protocols in the present study were compliant with the relevant research ethics guidelines of King Faisal University and the National Committee of Medical and Bioethics. Ethical approval for the study was obtained from the Research Ethics Committee (REC) at King Faisal University (KFU-REC/2021-02-17).

2.2. CC Seed Extraction and Analysis

CC fruits were obtained from different herbalists in Al-Ahsa, Saudi Arabia. Seeds were isolated manually from the pulp of the fruits, and then the collected seeds were dried and ground into a powder using a grinder (Moulinex Type LM201, Mayenne, France). The powder was then extracted three times using a mixture of 30% ethanol in distilled water. After that, the hydro-ethanol extracts were pooled and concentrated after removing ethanol residues by a vacuum evaporator at a maximum temperature of 50 °C [28]. The extracted CC seeds (ECCs) were kept as a powder in a clean, dark bottle at 4 °C until use in the experiment.

The chemical analysis of ECCs was performed following the methods of AOAC International [34] for caloric content (method 971.10), dry matter (method 930.15), crude fiber (method 920.169), total fat (method 950.54), crude protein (method 954.01), total ash (method 942.05), and calcium, potassium, magnesium, phosphorus, iron, and sodium minerals (method 985.35), and the results are presented in Table 1. In addition, ECCs were subjected to a qualitative phytochemical screening for the detection of various active chemical constituents according to methods described by Karumi et al. [35]. The presence of saponins was tested by foam appearance and persistence for at least 15 min after vigorous shaking of ECCs with distilled water. The presence of tannins was tested by the appearance of a green precipitate after adding ferric chloride reagent to a filtrated mixture of ECCs in distilled water. The presence of flavonoids was examined by Shinoda's test (cyanidin reaction) based on red/orange coloration after treating ECCs with metallic magnesium and concentrated hydrochloric acid. The reducing compounds were tested by observing brick-red precipitation resulting from boiling ECCs in Fehling's solution. The presence of alkaloids was tested by the appearance of turbidity or precipitation after treating the ECCs with Dragendorff's and Mayer's reagents. The presence of terpenoids was examined by the Liebermann–Burchard test based on the observation of a dark-green color when ECCs were treated with chloroform, acetic anhydride, and a few drops of sulfuric acid. The test for quinones was performed by treating a small amount of ECCs with concentrated hydrochloric acid and observing the formation of a yellow-colored precipitate. Finally, coumarins were examined by using the Erlich test based on the development of an orange color when ECCs were treated with a solution of dimethylamino-benzaldehyde (5% ethanol) and then acidified by hydrochloric acid. The results of phytochemical analysis revealed the presence of saponins, catechic tannins, flavonoids, and coumarins in the ECCs (Table 1).

Table 1. The chemical analysis and phytochemical screening of extracted *Citrullus colocynthis* seeds (ECCs).

Chemical Analysis	Result [1]	Phytochemical Screening	Result [2]
Dry matter (DM)	94.09%	Saponins	+
Caloric content	41.33 MJ	Tannins	+
Crude fiber	453.7 g	Flavonoids	+
Total fat	172.9 g	Reducing compounds	−
Crude protein	129.5 g	Alkaloids	−
Total ash	25.8 g	Terpenoids	−
Calcium	7.6 g	Quinones	−
Potassium	6.9 g	Coumarins	+
Magnesium	2.6 g		
Phosphorus	0.3 g		
Iron	0.2 g		
Sodium	0.1 g		

[1] Results of chemical analyses calculated per kg DM. [2] Positive test (+); negative test (−).

2.3. Birds and Experimental Protocol

Two hundred sixteen commercial Hy-Line brown laying hens (40-weeks-old) were settled in cages (3 layers/cage) in an open housing system. The hens were kept in standard environmental and hygienic conditions, with a photoperiod of 16 h light and 8 h darkness, and were provided with ad libitum water and feed throughout the experimental period. The hens were randomly allocated into four equal groups (18 replicates × 3 hens each per group), including the control group, which received a basal diet without supplementation, and the experimental groups 1, 2, and 3, which received a basal diet supplemented with 0.5, 1.0, and 2.0 g of ECCs per kg diet, respectively. The basal diet was formulated in grist form (Table 2) to meet the nutritional recommendations for brown commercial layers under the management guide of Hy-Line International (available at https://www.hyline.com/varieties/brown. Accessed on 23 August 2021). ECCs powder was mixed daily with the basal diet before introducing it to the laying hens. The trials were continued for 12 consecutive weeks (from 40 to 52 weeks of age).

Table 2. The nutrient ingredients and calculated and determined chemical analyses of the basal diet.

Ingredients [1]	Content (g/kg)
Yellow corn	565.5
Soybean meal (44%)	276.0
Wheat bran	10.0
Soybean oil	30.0
Bone meal	30.0
Limestone	80.0
Salt (NaCl)	4.0
Premix *	3.0
DL-methionine	1.5
Calculated chemical analysis [2]	
Metabolizable energy (MJ/kg)	1.26
Crude protein (g/kg)	174.7
Calcium (g/kg)	40.2
Available phosphorus (g/kg)	5.2
Lysine (g/kg)	9.5
Methionine (g/kg)	4.2
Linoleic acid (g/kg)	28.8
Determined chemical analysis [2]	
Dry matter (g/kg)	890.0
Crude protein (g/kg)	167.5
Crude fat (g/kg)	66.0
Crude fiber (g/kg)	47.0
Total ash (g/kg)	129.0
Calcium (g/kg)	42.2
Available phosphorus (g/kg)	4.2

[1] Ingredients were formulated as g per kg of the basal diet. [2] Calculated and determined chemical analyses were expressed per kg of the basal diet. * Premix provided the following vitamins and minerals per kg of the basal diet: 8000 IU of vitamin A, 1500 IU of vitamin D, 4 mg of riboflavin, 10 μg of cobalamin, 15 mg of vitamin E, 2 mg of vitamin K, 500 mg of choline, 25 mg of niacin, 60 mg of manganese, and 50 mg of zinc.

2.4. Productive Performance

The total egg number (EN), egg weight (EW), and feed intake (FI) per cage were recorded daily during the whole experimental period. The average hen–day egg production (%), average EW (g), and average FI (g) were then calculated for the entire experimental period per treatment group. The feed conversion ratio was calculated per cage on the basis of the total feed consumption per total egg mass (EM = EN × EW) during the whole experimental period.

2.5. Blood Samples

At the end of the experimental period (52 weeks of age), blood samples were collected from the brachial vein of 18 hens per treatment group (6 replicates per group) and transferred into heparinized tubes. To avoid handling stress, the birds were quickly bled within 3 min during the night [36]. Blood samples were centrifuged at 2000× g for 10 min at 4 °C, and then the plasma was separated and stored at −20 °C until further assay of physiological parameters and stress indicators.

2.5.1. Physiological Parameters

Plasma total protein (TP), alanine amino transferase (ALT), and aspartate amino transferase (AST) concentrations were determined for each treatment group using available colorimetric kits (ab102535, ab241035, and ab105135, respectively; Abcam, Waltham, MA, USA). The concentration of triiodothyronine (T_3) was measured using ELISA diagnostic kits specific for chickens (MBS269454; MyBioSource, San Diego, CA, USA). The intra-assay and inter-assay coefficients of variability (CVs) for T_3 analysis were less than 8% and 12%, respectively, with a detection range of 0.3–20 µM/mL.

2.5.2. Stress Indicators

Malondialdehyde (MDA) levels in the plasma were determined using quantitative colorimetric assay kits (ab118970; Abcam). Plasma corticosterone (CORT), tumor necrosis factor alpha (TNFα), and heat shock protein 70 (HSP70) levels were quantified using ELISA kits specific for chickens (MBS701668, MBS2509660, and MBS2702636, respectively; MyBioSource). The intra-assay and inter-assay CVs were less than 8% and 10% for CORT, 5.57% and 5.89% for TNFα, and 10% and 12% for HSP70, respectively. The detection ranges were 0.5–20 ng/mL, 31.25–2000 pg/mL, and 1.56–100 ng/mL for CORT, TNFα, and HSP70, respectively.

2.6. Lipid Profile

At the end of the experiment (52 weeks of age), 18 hens per treatment group (3 hens per replicate × 6 replicates per group) were fasted for 12 h, then blood samples were obtained from the brachial vein into heparinized tubes, and plasma was separated, as mentioned above. Plasma triglycerides (TGs) were analyzed using colorimetric techniques of enzymatic diagnostic kits (ab65336, Abcam), while total plasma cholesterol (CH), high-density lipoprotein cholesterol (HDL-CH), and low-density lipoprotein cholesterol (LDL-CH) were separated and quantified according to the kits' protocols (ab65390, Abcam).

Livers were also removed from 18 hens per treatment group (3 hens per replicate × 6 replicates per group) after slaughtering at the end of the experiment (52 weeks of age) and were immediately stored at −20 °C until further processing. In addition, 18 eggs per group were gathered randomly in the last week of the experiment (3 eggs per replicate × 6 replicates per group) and broken to separate yolks for further analysis. Cholesterol was measured in the liver and egg yolk samples using cholesterol/cholesteryl ester detection kits (ab102515, Abcam) following the manufacturer's protocol with a minor modification. Briefly, 1 g of the liver/yolk sample was homogenized with 20 mL of chloroform:isopropanol:NP-40 lysis buffer (7:11:0.1) and then centrifuged for 5 min at 15,000× g. The liquid phase was transferred to a new tube and dried under vacuum at 50 °C for 30 min. The samples were then dissolved in 200 µL of cholesterol assay buffer and vigorously shaken using a vortex until homogeneous. The cholesterol content in 50 µL of reaction mixtures was determined at 450 nm against a standard curve generated using 0.25 mg/mL of cholesterol standard solution.

2.7. Statistical Analysis

The experimental unit for statistical analysis was the replicate (cage). Six replicates per group were statistically analyzed for the lipid profile, stress indicators, and physiological parameters, considering the promediate value of 3 samples per cage as a value of the replicate ($n = 6$), while the productive performance data of 3 hens per cage was averaged for the entire experimental period and used as a replicate for the statistical analysis ($n = 18$ replicates per group). All data were statistically analyzed with the one-way analysis of variance (ANOVA) procedure with respect to the ECCs supplementation levels as independent variables. Orthogonal polynomial contrast and regression analyses were also performed to test the linear and quadratic trends in the increasing ECCs doses applied in the present study. Results were expressed as means ± standard deviation (SD), and the differences between treatment groups were determined using Duncan's post hoc test. The level of statistical significance was set at $p < 0.05$. All statistical analyses were performed using the SPSS software package (version 22.0; IBM Corp., Armonk, NY, USA, 2013).

3. Results

3.1. Lipid Profile

The effect of ECCs treatment on the lipid profile of laying hens is shown in Table 3. Plasma TG, CH, and LDL-CH were significantly ($p < 0.05$) reduced, while plasma HDL-CH was significantly increased by ECCs inclusion in the diets of laying hens. Results showed that there was a significant linear effect of increasing doses of ECCs on the plasma lipid profile (R^2 = 97.3, 97.0, 96.4, and 80.4% for TG, CH, LDL-CH, and HDL-CH, respectively; $p < 0.05$). It was found that ECCs at 0.5 g/kg feed significantly ($p < 0.05$) decreased the liver and egg yolk CH compared to the control group. The higher doses of ECCs (1.0–2.0 g/kg) significantly decreased the liver and egg yolk CH when compared to the lower doses (0–0.5 g/kg), showing a linear trend (R^2 = 75.3 and 69.2%, respectively) and a quadratic trend (R^2 = 22.6 and 21.6%, respectively).

Table 3. Effect of dietary extracted *Citrullus colocynthis* seeds (ECCs) supplementation on the lipid profile of laying hens.

Dietary ECCs Groups	Plasma TG (mg/dL)	Plasma CH (mg/dL)	Plasma HDL-CH (mg/dL)	Plasma LDL-CH (mg/dL)	Liver CH (mg/g)	Egg Yolk CH (mg/g)
Control	207.4 ± 1.59 [a]	152.3 ± 1.54 [a]	40.8 ± 2.27 [c]	113.2 ± 2.11 [a]	5.4 ± 0.18 [a]	12.3 ± 0.43 [a]
Group 1	200.0 ± 1.83 [b]	146.8 ± 1.62 [b]	42.3 ± 2.18 [c]	106.0 ± 2.82 [b]	4.2 ± 0.11 [b]	11.1 ± 0.29 [b]
Group 2	188.6 ± 2.53 [c]	135.5 ± 2.69 [c]	47.3 ± 1.06 [b]	89.8 ± 3.05 [c]	3.3 ± 0.03 [c]	10.4 ± 0.07 [c]
Group 3	173.1 ± 2.32 [d]	120.3 ± 1.68 [d]	54.1 ± 4.17 [a]	67.6 ± 4.08 [d]	3.2 ± 0.09 [c]	10.2 ± 0.27 [c]
Polynomial contrast test; sum of squares (p-value)						
Combined	4008.929 (<0.001)	3581.897 (<0.001)	647.676 (<0.001)	7356.190 (<0.001)	18.601 (<0.001)	17.142 (<0.001)
Linear term	3985.304 (<0.001)	3549.191 (<0.001)	635.222 (<0.001)	7273.272 (<0.001)	14.206 (<0.001)	13.065 (<0.001)
Quadratic term	2.613 (0.451)	0.454 (0.732)	3.101 (0.516)	5.839 (0.444)	4.263 (<0.001)	4.065 (<0.001)
Regression analysis; R^2 (p-value)						
Linear effect	0.973 (<0.001)	0.970 (<0.001)	0.804 (<0.001)	0.964 (<0.001)	0.753 (<0.001)	0.692 (<0.001)
Quadratic effect	0.000 (0.486)	0.001 (0.769)	0.004 (0.519)	0.000 (0.507)	0.226 (<0.001)	0.216 (<0.001)

Data are presented as means ± SD for six replicates per group (n = 6). Means within the same column with different superscripts are significantly different ($p < 0.05$). R-squared (R^2) represents the linear and quadratic regression effects of increasing ECCs doses in diets on the lipid profile parameters of laying hens. Dietary ECCs groups: the control group received a basal diet without supplementation, while experimental groups 1, 2, and 3 received a basal diet supplemented with 0.5, 1.0, and 2.0 g of ECCs per kg feed, respectively. TG, triglyceride; CH, cholesterol; HDL-CH, high-density lipoprotein cholesterol; LDL-CH, low-density lipoprotein cholesterol.

3.2. Stress Indicators

The effect of ECCs treatment on the plasma stress indicators of laying hens is presented in Table 4. A significant ($p < 0.05$) decrease was observed in the plasma MDA of laying hens supplemented with 1–2 g/kg of ECCs compared to those hens supplemented with 0–0.5 g/kg of ECCs. The plasma CORT, TNFα, and HSP70 were significantly lowered ($p < 0.05$) by 0.5–2 g/kg of ECCs compared to the control group. Moreover, there was a significant ($p < 0.05$) linear effect of increasing doses of ECCs on the plasma stress indicators, whereas the quadratic effect was found to be statistically insignificant ($p > 0.05$). Results also showed that increasing doses of ECCs were highly correlated with plasma CORT and MDA (R^2 = 88.1 and 73.8%, respectively) than with TNFα and HSP70 (R^2 = 52.3 and 52.9%, respectively) of laying hens.

Table 4. Effect of dietary extracted *Citrullus colocynthis* seeds (ECCs) supplementation on some stress indicators of laying hens.

Dietary ECCs Groups	MDA	CORT	TNFα	HSP70
	(µM/mL)	(ng/mL)	(pg/mL)	(ng/mL)
Control	2.84 ± 0.174 [a]	4.88 ± 0.129 [a]	94.62 ± 0.515 [a]	24.13 ± 0.897 [a]
Group 1	2.64 ± 0.386 [a]	4.35 ± 0.288 [b]	93.37 ± 0.683 [b]	23.39 ± 1.188 [b]
Group 2	2.14 ± 0.190 [b]	3.72 ± 0.338 [c]	93.25 ± 0.649 [b]	22.07 ± 1.069 [b]
Group 3	1.79 ± 0.167 [c]	3.01 ± 0.243 [d]	92.34 ± 1.038 [c]	21.34 ± 1.006 [c]
Polynomial contrast test; sum of squares (*p*-value)				
Combined	4.112 (<0.001)	11.726 (<0.001)	15.812 (<0.001)	28.484 (<0.001)
Linear term	3.931 (<0.001)	11.529 (<0.001)	14.095 (<0.001)	26.611 (<0.001)
Quadratic term	0.052 (0.367)	0.158 (0.143)	0.813 (0.242)	1.156 (0.316)
Regression analysis; R^2 (*p*-value)				
Linear effect	0.738 (<0.001)	0.881 (<0.001)	0.523 (<0.001)	0.529 (<0.001)
Quadratic effect	0.010 (0.379)	0.012 (0.139)	0.030 (0.248)	0.023 (0.311)

Data are presented as means ± SD for six replicates per group ($n = 6$). Means within the same column with different superscripts are significantly different ($p < 0.05$). R-squared (R^2) represents the linear and quadratic regression effects of increasing ECCs doses in diets on the stress indicators of laying hens. Dietary ECCs groups: the control group received a basal diet without supplementation, while experimental groups 1, 2, and 3 received a basal diet supplemented with 0.5, 1.0, and 2.0 g of ECCs per kg feed, respectively. MDA, plasma malondialdehyde; CORT, plasma corticosterone; TNFα, plasma tumor necrosis factor alpha; HSP70, plasma heat shock protein 70.

3.3. Physiological Parameters

The effect of ECCs treatment on the physiological parameters of laying hens is shown in Table 5. Supplementing the diet with ECCs at 1.0 g/kg significantly ($p < 0.05$) increased the plasma TP and T_3 by 17.5% and 46.2%, respectively, compared to the control group. Increasing the level of ECCs in the diet to 2.0 g/kg induced a linear increase in the plasma TP and T_3 of laying hens ($R^2 = 69.3\%$ and 85%, respectively; $p < 0.05$). In contrast, supplementing the diets with 1.0–2.0 g/kg of ECCs significantly ($p < 0.05$) decreased the plasma ALT and AST of laying hens compared to the hens supplemented with 0–0.5 g/kg of ECCs. Moreover, increasing the ECCs doses had significant ($p < 0.05$) linear and quadratic effects that explained 57.5% and 10.5% of the total variance for ALT and 67.2% and 15.4% of the total variance for AST, respectively.

3.4. Productive Performance

Results of the productive performance of laying hens that fed on diets with or without ECCs supplementation are represented in Table 6. The ECCs treatment at 1.0–2.0 g/kg significantly ($p < 0.05$) increased the hen–day egg production by at least 2.5 p.p. compared to the ECCs treatment at 0–0.5 g/kg. The average egg weight was significantly ($p < 0.05$) increased in all experimental groups by approximately 0.6, 1.3, and 2.2 g in the groups treated with 0.5, 1.0, and 2.0 g/kg of ECCs, respectively, compared to the control group. In contrast, the feed intake of laying hens was significantly ($p < 0.05$) reduced by 3.1% and 4.2% in the groups treated with 0.1 and 2.0 g/kg of ECCs, respectively, compared to the control group. The feed conversion ratio of hens treated with 1.0 and 2.0 g/kg of ECCs was significantly (P < 0.05) improved by approximately 7.9% and 10.8%, respectively, compared to the control group. Furthermore, it was found that the egg production and egg weight of laying hens were significantly increased in a linear trend with increasing doses of ECCs in the diets ($R^2 = 39.5\%$ and 70.2%, respectively; $p < 0.05$). In contrast, feed intake and feed conversion ratio were significantly decreased in a linear trend with increasing doses of ECCs in the diets ($R^2 = 72.9\%$ and 75.9%, respectively; $p < 0.05$), while the quadratic effect of ECCs doses on the productive performance was not significant ($p > 0.05$).

Table 5. Effect of dietary extracted *Citrullus colocynthis* seeds (ECCs) supplementation on some physiological parameters of laying hens.

Dietary ECCs Groups	TP	T_3	ALT	AST
	(g/dL)	(µM/mL)	(U/mL)	(U/mL)
Control	4.63 ± 0.579 [c]	5.69 ± 0.859 [c]	13.71 ± 1.344 [a]	30.88 ± 2.635 [a]
Group 1	4.99 ± 0.122 [bc]	6.15 ± 0.544 [c]	12.41 ± 1.421 [a]	25.46 ± 1.551 [b]
Group 2	5.44 ± 0.298 [b]	8.32 ± 0.619 [b]	10.10 ± 0.894 [b]	21.99 ± 1.337 [c]
Group 3	6.05 ± 0.379 [a]	10.05 ± 0.627 [a]	9.92 ± 0.459 [b]	20.73 ± 2.133 [c]
Polynomial contrast test; sum of squares (p-value)				
Combined	6.713 (<0.001)	73.422 (<0.001)	61.039 (<0.001)	371.368 (<0.001)
Linear term	6.671 (<0.001)	70.065 (<0.001)	48.976 (<0.001)	302.328 (<0.001)
Quadratic term	0.026 (0.680)	0.046 (0.754)	8.970 (0.013)	69.039 (<0.001)
Regression analysis; R^2 (p-value)				
Linear effect	0.693 (<0.001)	0.850 (<0.001)	0.575 (<0.001)	0.672 (<0.001)
Quadratic effect	0.003 (0.673)	0.000 (0.784)	0.105 (0.016)	0.154 (<0.001)

Data are presented as means ± SD for six replicates per group (n = 6). Means within the same column with different superscripts are significantly different ($p < 0.05$). R-squared (R^2) represents the linear and quadratic regression effects of increasing ECCs doses in diets on the physiological parameters of laying hens. Dietary ECCs groups: the control group received a basal diet without supplementation, while experimental groups 1, 2, and 3 received a basal diet supplemented with 0.5, 1.0, and 2.0 g of ECCs per kg feed, respectively. TP, plasma total protein; T_3, plasma triiodothyronine; ALT, plasma alanine amino transferase; AST, plasma aspartate amino transferase.

Table 6. Effect of dietary extracted *Citrullus colocynthis* seeds (ECCs) supplementation on the productive performance of laying hens.

Dietary ECCs Groups	Egg Production	Egg Weight	Feed Intake	Feed Conversion
	(%)	(g)	(g/d)	(kg FI/kg EM)
Control	91.0 ± 1.51 [b]	62.1 ± 0.48 [d]	114.3 ± 0.88 [a]	2.03 ± 0.044 [a]
Group 1	91.5 ± 2.36 [b]	62.7 ± 0.67 [c]	114.7 ± 0.90 [a]	2.00 ± 0.061 [a]
Group 2	93.3 ± 1.08 [a]	63.4 ± 0.57 [b]	110.8 ± 1.07 [b]	1.87 ± 0.032 [b]
Group 3	94.3 ± 1.09 [a]	64.3 ± 0.47 [a]	109.5 ± 0.53 [c]	1.81 ± 0.028 [c]
Polynomial contrast test; sum of squares (p-value)				
Combined	126.775 (<0.001)	51.092 (<0.001)	359.014 (<0.001)	0.596 (<0.001)
Linear term	118.734 (<0.001)	50.518 (<0.001)	298.909 (<0.001)	0.549 (<0.001)
Quadratic term	1.819 (0.402)	0.386 (0.266)	0.721 (0.331)	0.006 (0.077)
Regression analysis; R^2 (p-value)				
Linear effect	0.395 (<0.001)	0.702 (<0.001)	0.729 (<0.001)	0.759 (<0.001)
Quadratic effect	0.006 (0.407)	0.006 (0.264)	0.001 (0.505)	0.008 (<0.120)

Data are presented as means ± SD for 18 replicates per group (n = 18). Means within the same column with different superscripts are significantly different ($p < 0.05$). R-squared (R^2) represents the linear and quadratic regression effects of increasing ECCs doses in diets on the productive performance of laying hens. Dietary ECCs groups: the control group received a basal diet without supplementation, while experimental groups 1, 2, and 3 received a basal diet supplemented with 0.5, 1.0, and 2.0 g of ECCs per kg feed, respectively. FI, feed intake; EM, egg mass. All productive parameters were calculated as an average of the entire experimental period (12 weeks).

4. Discussion

The doses of ECCs used in the current study were determined based on a preliminary study on a group of laying hens. It was found that notable effects of ECCs on the egg productivity of laying hens are obtained at levels of 0.5–2.0 g per kg of diet, while adverse effects are observed at higher levels. These levels were used in the present study, and they were safe and did not display any signs of toxicity on the laying hens during the experimental period.

Results demonstrated that basal diets supplemented with ECCs lower the lipid profile by decreasing plasma TG, CH, and LDL-CH, as well as decreasing liver and egg yolk CH, while increasing plasma HDL-CH. These results may present the most important advantage of ECCs inclusion into layer diets in view of producing healthier eggs that contain lower yolk CH for humans [3], especially the elderly, who are more susceptible to the chronic inflammatory disease of the arteries, atherosclerosis [2]. Such antilipidemic activity of CC seeds has been widely demonstrated in previous studies on both animals and humans [33]. It was found that the ethanolic extract of CC plants lowers the TG and CH levels in hyperlipidemic rabbits [28] and that administration of CC seed powder to hyperlipidemic non-diabetic patients also lowers the TG and CH levels [27]. Furthermore, a significant reduction in hyperlipidemia was induced by the administration of CC extracts to diabetic rats [23]. The hypotriglyceridemic and hypocholesterolemic effects of ECCs may be due to the catabolism or removal of lipoproteins in the liver [37] and the inhibition of specific hepatic enzymes responsible for lysosomal lipid hydrolysis [38]. Dallak et al. [39] also reported that CC pulp extracts depress the hepatic gluconeogenesis and lipogenesis in diabetic rats through downregulation of gluconeogenetic enzymes and increasing of insulin secretion. The phytochemical screening of ECCs in the present study revealed the presence of saponins and catechic tannins, which are bioactive components that contribute to hypolipidemic activity [25]. It is known that both components can participate in the suppression of lipid and cholesterol absorption in the gut via dietary fat β-oxidation, enzymatic CH biosynthesis inhibition, and/or interference with CH–micellar affinity [40,41]. Due to the intensive system of modern egg production and the continuous selection for high egg yield, commercial laying hens commonly are stressed and sensitive to any unfavorable conditions [12]. Plasma MDA levels are the principal measurement for oxidative stress status in animal plasma and tissues [13]. The lower MDA levels in the hens treated with ECCs may be attributed to the direct free-radical-scavenging activity of polyphenols and flavonoids in the ECCs [25,42]. In birds, CORT is the main end product of the hypothalamic–pituitary–adrenal (HPA) axis response to stress [11]. So, lower plasma CORT levels in laying hens treated with ECCs suggests that ECC treatment can alleviate the stress induced by intensive production management in laying hens [43]. However, lower blood CORT in the ECC-treated groups can also be associated more with the lower blood cholesterol levels shown in the same groups [44]. Moreover, the TNFα and HSP70 in laying hens were significantly decreased by ECC treatment, whereas TNFα, as a pro-inflammatory cytokine, plays a critical role in inflammation and immune responses during stress [45] and HSP70 is mainly synthesized to protect stressed cells [46]. The reduction in plasma TNFα in ECC-treated groups may be explained by the decrease in CORT in the ECC-treated groups, as this correlation between CORT secretion and TNFα expression has been previously reported in stressed mice [47,48]. In contrast, HSP70 expression can be promoted more as a consequence of the elevated levels of MDA in order to protect cells against peroxidation damage [49], and this may explain the lower levels of both HSP70 and MDA in the ECC-treated groups in the present study.

The reduction in plasma ALT and AST levels could indicate that ECCs can maintain the liver functions of layers. Similar effects of CC extracts were obtained in rats [50]; moreover, CC treatment can ameliorate the deleterious effects of heat stress on liver tissues of broilers [51]. The increased plasma TP in ECC-treated groups may be also due to such hepatoprotective properties of CC extracts. It was reported that a healthy liver orchestrates the metabolism of proteins and amino acids [52]. It was further reported that the hepatoprotective activity of medicinal fungus in rats increases plasma albumin, which is the major plasma protein produced by the liver and then circulated in the bloodstream [16]. Moreover, ECC treatment increased plasma T_3 in the present study, thereby increasing metabolism [17] and, consequently, improving the performance of laying hens [53].

In general, layer productive performance was enhanced in a linear trend by increasing the doses of ECCs. The positive effects of ECCs on layer egg production and egg weight could be correlated to the availability of protein, essential amino acids, and fatty acids in CC extracts [54]. As mentioned earlier, the bioactive components of ECCs alleviated stress; therefore, birds treated with ECCs suppressed the HPA axis pathway and displayed low levels of CORT. Consequent to the opposite effect of CORT on FI [55], it was expected that FI would increase in these groups; instead, it significantly decreased (Table 6). The decreased FI in the ECC-treated groups may be due to the bitter taste of CC seeds, which may affect the palatability of the diet [56]. In contrast, the gastrointestinal tract in the ECC-treated groups seems to be more efficient in the digestibility and bioavailability of nutrients due to the possible antimicrobial effect of tannins and flavonoids presented in the CC extracts [57]. However, neither the cecal microbiota nor the antimicrobial effect were analyzed in the present study. Furthermore, while activation of the HPA-CORT pathway increases the energy expenditure to maintain the immunity and survival of laying hens, the extremely inhibitory effect of ECC treatment on CORT may trigger energy investment from maintenance toward productivity [58]. Taking together, the positive events displayed as a result of ECC treatment in the present study, such as alleviating stress and improving the physiological status, could be considered a great suppressor for intestinal and reproductive pathologies, thus enhancing the performance of laying hens [59].

5. Conclusions

The dietary supplementation of ECCs at the level of 0.5 g/kg had limited effects on layer performance. However, higher levels of ECCs at 1.0 and 2.0 g/kg showed a significant improvement in the productive and physiological performance of laying hens. In addition, there was a significant hypolipidemic effect and stress alleviation in laying hens after inclusion of ECCs into diets at the three levels, compared to the control group, whereas the highest levels were observed with 2.0 g/kg of ECCs. Furthermore, a high linear correlation was noticed between increasing doses of ECCs and MDA, liver CH, and egg yolk CH concentrations and egg weight, feed intake, and feed conversion ratio; moreover, the correlation was extremely high with the levels of TG, CH, LDL-CH, HDL-CH, and CORT. These results indicate that dietary supplementation with 2.0 g/kg of ECCs could be a promising nutritional approach to producing healthier, lower-cholesterol eggs for consumers, in addition to enhancing the physiological and productive performance of laying hens by alleviating the stress of intensive commercial production.

Author Contributions: Conceptualization, M.I.A., A.A.A., A.O.A. and G.M.K.M.; methodology, M.I.A., A.A.A. and A.O.A.; validation, G.F.G., G.M.K.M. and E.S.M.; formal analysis, F.S.N., G.M.K.M., G.F.G. and E.S.M.; investigation, M.I.A., A.A.A., A.O.A., G.M.K.M., G.F.G., H.K.A.E.-A. and E.S.M.; resources, M.I.A., A.A.A. and H.K.A.E.-A.; data curation, A.O.A., G.M.K.M. and H.K.A.E.-A.; writing—original draft, G.M.K.M. and A.O.A.; writing—review and editing, G.M.K.M., G.F.G. and E.S.M.; visualization, A.O.A. and H.K.A.E.-A.; supervision, A.O.A.; project administration, A.O.A.; funding acquisition, M.I.A., A.A.A., A.O.A. and G.M.K.M. All authors have read and agreed to the published version of the manuscript.

Funding: This research was supported by funds obtained from the College of Agricultural and Food Sciences at King Faisal University and from the General Scientific Research Department at Cairo University (GSRD-CU; http://gsrd.cu.edu.eg/. Accessed on 23 August 2021).

Institutional Review Board Statement: The study was conducted according to the guidelines of King Faisal University and the National Committee of Medical and Bioethics and approved by the Research Ethics Committee (REC) at King Faisal University (KFU-REC/2021-02-17).

Data Availability Statement: Not applicable.

Conflicts of Interest: The authors declare no conflict of interest. The funders had no role in the design of the study; in the collection, analyses, or interpretation of data; in the writing of the manuscript; or in the decision to publish the results.

References

1. Jeloka, T.K.; Pandit, M.; Dharmatti, G.; Jamdade, T. Are oral protein supplements helpful in the management of malnutrition in dialysis patients? *Indian J. Nephrol.* **2013**, *23*, 1–4. [CrossRef] [PubMed]
2. Lewington, S.; Whitlock, G.; Clarke, R.; Sherliker, P.; Emberson, J.; Halsey, J.; Qizilbash, N.; Peto, R.; Collins, R. Blood cholesterol and vascular mortality by age, sex, and blood pressure: A meta-analysis of individual data from 61 prospective studies with 55000 vascular deaths. *Lancet* **2007**, *370*, 1829–1839. [CrossRef] [PubMed]
3. Weggemans, R.M.; Zock, P.; Katan, M.B. Dietary cholesterol from eggs increases the ratio of total cholesterol to high-density lipoprotein cholesterol in humans: A meta-analysis. *Am. J. Clin. Nutr.* **2001**, *73*, 885–891. [CrossRef] [PubMed]
4. Law, M.R.; Wald, N.J.; Rudnicka, A. Quantifying effect of statins on low density lipoprotein cholesterol, ischaemic heart disease, and stroke: Systematic review and meta-analysis. *BMJ* **2003**, *326*, 1423. [CrossRef] [PubMed]
5. Keast, D.R.; Fulgoni, V.L.; Nicklas, T.A.; O'Neil, C.E. Food sources of energy and nutrients among children in the United States: National Health and Nutrition Examination Survey 2003–2006. *Nutrients* **2013**, *5*, 283–301. [CrossRef] [PubMed]
6. O'Neil, C.E.; Keast, D.R.; Fulgoni, V.L.; Nicklas, T.A. Food sources of energy and nutrients among adults in the US: NHANES 2003–2006. *Nutrients* **2012**, *4*, 2097–2120. [CrossRef] [PubMed]
7. Fernandez, M.L.; Webb, D. The LDL to HDL cholesterol ratio as a valuable tool to evaluate coronary heart disease risk. *J. Am. Coll. Nutr.* **2008**, *27*, 1–5. [CrossRef]
8. Mozaffarian, D.; Ludwig, D. The 2015 US dietary guidelines. *JAMA* **2015**, *313*, 2421–2422. [CrossRef]
9. Downing, J.A.; Bryden, W.L. A non-invasive test of stress in laying hens. *RIRDC Publ.* **2002**, *1–118*.
10. Lara, L.; Rostagno, M.H. Impact of heat stress on poultry production. *Animals* **2013**, *3*, 356–369. [CrossRef]
11. Charmandari, E.; Tsigos, C.; Chrousos, G. Endocrinology of the stress response. *Annu. Rev. Physiol.* **2005**, *67*, 259–284. [CrossRef]
12. Singh, R.; Cook, N.; Cheng, K.M.; Silversides, F.G. Invasive and noninvasive measurement of stress in laying hens kept in conventional cages and in floor pens. *Poult. Sci.* **2009**, *88*, 1346–1351. [CrossRef]
13. Pham-Huy, L.A.; He, H.; Pham-Huy, C. Free radicals, antioxidants in disease and health. *Int. J. Biomed. Sci.* **2008**, *4*, 89–96.
14. Bahoran, T.; Soobrattee, M.; Luximon-Ramma, V.; Aruoma, O. Free radicals and antioxidants in cardiovascular health and disease. *Internet J. Med. Update* **2007**, *1*, 25–41. [CrossRef]
15. Mumma, J.O.; Thaxton, J.P.; Vizzier-Thaxton, Y.; Dodson, W.L. Physiological stress in laying hens. *Poult. Sci.* **2006**, *85*, 761–769. [CrossRef] [PubMed]
16. Chiu, H.-W.; Hua, K.-F. Hepatoprotective effect of wheat-based solid-state fermented *Antrodia cinnamomea* in carbon tetrachloride-induced liver injury in rat. *PLoS ONE* **2016**, *11*, e0153087. [CrossRef]
17. Al-Ardi, M.H. The uses of gold nanoparticles and *Citrullus colocynthis* L. nanoparticles against *Giardia lamblia* in vivo. *Clin. Epidemiol. Glob. Health* **2020**, *8*, 1282–1286. [CrossRef]
18. Alaqil, A.; Abbas, A.; El-Beltagi, H.; El-Atty, H.; Mehaisen, G.; Moustafa, E. Dietary supplementation of probiotic *Lactobacillus acidophilus* modulates cholesterol levels, immune response, and productive performance of laying hens. *Animals* **2020**, *10*, 1588. [CrossRef]
19. Abbas, A.; Alaqil, A.; El-Beltagi, H.; El-Atty, H.A.; Kamel, N. Modulating laying hens productivity and immune performance in response to oxidative stress induced by *E. coli* challenge using dietary propolis supplementation. *Antioxidants* **2020**, *9*, 893. [CrossRef] [PubMed]
20. Hazrati, S.; Rezaeipour, V.; Asadzadeh, S. Effects of phytogenic feed additives, probiotic and mannan-oligosaccharides on performance, blood metabolites, meat quality, intestinal morphology, and microbial population of Japanese quail. *Br. Poult. Sci.* **2019**, *61*, 132–139. [CrossRef]
21. Gurudeeban, S.; Satyavani, K.; Ramanathan, T. Bitter apple (*Citrullus colocynthis*): An overview of chemical composition and biomedical potentials. *Asian J. Plant Sci.* **2010**, *9*, 394–401. [CrossRef]
22. Qureshi, R.; Raza Bhatti, G.; Memon, R.A. Ethnomedicinal uses of herbs from northern part of Nara desert, Pakistan. *Pak. J. Bot.* **2010**, *42*, 839–851.
23. Dallak, M. In vivo, hypolipidemic and antioxidant effects of Citrullus colocynthis pulp extract in alloxan-induced diabetic rats. *Afr. J. Biotechnol.* **2011**, *10*, 9898–9903. [CrossRef]
24. Marzouk, Z.; Marzouk, B.; Mahjoub, M.A.; Haloui, E.; Mighri, Z.; Aouni, M.; Fenina, N. Screening of the antioxidant and the free radical scavenging potential of Tunisian *Citrullus colocynthis* Schrad. from Mednine. *J. Food Agric. Environ.* **2010**, *8*, 261–265.
25. Kumar, S.; Kumar, D.; Jusha, M.; Saroha, K.; Singh, N.; Vashishta, B. Antioxidant and free radical scavenging potential of *Citrullus colocynthis* (L.) Schrad. methanolic fruit extract. *Acta Pharm.* **2008**, *58*, 215–220. [CrossRef] [PubMed]
26. Huseini, H.F.; Darvishzadeh, F.; Heshmat, R.; Jafariazar, Z.; Raza, M.; Larijani, B. The clinical investigation of *Citrullus colocynthis* (L.) schrad fruit in treatment of Type II diabetic patients: A randomized, double blind, placebo-controlled clinical trial. *Phytother. Res.* **2009**, *23*, 1186–1189. [CrossRef]
27. Rahbar, A.R.; Nabipour, I. The hypolipidemic effect of *Citrullus colocynthis* on patients with hyperlipidmia. *Pak. J. Biol. Sci.* **2010**, *13*, 1202–1207. [CrossRef] [PubMed]
28. Daradka, H.; Almasad, M.M.; Qazan, W.S.; El-Banna, N.M.; Samara, O.H. Hypolipidaemic effects of *Citrullus colocynthis* L. in rabbits. *Pak. J. Biol. Sci.* **2007**, *10*, 2768–2771. [CrossRef] [PubMed]
29. Najafi, S.; Sanadgol, N.; Nejad, B.S.; Beiragi, M.A.; Sanadgol, E. Phytochemical screening and antibacterial activity of *Citrullus colocynthis* (Linn.) schrad against Staphylococcus aureus. *J. Med. Plants Res.* **2010**, *4*, 2321–2325. [CrossRef]

30. Ali, A.A.; Alian, M.A.; Elmahi, H.A. Phytochemical analysis of some chemical metabolites of colocynth plant (*Citrullus colocynths* L.) and its activities as antimicrobial and antiplasmodial. *J. Basic Appl. Sci. Res.* **2013**, *3*, 228–236.
31. Aly, A.M.; Naddaf, A. Anti-inflammatory activities of Colocynth topical gel. *J. Med. Sci.* **2006**, *6*, 216–221.
32. Marzouk, B.; Marzouk, Z.; Haloui, E.; Fenina, N.; Bouraoui, A.; Aouni, M. Screening of analgesic and anti-inflammatory activities of *Citrullus colocynthis* from southern Tunisia. *J. Ethnopharmacol.* **2010**, *128*, 15–19. [CrossRef] [PubMed]
33. Hussain, A.I.; Rathore, H.; Sattar, M.Z.; Chatha, S.A.S.; Sarker, S.D.; Gilani, A.H. *Citrullus colocynthis* (L.) Schrad (bitter apple fruit): A review of its phytochemistry, pharmacology, traditional uses and nutritional potential. *J. Ethnopharmacol.* **2014**, *155*, 54–66. [CrossRef] [PubMed]
34. AOAC Association of Official Analysis Chemists International. *Official Methods of Analysis of AOAC International*; AOAC International: Washington, DC, USA, 2005.
35. Karumi, Y.; Onyeyili, P.A.; Ogugbuaja, V.O. Identification of active principles of *M. balsamina* (balsam apple) leaf extract. *J. Med. Sci.* **2004**, *4*, 179–182.
36. Romero, L.M.; Reed, J.M. Collecting baseline corticosterone samples in the field: Is under 3 min good enough? *Comp. Biochem. Physiol. A Mol. Integr. Physiol.* **2005**, *140*, 73–79. [CrossRef]
37. Begley, M.; Hill, C.; Gahan, C.G.M. Bile salt hydrolase activity in probiotics. *Appl. Environ. Microbiol.* **2006**, *72*, 1729–1738. [CrossRef]
38. Schlager, S.; Vujic, N.; Korbelius, M.; Duta-Mare, M.; Dorow, J.; Leopold, C.; Rainer, S.; Wegscheider, M.; Reicher, H.; Ceglarek, U.; et al. Lysosomal lipid hydrolysis provides substrates for lipid mediator synthesis in murine macrophages. *Oncotarget* **2017**, *8*, 40037–40051. [CrossRef]
39. Dallak, M.; Bashir, N.; Abbas, M.; Elessa, R.; Haidara, R.E.M.; Khalil, M.; Al-Khateeb, M.A. Concomitant down regulation of glycolytic enzymes, upregulation of gluconeogenic enzymes and potential hepato-nephro-protective effects following the chronic administration of the hypoglycemic, insulinotropic citrullus colocynthis pulp extract. *Am. J. Biochem. Biotechnol.* **2009**, *5*, 153–161. [CrossRef]
40. Raederstorff, D.G.; Schlachter, M.F.; Elste, V.; Weber, P. Effect of EGCG on lipid absorption and plasma lipid levels in rats. *J. Nutr. Biochem.* **2003**, *14*, 326–332. [CrossRef]
41. Zamani, M.; Rahimi, A.O.; Mahdavi, R.; Nikbakhsh, M.; Jabbari, M.V.; Rezazadeh, H.; Delazar, A.; Nahar, L.; Sarker, S.D. Assessment of anti-hyperlipidemic effect of *Citrullus colocynthis*. *Rev. Bras. Farm.* **2007**, *17*, 492–496. [CrossRef]
42. Tannin-Spitz, T.; Bergman, M.; Grossman, S. Cucurbitacin glucosides: Antioxidant and free-radical scavenging activities. *Biochem. Biophys. Res. Commun.* **2007**, *364*, 181–186. [CrossRef]
43. Caulfield, M.P.; Padula, M.P. HPLC MS-MS analysis shows measurement of corticosterone in egg albumen is not a valid indicator of chicken welfare. *Animals* **2020**, *10*, 821. [CrossRef] [PubMed]
44. Pavlík, A.; Sláma, P.; Mazalová, L.; Kabourková, E. Total cholesterol and corticosterone concentration relationship in blood plasma of laying hens. *J. Microbiol. Biotechnol. Food Sci.* **2016**, *05*, 17–19. [CrossRef]
45. Mocellin, S.; Rossi, C.; Pilati, P.; Nitti, D. Tumor necrosis factor, cancer and anticancer therapy. *Cytokine Growth Factor Rev.* **2005**, *16*, 35–53. [CrossRef]
46. Al-Aqil, A.; Zulkifli, I. Changes in heat shock protein 70 expression and blood characteristics in transported broiler chickens as affected by housing and early age feed restriction. *Poult. Sci.* **2009**, *88*, 1358–1364. [CrossRef] [PubMed]
47. Hayley, S.; Kelly, O.; Anisman, H. Corticosterone changes in response to stressors, acute and protracted actions of tumor necrosis factor-α, and lipopolysaccharide treatments in mice lacking the tumor necrosis factor-α p55 receptor gene. *Neuroimmunomodulation* **2004**, *11*, 241–246. [CrossRef] [PubMed]
48. Villar, S.R.; Ronco, M.T.; Bussy, R.F.; Roggero, E.; Lepletier, A.; Manarin, R.; Savino, W.; Pérez, A.R.; Bottasso, O. Tumor necrosis factor-α regulates glucocorticoid synthesis in the adrenal glands of trypanosoma cruzi acutely-infected mice. The role of TNF-R1. *PLoS ONE* **2013**, *8*, e63814. [CrossRef]
49. Kregel, K.C. Invited review: Heat shock proteins: Modifying factors in physiological stress responses and acquired thermotolerance. *J. Appl. Physiol.* **2002**, *92*, 2177–2186. [CrossRef]
50. Al-Ghaithi, F.; El-Ridi, M.R.; Adeghate, E.; Amiri, M.H. Biochemical effects of *Citrullus colocynthis* in normal and diabetic rats. *Mol. Cell. Biochem.* **2004**, *261*, 143–149. [CrossRef]
51. Alzarah, M.; Althobiati, F.; Abbas, A.; Mehaisen, G.; Kamel, N. *Citrullus colocynthis* seeds: A potential natural immune modulator source for broiler reared under chronic heat stress. *Animals* **2021**, *11*, 1951. [CrossRef]
52. Charlton, M.R. Protein metabolism and liver disease. *Baillieres. Clin. Endocrinol. Metab.* **1996**, *10*, 617–635. [CrossRef]
53. Elnagar, S.; Scheideler, S.; Beck, M. Reproductive hormones, hepatic deiodinase messenger ribonucleic acid, and vasoactive intestinal polypeptide-immunoreactive cells in hypothalamus in the heat stress-induced or chemically induced hypothyroid laying hen. *Poult. Sci.* **2010**, *89*, 2001–2009. [CrossRef]
54. Al-Snafi, A.E. Chemical constituents and pharmacological effects of Citrullus colocynthis—A review. *Asian J. Pharm. Res.* **2016**, *6*, 57–67.
55. Deng, W.; Dong, X.F.; Tong, J.M.; Zhang, Q. The probiotic Bacillus licheniformis ameliorates heat stress-induced impairment of egg production, gut morphology, and intestinal mucosal immunity in laying hens. *Poult. Sci.* **2012**, *91*, 575–582. [CrossRef]
56. Idoko, A.S.; Oladiji, A.T.; Ilouno, L.E. Growth performance of rats maintained on citrullus colocynthis seed coatbased diet. *IOSR J. Biotechnol. Biochem.* **2015**, *1*, 9–14.

57. Marzouk, B.; Marzouk, Z.; Décor, R.; Edziri, H.; Haloui, E.; Fenina, N.; Aouni, M. Antibacterial and anticandidal screening of Tunisian *Citrullus colocynthis* Schrad. from Medenine. *J. Ethnopharmacol.* **2009**, *125*, 344–349. [CrossRef]
58. Wang, X.; Liu, L.; Zhao, J.; Jiao, H.; Lin, H. Stress impairs the reproduction of laying hens: An involvement of energy. *World's Poult. Sci. J.* **2017**, *73*, 845–856. [CrossRef]
59. Shini, S.; Shini, A.; Blackall, P. The potential for probiotics to prevent reproductive tract lesions in free-range laying hens. *Anim. Prod. Sci.* **2013**, *53*, 1298–1308. [CrossRef]

 *agriculture*

 MDPI

Article

Preliminary Study of Slaughter Value and Meat Characteristics of 18 Months Ostrich Reared in Hungary

Lili Dóra Brassó [1,2,*], Vanessza Szabó [1], István Komlósi [1], Tünde Pusztahelyi [3] and Zsófia Várszegi [1]

[1] Department of Animal Husbandry, Faculty of Agricultural and Food Sciences and Environmental Management, University of Debrecen, 4032 Debrecen, Hungary; kylee6992@gmail.com (V.S.); komlosi@agr.unideb.hu (I.K.); varszegi@agr.unideb.hu (Z.V.)

[2] Doctoral School of Animal Science, University of Debrecen, 4032 Debrecen, Hungary

[3] Central Laboratory of Agricultural and Food Products, 4032 Debrecen, Hungary; pusztahelyi@agr.unideb.hu

[*] Correspondence: brasso.dora@agr.unideb.hu

Citation: Brassó, L.D.; Szabó, V.; Komlósi, I.; Pusztahelyi, T.; Várszegi, Z. Preliminary Study of Slaughter Value and Meat Characteristics of 18 Months Ostrich Reared in Hungary. *Agriculture* **2021**, *11*, 885. https://doi.org/10.3390/agriculture11090885

Academic Editor: Vincenzo Tufarelli

Received: 11 August 2021
Accepted: 13 September 2021
Published: 15 September 2021

Publisher's Note: MDPI stays neutral with regard to jurisdictional claims in published maps and institutional affiliations.

Abstract: The study aimed to evaluate the slaughter value and meat characteristics of ten ostrich females reared and slaughtered at the age of 18 months in Hungary. The ratio of selected body parts, the main organs and the lean meat parts were examined. The nutritive composition, the colour, the technological and organoleptic characteristics of five valuable meat parts (outside strip—M. flexor cruris lateralis, oyster—M. iliofemoralis externus, tip—M. femorotibialis medius, outside leg—M. gastrocnemius pars externa, medal—M. ambiens) and the amino acid, fatty acid and mineral composition of outside strip (M. flexor cruris lateralis) were also evaluated. The ratio of body parts and the main organs as the percentage of live weight, and the lean meat part as the percentage of carcass weight showed 16.74 ± 0.01%, 6.16 ± 0.01% and 57.29 ± 0.59%, respectively. The dry matter content of the examined valuable meat parts ranged between 24.89 ± 0.08 and 26.23 ± 0.13%, the protein ratio took on values between 18.40 ± 0.09 and 20.62 ± 0.16%, the fat content showed values between 2.36 ± 0.07 and 4.50 ± 1.09% and the hydroxyproline content ranged between 0.01 ± 0.001 and 0.08 ± 0.001%. The amino acid content of the outside strip showed a range between 0.15 and 3.33%. The ratio of SFA, MUFA and PUFA was 35.10 ± 0.53, 37.37 ± 1.52 and 27.54 ± 1.01. The n-6/n-3 ratio showed 3.91 ± 0.43 and the SFA/UFA ratio was 0.54 ± 0.02. Among the examined minerals, the content of Ca, K, Mg, Na and P was the highest in the meat. In the case of the colour, regarding L* value, we could reveal no significant difference between the examined meat parts. For a* and b* values, the outside leg had the lowest data of all. We could not reveal a significant difference between the pH values of the meat parts. Regarding technological parameters, meat differed only in thawing loss. The significantly lowest thawing loss could be detected in the outside leg (2.72 ± 0.01%) and in the medal (2.32 ± 0.01%). The results of the organoleptic evaluation showed that the outside strip and the tip had the best flavour and tenderness. In comparison with the younger birds (10–14 months of age) in the literature, the 18-month-old ostriches in our study showed similar or slightly lower slaughter weight, skin weight and head ratio, greater liver weight, lighter meat, lower protein and higher fat content, higher essential amino acid and lower non-essential amino acid content and higher SFA content in some cases. However, data on nutrition and population size were not always available. In comparison with other ratites (emu and rhea), ostrich meat has lower dry matter and protein, but higher fat, SFA, MUFA and PUFA content and lower n-6/n-3 ratio.

Keywords: ostrich; meat production; meat nutritive composition; amino acids; fatty acids; minerals; technological and organoleptic characteristics

1. Introduction

Alternative food sources that support healthy human nutrition are in heavy demand. Ostrich meat, as lean meat with low intramuscular fat (0.5%) and cholesterol content [1], is suitable for this purpose. Due to its status as a gourmet product and its high price, ostrich meat is usually purchased by the wealthy. Compared to beef and poultry, ostrich meat has a favourable fatty acid profile and low fat content [2], and it is characterised by a beef-like taste and tenderness [3]. According to Balog and Almeida [4], consumers prefer ostrich to beef because of its low intramuscular fat content. Although the prime product of the ostrich industry is meat, skin, oil and feathers are also processed. Ostrich meat production ranges from 12 to 15 thousand tonnes, 60% of which comes from South Africa. The major producers are South Africa, the USA, Australia, Spain, Poland and the Middle East. The ostrich meat industry produces mostly fresh meat, including steak and hamburger meat, as well as dry goods. [5].

Hungary exports live ostriches and ostrich meat (approximately three thousand slaughtered birds a year) to France. Ostrich meat is also purchased by local consumers and restaurants in the capital. The outside strip, the oyster, the tip, the outside leg, the medal and the fan are considered to be the most valuable cuts of the lean meat parts. In Hungary, the outside strip and the fan are sold vacuum packed as raw or smoked meat or sliced prosciutto crudo. The oyster is used for preparing pate, while the tip and the outside leg are used for sausages and salami.

As with all ratites, the largest amount of meat can be found on the thigh, while the smallest is on the breast, wings and back. Ostriches reach the optimal slaughter age at 12–14 months [6], although birds of the Zimbabwean blue neck genotype provide a satisfying amount of meat at the age of 10–12 months [7]. The *Struthio camelus* species can be characterized with a 57–58% slaughter yield and 62.5% lean meat [8,9]. For slaughter animals, maintainer feed is usually provided from 10 months of age until slaughter with 8–10 MJ/kg energy, 12–14% protein, 2.5% fat, 0.9–1.4% Ca and 0.6 g/kg lysine content [10]. Due to its high iron content, ostrich meat has dark red pigmentation but the iron concentration may vary among meat parts [11]. There has been no scientific result on raw meat smell, but packaging technology and storage time influence this parameter [12]. The smell is determined by water-soluble molecules, whereas flavour is mainly influenced by fat-soluble compounds [13]. For consumers, the other significant meat characteristic is tenderness. This parameter is affected by the cooking technology employed and also by the cooking temperature [14]. When compared to beef, ostrich meat is stiffer and drier because it has lower connective tissue and collagen content. Juiciness and tenderness are closely linked and consumers consider juicier meat to be more tender, too [13]. Taylor et al. [14] also report a relation between low fat content, juiciness and texture. Under 15 months of age, ostrich meat is more tender [15]. Technological characteristics of meat, such as thawing, dripping and cooking loss are also relevant from the market point of view. These characteristics, including texture and tenderness, are influenced by the water-holding capacity of the meat [16]. Cooking loss increases with longer storage time [17]. Higher pH value improves water-holding capacity, thus it decreases drip loss [18]. The mean pH value of ostrich meat is 7.2, but it increases after 24 h and facilitates meat processing [19]. Ostrich meat has a regular (<5.8) to high (>6.2) 24 h postmortem pH [1].

In this preliminary study, we aimed to assess the slaughter value and meat characteristics of ostriches reared in Hungary. The slaughtered animals were six months older than the optimal slaughter age (12 months) since the COVID-19 pandemic substantially decreased the capacity of the HORECA (hotel/restaurant/catering) sector. Zimbabwean ostriches are known to have constant weight over the age of one year [20]. As regards the age of the birds, we wanted to assess the slaughter value and meat characteristics of older animals and to compare them with those of the ostriches that were slaughtered at the optimal age (10–14 months). Although extensive literature is available on this topic, none of the studies included a wide range of parameters examined similar to ours. In Hungary, no research findings have been published on this species so far. The present article is the first one to evaluate the whole spectrum of meat production characteristics of ostriches. In this respect, our findings can be considered to be gap filling. However, regarding the small number of the evaluated birds, the present research can offer only preliminary results.

2. Materials and Methods

2.1. The Animals and the Slaughter Process

The ten female ostriches used in our examination came from a farm in Jászberény in the Central Hungary region with a continental climate. Polipor-98 Ltd. is the largest company producing slaughter birds for export. Semi-intensive, grain-based fattening technology was applied with coarse feed supplementation (alfalfa). The birds received starter feed (0.02–0.12 kg/day/bird) from week 0 to 8, then starter + grower (0.55–0.65 kg/day/bird) from week 9 to 11. After that, grower (0.75–1.4 kg/day/bird) feed was provided for them from week 12 to 24, then maintainer (1.5 kg/day/bird) from week 25 until slaughter. The starter feed was composed of maize, wheat, non-GMO soybean meal, alfalfa pellets, full-fat soy, sunflower meal, sugar beet pellets and vegetable oil. The starter feed had 12 MJ/kg energy, 21.00% raw protein, 4.00% fat, 6.70% fibre, 7.5% ash and 1.15% lysine content. The grower feed included maize, wheat, alfalfa meal, sunflower meal, soybean meal, FF soybean, dried beet pulp and non-GMO soybean meal. The grower feed contained 10 MJ/kg energy, 18.00% protein, 3.70% fat, 8.20% fibre, 7.52% ash and 0.95% lysine. The maintainer feed comprised maize, alfalfa pellets, wheat, sunflower meal, full-fat soybean, beet pellets and vegetable oil. The maintainer had 8 MJ/kg energy, 16.20% protein, 3.80% fat, 9.40% fibre, 6.92% ash and 0.80% lysine content. The feed was purchased from Agrofeed Ltd.

Chicks were nursed on a 2 m × 0.8 m × 1 m sized battery until the age of two weeks with a total of twenty birds of mixed gender. After that, they were moved to deep litter until three months of age with 5 m^2/bird stocking density and mixed gender. From the age of three months until slaughter, the animals were kept in a 300 m^2 pen with a stocking density of 10 m^2/bird and mixed gender. The fattening process was accomplished on the same farm; the pens were placed next to each other, so the environmental conditions were considered equal for each animal. The keeping conditions met the requirements of the Model Code of Practice for the Welfare of Animals [21].

After the birds were stunned and bled on the farm, they were processed three hours later in a slaughterhouse in Tokaj. The birds were 18 months of age with a 100.9 ± 4.48 kg live and 51.75 ± 2.18 kg carcass weight. In the slaughterhouse, we measured the weight of skin with feathers, the head, the feet, the diaphragm and the internal organs (heart, lung, liver and gizzard) with a two-decimal digital balance. The carcass was cooled to 4 °C and divided into meat parts 24 h after slaughter. The weight of all meat parts located on the left leg and the back was measured.

2.2. pH and Colour Determinations

We recorded the colour and the pH value of five valuable meat parts (outside strip, oyster, tip, outside leg and medal) all used in further evaluations. The pH value was measured 24 h after slaughter by Testo AG Germany 205 pH value gauge. The pH value gauge was immersed in buffer solution before measurement. For meat colour (L*, a*, b*) examination, the Konica Minolta CR-410 Chroma Meter was applied 24 h after slaughter with a 21 min blooming time. The Chroma Meter was calibrated with the use of a white calibration plate before the analysis, setting the Y, x and y illuminant coordinates (Y = 93.7, x = 0.3144, y = 0.3204). Regarding meat colour, L*, a* and b* values represent a colour space defined by CIE. L* value expresses lightness, a* and b* refer to the greenness and redness, blueness and yellowness of meat [22].

2.3. Kjeldahl-Method for N Analysis

From the prepared (ground) samples, 1 g was measured on nitrogen-free paper and folded into it. The digestion of these samples was initiated with two 3.5 g Se-Kjeldahl tablets (VWR International) and 14 mL conc. H_2SO_4. The digestion was performed at 420 °C. After cooling, the samples were distilled in UDK 149 distiller (VELP Scientifica Srl, Usmate, Italy) and ammonium was released by 33 (*m/m*)% NaOH addition and loaded into 4% (w/w)% boric acid solution. The N was determined by titrimetric analysis with 0.2 N H_2SO_4 in TITROLINE 5000 automatic titrator (VELP Scientifica Srl, Usmate, Italy).

2.4. Total Fat Content Assessment

The measurement was performed according to the MSZ ISO 1443:2002 standard. The ground samples (3–5 g) were measured and treated with 50 mL conc. HCl for 1 h; then 150 mL hot water was added to it. The solution was filtered through a filter paper and washed with hot water until the pH was neutral. The filtrate was dried at $103 \pm 2\,^\circ$C for 1 h. The filter paper with the filtrate was inserted into TECATOR Soxtec fat extractor and extracted with petrol ether (40–60 $^\circ$C). The extracted fat was dried and measured. The fat was further analysed for fatty acid composition.

2.5. The Examination of Fatty Acid Composition

The fatty acid composition was determined as methylated fatty acid composition based on the ISO 12966-2:217 standard sample preparation method. The fat (Section 2.4) was dissolved in 6 mL hexane and 12 mL NaOH:MeOH and was treated at 80 $^\circ$C for 10 min for saponification. Afterwards, the solution was diluted with water and the non-saponified compound was extracted with hexane. The remaining solution was acidified with 0.5 mL 6 M H_2SO_4 and with hexane again; then the fatty acids were extracted and separated. After the addition of 2 mL BF3:MeOH, it was heated at 80 $^\circ$C for 30 min. Then, the solution was treated with conc. NaCl solution and the supernatant was applied into GC-FID FAME (Varian GC 3800, San Diego, CA, USA) detection with CP-Select for FAME column (100 m $\times$ 0.25 mm). FAME standard (Sigma-Aldrich) was applied as a reference. The total SFA content was calculated as the sum of C4:0, C6:0, C8:0, C10:0, C12:0, C14:0, C15:0, C16:0, C18:0, C20:0, C21:0, C22:0, C23:0 and C24:0. The total MUFA content was given as the sum of C15:1, C16:1, C17:1, C18:1, C20:1, C20:3, C22:1, C23:1 and C24:1. For calculating the total PUFA content, the sum of 18:2n6, 18:3n6, 18:3n3, C20:3n6, C20:3n3 and C22:6n3 was used.

2.6. The Preparation of Meat Samples for Elemental Analysis (HNO_3-H_2O_2 Digestion)

From the prepared (ground) meat samples, 1 g was loaded into digester tubes. Ten ml of distilled conc. HNO_3 was added and heated at 60 $^\circ$C for 30 min, then 3 mL 30% (v/v) H_2O_2 (Scharlab) was added and the samples were digested further at 120 $^\circ$C for 90 min. After the digestion, all samples were washed into 50 mL volumetric flasks with distilled water, homogenised and filtered (MN 640 W paper; Macherey-Nagel). The ICP-OES technique was applied on iCAP 7000 spectrophotometer (Thermo Scientific, Cambridge, UK). For the calibration, a multielement standard solution was applied.

2.7. Amino Acid Analysis

For total amino acid analysis, 5 mg homogenised samples were hydrolysed with 6 M HCl at 110 $^\circ$C for 23 h. After filtration and dilution to even protein content, the samples were loaded to ion-exchange chromatography on AAA500 (INGOS Ltd., Prague, Czech Republic) semi-automated amino acid analysing system in five buffer systems (citrate -sodium citrate, pH 2.2; 2.7; 3; 4.25; 8) with ninhydrin detection on two wavelengths (440 and 570 nm). Cya, Asp, MetS, Thr, Ser and Glu at pH 2.7; Pro, Gly, Ala, Cys and Val at pH 3; Ile and Leu at pH 4.25; and Tyr, Phe, His, Lys and Arg at pH 8 were eluted. Reference amino acid mixture was applied (INGOS Ltd., Prague, Czech Republic).

2.8. Hydroxyproline Measurement

The measurement was performed according to the MSZ ISO 3496:2000 standard. Frozen meat was sliced into cubes and wrapped in plastic. It was taken to 70 $^\circ$C for 30 min. It was cooled and homogenised. Then, 4 g of the sample was added to a flask and 30 mL conc. H_2SO_4 was added for hydrolysis. The sample was left for 16 h at 105 $^\circ$C. The hydrolysed sample was filtered into a 250 mL measuring flask with 3 $\times$ 10 mL hot conc. H_2SO_4 and diluted up to the 250 mL mark with distilled water. Four millilitres of the solution was mixed with 2 mL chloramine T reagent (1.41 $(m/v)\%$ sodium-N-chlor-*p*-toluol-sulphonamide-trihydrate in citrate-sodium acetate buffer, pH 6.8) and left for 20 min at RT. Twenty millilitres of the 10 $(m/v)\%$ *p*-dimethyl-amino-benzaldehyde in 60 $(m/m)\%$ perchlorate was added to the solution, mixed thoroughly, closed with a loosened cap and put into a water bath at 60 °C for 20 min. After heating, the solution was cooled and left at RT for 30 min. The absorbance was determined at 560 nm in glass cuvettes with water as a reference. The hydroxyproline concentration was determined with a hydroxyproline reference curve (0.5 µg/mL–2 µg/mL).

2.9. Examination of Selected Technological Parameters and Shear Force of Meat

The measured technological characteristics included drip loss (%), thawing loss (%), cooking loss (%) and shear force (N/mm). For drip loss, meat pieces of 50 ± 5 g and 1 cm thick were cut from each meat part. Pieces were packed in an inflated nylon bag, then hung up in the fridge at 4 °C for 24 h and weighed again [23]. The meat samples for cooking loss measurement were cut into 100 ± 5 g and 1 cm thick pieces and weighed. For the evaluation of cooking loss, pieces were packed in nylon bags and cooked for half an hour until core temperature reached 75 °C. The samples were cooled under running water, wiped with a paper towel, and then weighed. After cooking, meat pieces were chilled at 4 °C overnight. The temperature of the samples was 4 °C during the measurement. Pieces were cut across the muscle fibres, two times each. XT + Texture Analyser (Warner-Blatzer) was used to measure the shear force (N/mm) on cooked samples, with shear blade set using 25 kg load cell, with 1.5 mm/s test speed from 40 mm distance. Once the trigger force is attained, the blade proceeds to shear through the sample. The maximum force denotes the point at which the sample completely fills the triangular cavity of the blade and cuts through the sample surface. After this point, shearing continues through the whole sample until the blade passes through the base plate slot. The blade then returns to its starting position. Curves were evaluated to obtain shear force data.

2.10. Organoleptic Evaluation of the Different Meat Parts

After freezing at –20 °C for three months, the five samples from the ten birds were cooked for organoleptic evaluation. The cooking procedure was carried out in a pot at 90 °C until reaching core temperature at 75 °C [24]. The samples were cut into four pieces of 2 cm × 2 cm × 2 cm cubes and kept at 45–55 °C until analysis, which took place immediately after. An experienced panel of eight men and thirteen women aged between 24 and 70 were included in the examination. Four pieces of the five meat parts were placed on each plate but the different meat parts were separated. Each meat part was tasted by four people, so we had two hundred pieces of meat in total. Fresh water and bread were also served to the panellists, who also received a questionnaire with questions about their personal demographical specificities and habits of meat consumption. Having filled in the questionnaire, the panellists commenced the organoleptic evaluation using a 5-point hedonic scale (1—unfavourable, 5—excellent). For organoleptic characteristics, smell, flavour, juiciness, tenderness, texture and the presence of an aftertaste were scored.

2.11. Statistical Analysis

The results are presented as mean values and the standard error of the mean. The statistical analysis was carried out with the IBM SPSS Statistics 23.0 and Microsoft Office Excel 2016 programs. Mean values were compared using univariate analysis of variance and a Tukey test with a significance level of $p < 0.05$.

3. Results

3.1. The Ratio of Body Parts, Meat Parts and the Main Organs

The ratio of the main body parts is shown in Table 1. The skin with feathers made up the greatest part of live weight among the examined body parts (Table 1). The weight of the diaphragm and the feet contributed equally to the live weight. The neck was significantly heavier than the head, but lighter than the other body parts. The head made the lowest percentage of all the measured body parts. The total ratio of the evaluated body parts was $16.74 \pm 0.01\%$ of live weight.

Table 1. The ratio of body parts expressed in the percentage of live weight ($n = 10$).

Body Parts	Ratio (%)
Skin with feathers	8.06 [d] $\pm$ 0.01
Neck	2.12 [b] $\pm$ 0.01
Diaphragm	2.86 [c] $\pm$ 0.01
Head	0.71 [a] $\pm$ 0.01
Feet	2.99 [c] $\pm$ 0.01
Total	16.74 $\pm$ 0.01

[a–d]: Different letters represent significant differences ($p < 0.05$).

The ratio of meat parts located on the thigh and back expressed as the percentage of carcass weight is presented in Table 2. The weights in descending order were the following: tip, outside thigh, fan, outside leg, leg, drumstick, other meat cuts, inside leg, oyster, outside strip, back tender, inside strip and medal. The total ratio of all examined lean meat parts was $57.29 \pm 0.59\%$ and we found significant differences among the muscles. The medal showed the lowest percentage of all, followed by the inside strip, the back tender and the outside strip making the second weight category. The oyster had a significantly higher percentage than the previous cuts. The ratio of the inside leg showed the next weight category, followed by the drumstick and the other meat cuts. The ratio of the fan, outside leg, leg and outside thigh was similar. The tip had the highest percentage of all. The ratio of the outside strip, oyster, tip, outside leg and medal made $19.16 \pm 0.01\%$ of all the examined meat parts. The owner did not allow us to evaluate the fan as it is the most valuable meat part and is sold whole and only a few birds are slaughtered annually.

Table 2. Percentage of individual valuable meat parts ($n = 10$).

Meat Part	Ratio (%)
Outside strip	2.32 [b] $\pm$ 0.23
Oyster	2.72 [c] $\pm$ 0.09
Fan	6.49 [fg] $\pm$ 0.21
Tip	6.83 [g] $\pm$ 0.26
Outside leg	6.34 [f] $\pm$ 0.13
Medal	0.94 [a] $\pm$ 0.03
Inside leg	3.28 [d] $\pm$ 0.14
Inside strip	2.05 [b] $\pm$ 0.13
Outside thigh	6.57 [fg] $\pm$ 0.21
Back tender	2.23 [b] $\pm$ 0.13
Leg	6.28 [f] $\pm$ 0.17
Drumstick	5.70 [e] $\pm$ 0.21
Other meat cuts	5.54 [e] $\pm$ 0.15
Total	57.29 $\pm$ 0.59

[a–g]: Different letters represent significant differences ($p < 0.05$).

The ratio of the main organs expressed in the percentage of live weight is shown in Table 3. The ratio of heart and lung was similar; neither the liver, nor the empty gizzard differed in the percentage of live weight. The total ratio of the main metabolic organs was $6.16 \pm 0.01\%$.

Table 3. Ratio of the main organs expressed in the percentage of live weight (*n* = 10).

Organs	Ratio (%)
Heart	0.88 [a] ± 0.01
Lung	0.98 [a] ± 0.01
Liver	2.02 [b] ± 0.01
Empty gizzard	2.28 [b] ± 0.01
Total	6.16 ± 0.01

[a,b]: Different letters represent significant differences (*p* < 0.05).

3.2. Chemical Composition of the Examined Meat Parts

The nutritive composition of the examined meat parts is given in Table 4. The dry matter content showed values between 23.84 ± 0.31 and 26.23 ± 0.13%. The oyster and the tip showed the lowest dry matter content, with the outside strip showing the highest. The protein content of meat parts took on values between 18.40 ± 0.09 and 20.62 ± 0.16%. The tip showed the lowest protein content, while the outside strip and the medal revealed the highest protein content. The fat content revealed values between 2.36 ± 0.07 and 4.50 ± 1.09%. The oyster had the lowest fat content, while the tip and the outside strip showed the highest fat content. The hydroxyproline content ranged between 0.01 ± 0.001 and 0.08 ± 0.001%. The lowest hydroxyproline content was found for the outside strip and the outside leg; conversely, the tip was the richest in this compound.

Table 4. Nutritive composition of the five examined valuable meat parts (*n* = 10).

Meat Parts/Nutrients	Dry Matter% (*w/w*)	Protein% (*w/w*)	Fat% (*w/w*)	Hydroxyproline% (*w/w*)
Outside strip	26.23 [d] ± 0.13	20.49 [cd] ± 0.08	4.41 [c] ± 0.12	0.01 [a] ± 0.001
Oyster	23.84 [a] ± 0.31	20.10 [c] ± 0.32	2.36 [a] ± 0.07	0.03 [c] ± 0.001
Tip	24.35 [ab] ± 0.15	18.40 [a] ± 0.09	4.50 [c] ± 1.09	0.08 [d] ± 0.001
Outside leg	24.51 [b] ± 0.15	19.49 [b] ± 0.11	3.64 [b] ± 0.14	0.01 [a] ± 0.001
Medal	25.54 [c] ± 0.13	20.62 [d] ± 0.16	3.57 [b] ± 0.11	0.02 [b] ± 0.001

[a–d]: Different letters represent significant differences (*p* < 0.05).

The amino acid content of the outside strip is presented in Table 5. The amino acid content ranged between 0.15 ± 0.01and 3.33 ± 0.04 g/100 g meat. Regarding the amino acid and fatty acid composition, we had the opportunity to assess only one meat part, so we chose the outside strip, which is the second most valuable one after the fan. Among the evaluated amino acids, the essential amino acids showed 8.68 ± 0.15 g/100 g meat, the non-essential amino acids made 4.69 ± 0.27 g/100 g meat and the quantity of conditional amino acids was 6.18 ± 0.02 g/100 g meat. The outside strip was the richest in glutamine and the poorest in cysteine. The quantity of isoleucine, valine and threonine was equal. The examined meat part contained similar concentrations of leucine and lysine. On amino acid content, only Sales's results [8] could be found expressed in g/100 g meat of unknown meat parts. The author evaluated 10- to 12-month-old ostriches and found lower values for the content of asparagine, threonine, serine, glutamine, alanine, valine, isoleucine, tyrosine, histidine and lysine, while he had no data on proline and cysteine.

In Table 6, we present the main fatty acids which were published in the literature to enable comparison. The total values were calculated using all examined fatty acids. The total saturated fatty acid content of the outside strip was 35.10 ± 0.53%. The concentration of palmitic acid (C16:0) was the highest of the whole fatty acid composition of all examined saturated fatty acids. Stearic acid (C18:0) was the second fatty acid present in large quantity. The content of monounsaturated fatty acids showed 37.37 ± 1.52%, of which the concentration of vaccenic acid (C18:1) should be highlighted. The total polyunsaturated fatty acid ratio was 27.54 ± 1.01%, among which the ratio of linoleic acid (C18:2n6) was the most significant. The total unsaturated fatty acid content was 64.90 ± 1.68%. Our results, compared to the literature demonstrated in Table 6, agree with the 27 to 39% range of fatty acid ratio published by other authors. The SFA/UFA ratio was below the results of Horbańczuk et al. (1998) [25] and Hoffman et al. (2005) [26], but above that of Horbańczuk et al. (2019) [27]. The n-6/n-3 ratio in our examination was almost half of the one found by Horbańczuk et al. [27]. Information on the diet is not available for drawing further conclusions from the difference.

Table 5. Amino acid composition of outside strip ($n = 10$) compared to the literature.

Amino Acids	Present Results (g/100 g Meat)	Sales [8] (g/100 g Meat)
ASP	1.94 [m] ± 0.02	1.90
THR	1.00 [i] ± 0.01	0.76
SER	0.83 [e] ± 0.01	0.59
GLU	3.33 [n] ± 0.04	2.51
PRO	0.88 [h] ± 0.04	-
GLY	0.82 [d] ± 0.01	0.82
ALA	1.18 [k] ± 0.01	1.06
CYS	0.15 [a] ± 0.01	-
VAL	1.00 [i] ± 0.01	0.97
MET	0.54 [b] ± 0.02	0.55
ILE	0.98 [i] ± 0.01	0.92
LEU	1.62 [l] ± 0.02	1.70
TYR	0.74 [c] ± 0.01	0.61
PHE	0.85 [f] ± 0.01	0.94
HIS	0.86 [g] ± 0.01	0.39
LYS	1.84 [l] ± 0.08	1.65
ARG	1.00 [j] ± 0.01	1.36

[a-n]: Different letters represent significant differences ($p < 0.05$).

Table 6. Fatty acid composition (%) in the estimated total sum of fatty acid of ostrich outside strip and that of several authors.

Fatty Acids	Present Results ($n = 10$)	Horbańczuk et al. (1998) [25] ($n = 6$)	Hoffman et al. (2005) [26]	Horbańczuk et al. (2019) [27] ($n = 8$)
		Saturated fatty acids (SFAs)		
C8:0	0.02 [a] ± 0.01	0.03 ± 0.00	-	-
C10:0	0.05 [a] ± 0.02	0.09 ± 0.01	1.67 ± 0.45	-
C12:0	1.03 [a] ± 0.39	0.14 ± 0.01	0.00 ± 0.03	-
C14:0	0.75 [a] ± 0.08	1.53 ± 0.18	0.75 ± 0.20	0.57 ± 0.12
C15:0	0.24 [a] ± 0.03	- [x]	0.11 ± 0.08	0.02 ± 0.00
C16:0	22.25 [c] ± 1.31	24.06 ± 0.29	21.95 ± 0.56	21.37 ± 0.21
C18:0	10.38 [b] ± 1.06	11.84 ± 0.32	14.08 ± 0.66	9.81 ± 0.08
Total (SFA)	35.10 ± 0.53	37.71 ± 0.39	39.73 ± 0.77	31.33 ± 0.21
		Monounsaturated fatty acids (MUFAs)		
C14:1	<0.01 *	-	-	0.08 ± 0.00
C15:1	<0.01 *	-	0.12 ± 0.07	0.17 ± 0.01
C16:1	5.33 [b] ± 0.52	3.79 ± 0.11	3.51 ± 0.42	7.90 ± 0.09
C18:1	31.85 [c] ± 1.20	33.25 ± 0.52	21.15 ± 0.78	29.96 ± 0.15
C20:1	<0.01 *	0.29 ± 0.01	1.96 ± 0.57	0.21 ± 0.02
Total (MUFA)	37.37 ± 1.52	33.49 ± 0.40	27.27 ± 1.13	38.46 ± 0.16

Table 6. *Cont.*

Fatty Acids	Present Results ($n = 10$)	Horbańczuk et al. (1998) [25] ($n = 6$)	Hoffman et al. (2005) [26]	Horbańczuk et al. (2019) [27] ($n = 8$)
	Polyunsaturated fatty acids (PUFAs)			
C18:2n6	21.19 [d] ± 0.61	15.01 ± 0.55	18.06 ± 0.84	18.70 ± 0.10
C18:3n3	1.49 [b] ± 0.11	6.50 ± 0.52	5.76 ± 0.36	1.98 ± 0.04
C18:3n6	<0.01 *	-	0.59 ± 0.16	-
C20:3n6	<0.01 *	-	-	0.57 ± 0.01
C20:3n3	4.85 [c] ± 1.10	5.30 ± 0.13	6.15 ± 0.77	5.44 ± 0.05
C22:6n3	<0.01 *	0.73 ± 0.05	1.22 ± 0.55	0.67 ± 0.02
Total (PUFA)	27.54 ± 1.01	28.79 ± 0.61	32.99 ± 1.22	28.48 ± 0.10
n-6/n-3 ratio	3.91 ± 0.43	-	-	7.55 ± 0.21
SFA/UFA ratio	0.54 ± 0.02	0.61	0.66	0.47

[a–d]: Different letters represent significant differences ($p < 0.05$); * values below the level of detection; [x] not presented in the study.

The mineral composition of the outside strip is shown in Table 7. The content of Ca, K, Mg, Na and P was significantly the greatest among all examined minerals. Our results for all the examined minerals except for Mn surpassed the literature data presented in Table 7. In comparison with the studies, differences in the content of Ca and K were the greatest. The authors also revealed remarkably different results in some cases.

Table 7. Mineral composition of ostrich outside strip ($n = 10$) compared to the literature.

Minerals	Present Results (g/100 g)	Majewska [28] (g/100 g)	Sales and Oliver-Lyons [29] (g/100 g)
Ca	0.33 [b] ± 0.007	0.05 ± 0.008	0.08
Cu	0.002 [a] ± 0.0001	0.001 ± 0.0003	0.001
Fe	0.04 [a] ± 0.0006	0.04 ± 0.005	0.02
K	3.45 [b] ± 0.05	2.38 ± 0.14	2.69
Mg	0.26 [b] ± 0.003	0.25 ± 0.01	0.22
Mn	0.0004 [a] ± 0.00001	0.0002 ± 0.00003	0.0006
Na	0.51 [b] ± 0.01	0.33 ± 0.03	0.43
P	2.48 [b] ± 0.01	2.28 ± 0.12	2.13
Zn	0.03 [a] ± 0.0008	0.02 ± 0.006	0.02

[a,b]: Different letters represent significant differences ($p < 0.05$).

3.3. Colour and pH Value of the Examined Valuable Meat Parts

Table 8 presents the colour and pH value of the five examined meat parts. No significant difference was found between the meat parts in L* value. However, the a* and b* planes were more sensitive ($p < 0.05$) to the colour determinations. Only the a* value of outside leg differed from a* value of the other meat parts and proved to be the smallest. The outside leg showed the smallest, the oyster the greatest b* value. The pH value of the examined meat parts ranged between 5.95 ± 0.02 and 6.01 ± 0.03 and we found no significant difference among the muscles.

Table 8. The colour and the pH [24h] value of the five examined valuable meat parts ($n = 10$).

Meat Parts/Parameters	L*	a*	b*	pH [24h]
Outside strip	36.76 ± 0.92	25.61 [b] ± 0.32	6.15 [bcd] ± 0.38	5.95 ± 0.02
Oyster	35.47 ± 0.93	24.85 [b] ± 0.47	6.59 [d] ± 0.32	6.00 ± 0.04
Tip	36.66 ± 0.73	25.38 [b] ± 0.46	5.75 [bcd] ± 0.38	6.01 ± 0.03
Outside leg	35.81 ± 0.82	22.60 [a] ± 0.58	3.36 [a] ± 0.51	5.97 ± 0.01
Medal	37.04 ± 0.67	24.75 [b] ± 0.35	5.29 [bc] ± 0.30	6.00 ± 0.02

Different letters represent significant differences ($p < 0.05$).

3.4. Selected Technological Parameters and Shear Force of the Five Examined Valuable Meat Parts

The values of the technological parameters and shear force are shown in Table 9. We found no significant difference between the results of drip loss; however, differences in thawing loss were significant. The outside leg and medal showed the smallest thawing loss, while the outside strip and the oyster showed the biggest. It can be stated that the outside leg and the medal were less negatively affected by thawing, and these meat parts could hold water more efficiently after the process. The cooking loss showed values between 36.63 ± 1.38 and $41.23 \pm 1.47\%$ and we could not detect significant differences among meat parts. The shear force took on values between 2.90 ± 0.24 and 3.42 ± 0.37 N/mm. There was no significant difference among the meat parts.

Table 9. Technological characteristics and shear force of the five valuable meat parts ($n = 10$).

Meat Parts	Drip Loss (%)	Thawing Loss (%)	Cooking Loss (%)	Shear Force (N/mm)
Outside strip	5.88 ± 0.02	$4.22^{b} \pm 0.01$	41.23 ± 1.47	2.99 ± 0.25
Oyster	4.02 ± 0.01	$4.48^{b} \pm 0.01$	37.32 ± 1.30	3.01 ± 0.24
Tip	4.00 ± 0.01	$3.50^{ab} \pm 0.01$	40.05 ± 1.23	3.28 ± 0.23
Outside leg	4.17 ± 0.01	$2.72^{a} \pm 0.01$	36.63 ± 1.38	3.42 ± 0.37
Medal	4.08 ± 0.01	$2.32^{a} \pm 0.01$	36.68 ± 1.30	2.90 ± 0.24

Different letters represent significant differences ($p < 0.05$).

3.5. Organoleptic Characteristics of the Examined Meat Parts

The results of the organoleptic evaluation are given in Table 10. Only the flavour and the tenderness revealed significant differences between the meat parts. The oyster and the medal were less favourable, while the outside strip and the tip had the best flavour and tenderness. Although we could not find significant differences in shear force among the meat parts, panellists found significant deviations in meat tenderness. Aftertaste (metallic, unfavourable, uncharacteristic of meat) was not detected. In summary, ostrich meat received an average score for every organoleptic parameter.

Table 10. Organoleptic characteristics of the examined valuable meat parts ($n = 10$).

Meat Parts	Smell	Flavour	Juiciness	Tenderness	Texture	Aftertaste
Outside strip	3.22 ± 0.14	$3.42^{b} \pm 0.14$	3.37 ± 0.13	$3.05^{b} \pm 0.14$	2.93 ± 0.14	1.90 ± 0.05
Oyster	2.85 ± 0.14	$2.88^{a} \pm 0.14$	3.37 ± 0.13	$2.85^{ab} \pm 0.14$	2.63 ± 0.14	1.88 ± 0.05
Tip	3.21 ± 0.14	$3.36^{b} \pm 0.14$	3.57 ± 0.12	$3.05^{b} \pm 0.14$	2.67 ± 0.14	1.93 ± 0.05
Outside leg	3.19 ± 0.14	$3.07^{ab} \pm 0.14$	3.41 ± 0.12	$2.52^{a} \pm 0.14$	2.45 ± 0.14	1.95 ± 0.05
Medal	3.00 ± 0.14	$2.98^{a} \pm 0.14$	3.33 ± 0.12	$2.71^{ab} \pm 0.14$	2.76 ± 0.14	1.91 ± 0.05

Different letters represent significant differences ($p < 0.05$).

4. Discussion

4.1. The Ratio of Body Parts, Meat Parts and the Main Organs

Although the feed given in our study contained the recommended quantity of energy, it was richer in protein, fat and lysine in comparison with the literature data [10]. The birds slaughtered at 18 months of age had similar live weight to that of the younger (10–14 months) animals. However, after 12 months of age, fattening becomes less economic [6,7]. The skin and feather represented a significant proportion of slaughter weight. These body parts are considered valuable products on the western market [30]. The head, the feet and the diaphragm are regarded as inedible body parts since they contain mostly bones. The neck is worthless from the consumers' point of view. It is usually purchased by zoos. The live weight of 12–14-month-old Zimbabwean blue neck ostriches was 103.72 kg in the study of Dijana et al. [31]. The ratios of the skin, neck and head of the slaughtered birds were similar to those of the 18-month-old birds in our study. Morris et al. [11] examined fourteen ostriches between 10 to 14 months of age with 95.54 kg live weight and drew similar conclusions. The authors presented a moderately higher proportion of head (1.78%)

and lower feet (2.51%). Data on genotype and nutrition were not available. Regarding meat parts, the weight of the outside strip, the oyster, the tip and the outside leg was 0.10–0.53, 0.08–0.29, 0.31–1.17 and 0.19–1.16 kg, respectively [32]. The lower values of the range stemmed from 7-month-old ostriches, the higher values represented 18-month-old birds. Results from the same age were far below what we experienced for the same meat parts. Data on genotype and nutrition were not published.

The weight of the empty gizzard, heart, liver and lung or lung with trachea of 12–14 months old ostriches ranged between 2.11–4.39%, 0.90–0.97, 0.54–0.56 and 0.42–1.29% [11,31]. When the organ weight of the 10-month-old and 14-month-old birds was compared, the biggest differences were observed in those of the lung and gizzard. In the present study, the weight of the same organs showed 2.28 ± 0.01, 0.88 ± 0.01, 2.02 ± 0.01 and $0.98 \pm 0.01\%$, respectively. Differences due to the age of the birds (12, 14 and 18 months) could be found in the weight of the liver since 18-month-old ostriches in our examination had four times heavier liver than the 12- to 14-month-old ones in the literature. The large size of the liver enables the sufficient metabolism of nutrients, vitamin and glycogen storage and the detoxification of the organism. The extensive ratio of the empty gizzard can be interpreted by its high capacity because ostriches swallow small pebbles to facilitate digestion.

4.2. Chemical Composition of the Examined Meat Parts

For the colour, L* defines black at 0 and white at 100. In the a* value, green is closer to −120 and red is closer to +120, in b*, blue takes the value of −120 and yellow shows +120 [22,31]. Species, breed, age, sex, nutrition, rearing, total haem and myoglobin content, storage time, pH and processing parameters also influence meat colour [33]. Meat with high pigmentation and after oxidation offers lower L*, a* and b* values [33]. The L*, a* and b* for the tip of ten 10- to 12-month-old Zimbabwean blue neck ostriches were 32.6 ± 2.46, 16.9 ± 1.84 and 11.4 ± 1.22 [28], compared to our results of 36.66 ± 0.73, 25.38 ± 0.46 and 5.75 ± 0.38 for the same meat part. The cited authors determined lower values for L* and a* and higher for b*, indicating a darker meat part. The birds in their research were younger; however, data on nutrition and rearing conditions were not available.

The final postmortem pH values of meat and drip loss are significantly and negatively correlated parameters. Higher pH (over 5.7) results in stronger water-holding capacity, but shorter shelf life. Higher pH value improves water-holding capacity and so decreases drip loss [18]. The mean pH value of ostrich meat is 7.2, but it decreases after 24 h and facilitates meat processing [19]. Generally speaking, ostrich meat has a regular (<5.8) to high (>6.2) 24 h postmortem pH [1]. Our results were slightly below the pH_{24} 6.12 for the inside leg and 6.11 ± 0.03 for an unknown meat part [1,34], but above the value of 5.81 for the fan [35]. In summary, ostrich meat has a naturally higher postmortem pH than the 5.7–5.9 generally detected in meat, which results in better water-holding capacity, but shorter shelf life. With chicken, pH_{24} significantly increases by bird age [36] but there are no data published on ostrich.

In comparison with our results on chemical composition, Sharaf et al. [37] demonstrated a 1.01% lower dry matter, a 1.33% higher protein and a 2.05% lower fat content and claimed that the dry matter, protein and fat content of ostrich fan from three 12-month-old ostriches were 23.88, 21.15 and 1.65%, respectively. The higher protein and lower fat content of the meat can be interpreted by the fact that they used six-month-younger birds in their study. However, no data are available on nutrition. The protein content of the inside leg, tip and fan deriving from seven 12- to 14-month-old ostriches showed 21.6 ± 0.49, 20.81 ± 0.72 and $21.0 \pm 0.58\%$ [38]. Our result for the tip was 2.41% lower than that observed by the authors. Their findings for all the examined meat parts show higher values than any meat part in our research. This fact can be interpreted by the age of the birds as younger ostriches were examined in their study. Data on nutrition were not available. The meat of rhea contains 25.9% dry matter, 22.5% protein and 1.6% fat [39–41], which indicates a higher dry matter and protein and a lower fat content when compared to our results. The same parameters in emu were 25.2, 22.3 and 1.8% [33]. Regarding the literature and our findings, it can be concluded that emu and rhea meat contain more dry matter and protein, but less fat than ostrich meat. The hydroxyproline content of meat determines meat tenderness. The lower the hydroxyproline content in meat, the more tender it is. The hydroxyproline content of ostrich meat is 0.09% [34], which is higher than the value we determined, providing more tender meat. Data on sample size, age of the bird and the examined meat parts were not available.

In respect of amino acid content, values in the research of Sales [8] were lower for threonine, serine, glutamine, isoleucine, alanine, valine, isoleucine, tyrosine, histidine and lysine for ostrich unknown meat parts as compared to our results on the outside strip. Sales and Hayes [38] examined the tip with 21.0% protein content and found essential and non-essential amino acid content to be 8.47 g/100 g meat and 10.09 g/100 g meat. We presented higher essential amino acid content (8.68 $\pm$ 0.15 g/100 g meat), but lower non-essential amino acid content (4.69 $\pm$ 0.27 g/100 g meat) when compared to the literature. Data on the amino acid composition of the outside strip were not available in the literature. Differences could be attributed to the different diets and ages of the animals. However, the description of the feed was not available in the literature. In addition, there was no information on the influence of ostrich age for this parameter.

Horbańczuk et al. [25] examined the fan and leg deriving from six Zimbabwean blue neck 12-month-old ostriches and reported SFA content being 37.71 $\pm$ 0.39% for the fan, and 39.37 $\pm$ 0.45% for the leg. The PUFA content of the fan and leg was 28.79 $\pm$ 0.61 and 23.78 $\pm$ 0.33% of total fatty acids. The authors did not find a significant difference between the meat parts. The SFA content was higher for both examined meat parts as compared to our results for the outside strip (35.10 $\pm$ 0.53%). The PUFA content of the fan was higher, while that of the leg was lower than it was found in our study for the outside strip (27.54 $\pm$ 1.01%). It can be argued that the meat from older birds shows lower SFA content than that of the younger ones. Conversely, differences between PUFA values are not obvious enough to draw further conclusions from the effect of the age of the bird on PUFA content. Data on nutrition were not available. The SFA content of the fan from ostriches of unknown ages was 39.73 $\pm$ 0.77 [26], which is 4% more than the value we established for the outside strip (35.10 $\pm$ 0.53%). The PUFA content was reported to be 5% higher (32.99 $\pm$ 1.22%) than our results for the outside strip (27.54 $\pm$ 1.01%), although in the study [26] unrefined fish oil diet supplementation was provided for the birds. Our research presented the n-6/n-3 ratio to be 3.91 $\pm$ 0.43, which is higher than the 3.02 for ostrich in general [42], but lower than the 7.55 $\pm$ 0.12 [27] in the fan. The saturated and unsaturated fatty acid composition of the ostrich fan is 48.0 $\pm$ 1.9 and 50.8 $\pm$ 1.9% [43]. The authors found higher saturated but lower unsaturated fatty acid content and SFA/UFA ratio in comparison with our results (35.10 $\pm$ 0.53%, 64.90 $\pm$ 1.68% and 0.54 $\pm$ 0.017). In their findings, the SFA/UFA ratio was 0.9 $\pm$ 0.4 for the ostrich fan. The meat was the richest in palmitic acid, oleic acid and linoleic acid. The SFA/UFA ratio was higher (0.9 $\pm$ 0.4) in their study for the fan than what we found in the ostrich outside strip (0.54 $\pm$ 0.02). The ostrich meat was the richest in palmitic acid, oleic acid and linoleic acid. The SFA/UFA ratio was higher in their study than our findings for the ostrich outside strip (0.54 $\pm$ 0.02). In another examination on 12-month-old rheas, the SFA, MUFA and PUFA content of the inside leg was 27.93 $\pm$ 0.7%, 42.36 $\pm$ 2.5 and 29.71 $\pm$ 1.9%, respectively [44]. The authors found lower SFA, MUFA and PUFA content; in addition, the n-6/n-3 ratio was also higher (31.30 $\pm$ 9.0) in comparison with our result on ostrich outside strip (3.91 $\pm$ 0.43). Nutrition data were not published. The PUFA content of ostrich meat is 8% higher than emu meat but 5% lower than rhea meat [45].

The Ca, P and Zn content of raw ostrich meat is 0.90 $\pm$ 0.003, 6.30 $\pm$ 0.07 and 0.10 $\pm$ 0.002 g/100 g [34]. Horbańczuk and Wierczbicka [42] claimed that the Ca content of emu and ostrich meat varies between 0.05 and 0.07 g/100 g, respectively. We found the same elements to be present in 0.33 $\pm$ 0.007, 2.48 $\pm$ 0.01 and 0.03 $\pm$ 0.0008 g/100 g in the outside strip. Neither the age nor the nutrition of the birds was mentioned in the study.

The Fe content of ratite meat is 0.04 g/100 g [42]. We experienced a higher value for this parameter in the outside strip (0.04 ± 0.0006 g/100 g). The Mg and Na content of ostrich meat was as considered 0.22 g/100 g and 0.43 g/100 g [29], respectively. Selenium showed a value of 0.0004 ± 0.00001 g/100 g. The content of Mg content of the outside strip was lower (0.26 ± 0.003 g/100 g) and the content of Na was higher (0.51 ± 0.01 g/100 g) in our research in comparison to the relevant literature.

4.3. Selected Technological Parameters, Shear Force and Organoleptic Characteristics of the Five Examined Valuable Meat Parts

The cooking loss and drip loss of ostrich meat of unknown origin were 21.18 and 2.85%, respectively [33]. In the study, there was no significant difference between the meat parts. The thawing loss (3.88 ± 0.42%) of tip was the greatest in every examined meat part [31]. Their examination revealed greater loss for the tip than our study. Cooking loss and shear force of the inside leg from ostriches of 12 months of age were 37.4% and 3.35 N/mm [46]. The cooking loss showed values closest to the oyster in our study (37.32 ± 1.30%), but higher than the inside leg and medal, and lower than the outside strip and tip. Even though birds in their study were half a year younger than in our research, the meat was firmer as compared to every meat part in our examination except for the outside leg (3.42 ± 0.3). For emus, the shear force is 2.95 N/mm, providing more tender meat as compared to ostriches [33]. The cooking loss of rhea meat is 41.9%, which is more than in ostrich [34].

Different meat parts from ostriches of unknown age were scored on a 9-point hedonic scale. The leg (outside leg, mid-leg and inside leg) was considered to have the most intense flavour [47]. In another experiment on a 9-point hedonic scale, meat flavour, tenderness and juiciness scored 6.80 ± 0.05, 7.17 ± 0.06 and 7.38 ± 0.12 [34], which meant that the judges liked ostrich meat moderately. The flavour of the meat was not significantly affected by the age of the birds [17]. Regarding our results, we found statistical differences in flavour and tenderness between the meat parts. The outside strip and tip had the best flavour (with scores 3.42 ± 0.14 and 3.36 ± 0.14) and tenderness (with scores 3.05 ± 0.14 and 3.05 ± 0.14).

In our analyses in respect to the organoleptic evaluation, the oyster and the medal were less favourable, while the outside strip and the tip had the best flavour and tenderness.

5. Conclusions

Ostrich meat can play a role as an alternative food resource in human nutrition. We aimed to conduct a study including meat parts that are considered valuable in Hungary. Concerning the amino acid composition, glutamine should be highlighted. Our results show that ostrich meat generally has low saturated fatty acid content, a low SFA/UFA and n-6/n-3 ratio and high polyunsaturated fatty acid content. Ostrich meat is dark red and we could prove that it is rich in minerals. Comparing the examined valuable meat parts, the protein content of the medal was the highest. Regarding L*a*b* meat colour, the meat parts significantly differed in a* and b* values. The outside leg had the lowest a* and b* values. As for technological characteristics, the outside strip and the oyster significantly had the greatest thawing loss. According to the panellists, cooked ostrich meat did not have an aftertaste and it received an average score for all the evaluated organoleptic parameters. In comparison with the younger birds (10–14 months of age) in the literature, the 18-month-old ostriches in our study showed similar or slightly lower slaughter yield, lighter meat, greater liver weight, lower protein and higher fat content, higher essential amino acid and lower non-essential amino acid content and higher SFA content in some cases.

Author Contributions: Conceptualisation, L.D.B., Z.V., I.K., T.P. and V.S.; methodology, Z.V., T.P., I.K. and L.D.B.; software, I.K. and L.D.B.; validation, L.D.B., I.K., Z.V. and T.P.; formal analysis, L.D.B.; writing—original draft preparation, L.D.B.; writing—review and editing, L.D.B., T.P., Z.V. and I.K.; visualisation, I.K., Z.V. and L.D.B.; supervision, I.K., Z.V. and T.P. All authors have read and agreed to the published version of the manuscript.

Funding: The publication is supported by the EFOP-3.6.3-VEKOP-16-2017-00008 project. The project is co-financed by the European Union and the European Social Fund.

Institutional Review Board Statement: Not applicable.

Informed Consent Statement: Not applicable.

Data Availability Statement: The data presented in this study are available on request from the corresponding author. The data are not publicly available due to privacy.

Conflicts of Interest: The authors declare no conflict of interest.

References

1. Poławska, E.; Marchewska, J.; Cooper, R.; Pomianowski, J.; Strzałkowska, N.; Horbańczuk, J.; Sartowska-Żygowska, K. The ostrich meat—An updated review. II. Nutritive value. *An. Sci. Pap. Rep.* **2011**, *29*, 89–97.
2. Al-Khalifa, H.; Al-Naser, A. Ostrich meat: Production, quality parameters, and nutritional comparison to other types of meats. *J. Appl. Poult. Res.* **2014**, *23*, 784–790. [CrossRef]
3. Harris, I.; Morris, C.; Jackson, T.; May, S.; Lucia, L.; Hale, D.; Miller, R.; Keeton, J.; Savell, J.; Acuff, G. Ostrich Meat Industry Development. Texas A&M University System, Texas. *Fin. Rep. Am. Ost. Ass.* **1993**, *1*, 1–44.
4. Balog, A.; Almeida, P.I.C.L. Ostrich (*Struthio camelus*) carcass yield and meat quality parameters. *Braz. J. Poult. Sci.* **2007**, *9*, 215–220. [CrossRef]
5. Al-Nasser, A.; Al-Khalaifa, H.; Holleman, K.; Al-Ghalaf, W. Ostrich production in the arid environment of Kuwait. *J. Arid. Environ.* **2003**, *54*, 219–224. [CrossRef]
6. Cooper, R.G. Critical factors in ostrich production: A focus on Southern Africa. *W. Poult. Sci. J.* **2000**, *56*, 247–265. [CrossRef]
7. Pollok, K.D.; Hale, D.S.; Miller, R.K.; Angel, R.; Blue-Mclendon, A.; Baltmanis, B.; Keeton, J.T. Ostrich slaughter and by-product yields. *Am. Ost.* **1997**, *4*, 31–35.
8. Sales, J. Ostrich meat research: An update. In Proceedings of the World Ostrich Congress, Warsaw, Poland, 26–29 September 2002; pp. 148–160.
9. Hoffman, L.; Botha, S.S.; Britz, T. Sensory properties of hot-deboned ostrich (*Struthio camelus* var. *domesticus*) Muscularis gastrocnemius, pars interna. *Meat Sci.* **2006**, *72*, 734–740. [CrossRef]
10. Aganga, A.A.B.; Aganga, A.O.; Omphile, U.J. Ostrich feeding and nutrition. *Pak. J. Nutr.* **2003**, *2*, 60–67.
11. Morris, C.A.; Harris, S.D.; May, S.G.; Jackson, T.C.; Hale, D.S.; Miller, R.K.; Keeton, J.T.; Acuff, G.R.; Lucia, L.M.; Savell, J. Ostrich Slaughter and Fabrication: 1. Slaughter Yields of Carcasses and Effects of Electrical Stimulation on Post-Mortem pH. *Poult. Sci.* **1995**, *74*, 1683–1687. [CrossRef]
12. Fernández-López, J.; Sayas-Barberá, E.; Muñoz, T.; Sendra, E.; Navarro, C.; Pérez-Alvarez, J.A. Effect of packaging conditions on shelf-life of ostrich steaks. *Meat Sci.* **2008**, *78*, 143–152. [CrossRef]
13. Váry, M.; Holló, G.; Ábrahám, C.; Csapó, J.; Seenger, J.; Holló, I.; Szűcs, E. Az íz szerepe a hús élvezeti értékében (*Irodalmi áttekintés*). *Acta Agr. Kap.* **2003**, *7*, 63–74.
14. Taylor, G.; Andrews, L.; Gillespie, J.; Schupp, A.; Prinyawiwatkul, W. How do ratite meats compare with beef? Implications for ratites industry. *J. Agribus.* **1998**, *16*, 97–114.
15. Hoffman, L.; Fisher, P. Comparison of meat quality characteristics between young and old ostriches. *Meat Sci.* **2001**, *59*, 335–337. [CrossRef]
16. Swatland, H.J. *On-Line Evaluation of Meat*; Technomic Pub. Co.: Lancaster, PA, USA, 1995; pp. 1–372. ISBN 978-1-56676-333-2.
17. Botha, S.S.C.; Hoffman, L.C.; Britz, T.C. Effect of hot-deboning on the physical quality characteristics of ostrich meat. *S. Afr. J. An. Sci.* **2006**, *36*, 197–208.
18. Thomas, A.R.; Gondoza, H.; Hoffman, L.; Oosthuizen, V.; Naudé, R.J. The roles of the proteasome, and cathepsins B, L, H and D, in ostrich meat tenderisation. *Meat Sci.* **2004**, *67*, 113–120. [CrossRef] [PubMed]
19. Fernández-López, J.; Jiménez, S.; Sayas-Barberá, E.; Sendra, E.; Pérez-Alvarez, J.A. Quality characteristics of ostrich (*Struthio camelus*) burgers. *Meat Sci.* **2006**, *73*, 295–303. [CrossRef]
20. Cooper, R.; Mahrose, K. Anatomy and physiology of the gastro-intestinal tract and growth curves of the ostrich (*Struthio camelus*). *An. Sci. J.* **2004**, *75*, 491–498. [CrossRef]
21. Csiro, P. *Model Code of Practice for the Welfare of Animals: Livestock at Slaughtering Establishments*; Csiro Publishing: Victoria, Australia, 2003. Available online: https://www.publish.csiro.au/book/2975 (accessed on 10 May 2021).
22. Hernández Salueña, B.; Sáenz Gamasa, C.; Diñeiro Rubial, J.M.; Alberdi Odriozola, C. CIELAB color paths during meat shelf life. *Meat Sci.* **2019**, *157*, 107–889. [CrossRef]
23. Honikel, K.O. Reference methods for the assessment of physical characteristics of meat. *Meat Sci.* **1998**, *49*, 447–457. [CrossRef]
24. Bejerholm, C.; Aaslyng, M.D. The influence of cooking technique and core temperature on results of a sensory analysis of pork—depending on the raw meat quality. *Food Qual. Prefer.* **2004**, *15*, 19–30. [CrossRef]
25. Horbańczuk, J.; Sales, J.; Celeda, T.; Celeda, A.; Ziêba, G.; Kawka, P. Cholesterol content and fatty acid composition of ostrich meat as influenced by subspecies. *Meat Sci.* **1998**, *50*, 385–388.

26. Hoffman, L.C.; Joubert, M.; Brand, T.S.; Manley, M. The effect of dietary fish oil rich in n-3 fatty acids on the organoleptic, fatty acid and physicochemical characteristics of ostrich meat. *Meat Sci.* **2005**, *70*, 45–53. [CrossRef]
27. Horbańczuk, O.K.; Moczkowska, M.; Marchewka, J.; Atanasov, A.G.; Kurek, M.A. The Composition of Fatty Acids in Ostrich Meat Influenced by the Type of Packaging and Refrigerated Storage. *Molecules* **2019**, *24*, 4128. [CrossRef]
28. Majewska, D.; Jakubowska, M.; Ligocki, M.; Tarasewicz, Z.; Szczerbińska, D.; Karamucki, T.; Sales, J. Physicochemical characteristics, proximate analysis and mineral composition of ostrich meat as influenced by muscle. *Food Chem.* **2009**, *117*, 207–211. [CrossRef]
29. Sales, J.; Oliver-Lyons, B. Ostrich meat: A review. *Food Aust.* **1996**, *48*, 504511.
30. Bitlisli, B.O.; Başaran, B.; Sari, Ö.; Ahmet, A.; Gokhan, Z. Some physical and chemical properties of ostrich skins and leathers. *Ind. J. Chem. Technol.* **2004**, *11*, 654–658.
31. Dijana, N.; Zlatko, P.; Lilić, S. Evaluation of the ostrich carcass reared and slaughtered in Macedonia. *Meat Technol. Mes.* **2010**, *51*, 143–148.
32. Engelbrecht, A. Establishing Genetic and Environmental Parameters for Ostrich (*Struthio camelus domesticus*) Growth and Slaughter Characteristics. Ph.D. Thesis, Stellenbosch University, Stellenbosch, South Africa, 2013.
33. Naveena, B.; Sen, A.; Muthukumar, M.; Girish, P.; Kumar, Y.P.; Kiran, M. Carcass characteristics, composition, physico-chemical, microbial and sensory quality of emu meat. *Br. Poult. Sci.* **2013**, *54*, 329–336. [CrossRef]
34. Akram, M.B.; Khan, M.I.; Khalid, S.; Shoaib, M.; Hassan, S.A. Quality and Sensory Comparison of Ostrich and Goat Meat. *SSR Inst. Int. J. Life Sci.* **2019**, *5*, 2168–2175. [CrossRef]
35. Yucel, B.; Taskin, T. *Animal Husbandry and Nutrition*; IntechOpen: London, UK, 2018; p. 202, ISBN 978-1-78923-420-6.
36. Li, J.; Yang, C.; Peng, H.; Yin, H.; Wang, Y.; Hu, Y.; Yu, C.; Jiang, X.; Du, H.; Li, Q.; et al. Effects of Slaughter Age on Muscle Characteristics and Meat Quality Traits of Da-Heng Meat Type Birds. *Animals* **2019**, *10*, 69. [CrossRef] [PubMed]
37. Sharaf, A. Chemical Characteristics of Ostrich Meat in Comparison with Beef and Chicken Meats. *Egypt J. Appl. Sci.* **2006**, *21*, 569–580.
38. Sales, J.; Hayes, J. Proximate, amino acid and mineral composition of ostrich meat. *Food Chem.* **1996**, *56*, 167–170. [CrossRef]
39. Wideman, N.; O'Bryan, C.; Crandall, P. Factors affecting poultry meat colour and consumer preferences—A review. *World's Poult. Sci. J.* **2016**, *1*, 1–14. [CrossRef]
40. Ma, D.; Kim, Y.H.B.; Cooper, B.; Oh, J.-H.; Chun, H.; Choe, J.-H.; Schoonmaker, J.P.; Ajuwon, K.; Min, B. Metabolomics Profiling to Determine the Effect of Postmortem Aging on Color and Lipid Oxidative Stabilities of Different Bovine Muscles. *J. Agric. Food Chem.* **2017**, *65*, 6708–6716. [CrossRef]
41. Watanabe, G.; Motoyama, M.; Nakajima, I.; Sasaki, K. Relationship between water-holding capacity and intramuscular fat content in Japanese commercial pork loin. *Asian-Australas. J. Anim. Sci.* **2018**, *31*, 914–918. [CrossRef] [PubMed]
42. Horbańczuk, O.K.; Wierzbicka, A. Technological and nutritional properties of ostrich, emu, and rhea meat quality. *J. Vet. Res.* **2016**, *60*, 279–286. [CrossRef]
43. Paleari, M.A.; Camisasca, S.; Beretta, G.; Renon, P.; Corsico, P.; Bertolo, G.; Crivelli, G. Ostrich meat: Physico-chemical characteristics and comparison with turkey and bovine meat. *Meat Sci.* **1998**, *48*, 205–210. [CrossRef]
44. Romanelli, P.F.; Trabuco, E.; Scriboni, A.B.; Visentainer, J.V.; De Souza, N.E. Chemical composition and fatty acid profile of rhea (*Rhea americana*) meat. *Arch. Latinoam. Nutr.* **2008**, *58*, 201–205.
45. Sales, J.; Navarro, J.; Martella, M.; Lizurume, M.; Manero, A.; Bellis, L.; Garcia, P. Cholesterol content and fatty acid composition of rhea meat. *Meat Sci.* **1999**, *53*, 73–75. [CrossRef]
46. Poławska, E.; Lisiak, D.; Jozwik, A.; Pierzchala, M.; Strzalkowska, N.; Pomianowski, J.; Wójcik, A. The effect of the diet supplementation with linseed and rapeseed on the physico-chemical and sensory characteristics of ostrich meat. *An. Sci. Pap. Rep.* **2012**, *30*, 65–72.
47. Cooper, R.G.; Horbańczuk, J.O. Anatomical and physiological characteristics of ostrich (*Struthio camelus* var. *domesticus*) meat determine its nutritional importance for man. *Anim. Sci. J.* **2002**, *73*, 167–173. [CrossRef]

agriculture

Article

Evaluation of Methionine Sources in Protein Reduced Diets for Turkeys in the Late Finishing Period Regarding Performance, Footpad Health and Liver Health

Jan Berend Lingens [1,*,†], Amr Abd El-Wahab [1,2,†], Juliano Cesar de Paula Dorigam [3], Andreas Lemme [3], Ralph Brehm [4], Marion Langeheine [4] and Christian Visscher [1]

[1] Institute for Animal Nutrition, University of Veterinary Medicine Hannover, Foundation, Bischofsholer Damm 15, D-30173 Hannover, Germany; amrwahab5@mans.edu.eg (A.A.E.-W.); christian.visscher@tiho-hannover.de (C.V.)

[2] Department of Nutrition and Nutritional Deficiency Diseases, Faculty of Veterinary Medicine, Mansoura University, Mansoura 35516, Egypt

[3] Evonik Operations GmbH, Rodenbacher Chaussee 4, D-63457 Hanau, Germany; juliano.dorigam@evonik.com (J.C.d.P.D.); andreas.lemme@evonik.com (A.L.)

[4] Institute for Anatomy, University of Veterinary Medicine Hannover, Foundation, Bischofsholer Damm 15, D-30173 Hannover, Germany; ralph.brehm@tiho-hannover.de (R.B.); marion.langeheine@tiho-hannover.de (M.L.)

* Correspondence: jan.berend.lingens@tiho-hannover.de

† These authors contributed equally to this work.

Citation: Lingens, J.B.; Abd El-Wahab, A.; de Paula Dorigam, J.C.; Lemme, A.; Brehm, R.; Langeheine, M.; Visscher, C. Evaluation of Methionine Sources in Protein Reduced Diets for Turkeys in the Late Finishing Period Regarding Performance, Footpad Health and Liver Health. *Agriculture* **2021**, *11*, 901. https://doi.org/10.3390/agriculture11090901

Academic Editor: István Komlósi

Received: 13 August 2021
Accepted: 16 September 2021
Published: 19 September 2021

Publisher's Note: MDPI stays neutral with regard to jurisdictional claims in published maps and institutional affiliations.

Abstract: Footpad dermatitis and hepatic lipidosis are health problems in fattening turkeys where a positive influence of higher methionine content in feed is discussed. The effects of the methionine supplements DL-methionine (DLM) and liquid methionine hydroxyl analogue free acid (MHA-FA) under the aspect of low protein diets were investigated in this study based on performance parameters, footpad health, liver health and oxidative stress. In this study, 80 female turkeys (B.U.T. Big 6) of 63 day-old, were randomly assigned to four groups characterising a 2 × 2 factorial design with five replicates each over five weeks. The groups were fed with diets differing in methionine source (DLM vs. MHA-FA, assuming a biological activity of MHA-FA of 65%) and crude protein content (15% vs. 18%) for 35 days. The results showed no significant interactions between the protein content and methionine source. Strong protein reduction significantly impaired water intake, feed intake, weight gain and feed conversation ratio, but improved footpad health. DLM and MHA-FA addition had no significant effect on weight gain, crude fat and protein contents in the liver, but DLM resulted in a significant increase in livers antioxidative capacity compared to MHA-FA. Although the protein reduction resulted in reduced performance, the study showed that MHA-FA can be replaced by DLM in a 100:65 weight ratio without compromising performance but with certain advantages in the antioxidative capacity of the liver.

Keywords: fattening turkey; protein; methionine; growth; hepatic lipidosis; oxidative stress

1. Introduction

In recent years, maintaining the competitiveness of agricultural enterprises has led to a further increase in the efficiency of animal production [1]. As a result of breeding advances, fattening turkeys with modern genetics have a high performance potential, which results in high expectations for nutrients and therefore requires better understanding of the feed composition [2]. The use of additives such as non-bound amino acids allows to adapt the diets to different feeding requirements [3]. Feed management practices, including those that substantially reduce dietary crude protein, are possibly the most important measure to reduce nitrogen excretion and the emission of nitrogen into the environment which is attained with the supplementation of non-bound amino acids [4,5].

In this context, methionine (Met) plays a central role, as it is the first limiting amino acid for poultry and serves as a building block for protein synthesis, being also a precursor for cysteine and a functional amino acid involved in methyl donation for glutathione to counter oxidative stress [4]. Basically, two products for balancing dietary methionine are commercially available: DL-methionine (DLM) and liquid methionine hydroxyl analogue-free acid (MHA-FA). Research suggests that they differ in their biological activity, which is confirmed in the assessment performed by EFSA in numerous publications as 75% on an equimolar basis [6,7]. Among other causes, a lack of methionine is discussed as a predisposing factor for footpad dermatitis (FPD), hepatic lipidosis and oxidative stress [8–11].

Various studies have shown a positive influence of Met on the footpad health of fattening turkeys [10,12,13]. Abd El-Wahab et al. [10] concluded that at almost identical litter dry matter (DM) contents, the FPD scores for young turkeys fed high level of Met in diet were lower than those fed dietary low Met level. It means that level of dietary Met plays an important role for health of skin rather than moisture content in the litter. Thus, it seems that Met has a specific function regards foot pad health (as known for skin and feathers) as mentioned by Abd El-Wahab et al. [14] via protein synthesis and continuous production of keratin. Furthermore, reduced protein intake also has a positive effect on footpad health by reducing excess excretion of uric acid [15]. A reduction in the protein content in feed is currently required in order to decrease the eutrophication potential of excremental ammonia as well as minimise fine dust and the nitrate intake in ground water [5,16,17]. Additionally, hepatic lipidosis is a metabolic disease of the liver whose pathogenesis has not been fully clarified yet, but low dietary protein and Met content as well as high starch content are discussed as predisposing factors [9,18,19].

A feed composition containing minimum protein requires a higher addition of synthetic amino acids such as Met [20]. While the biological availability of MHA-FA relative to DLM has been extensively studied in broilers, less information is available for turkeys [6,21–23]. Based on their literature surveys and experiments, Hoehler and Hooge [24] and Hoehler et al. [25] recommended a bioavailability of 65% for MHA-FA relative to DLM [25,26]. Indeed, other research studies came to different conclusions, but this concept has successfully been applied and validated [24,27,28]. Therefore, in the present study, the effects of supplementation of DLM and MHA-FA in protein reduced diets were examined on growth performance, footpad health liver health and parameters of oxidative stress in late fattening female turkeys.

2. Materials and Methods

The experiment was carried out in accordance with German regulations. The Ethics Committee of Lower Saxony for Care and Use of Laboratory Animals (LAVES) approved the experiment in accordance with § 4, paragraph 3 of the Animal Protection Act (reference number: 33.8-42502-05-18A313).

2.1. Birds and Housing

For the experiment, 80 female turkeys of a common fattening line were used (B.U.T. Big 6, Aviagen Turkeys Ltd., Tattenhall, UK). The birds had been grown on a commercial farm and received the Newcastle vaccine on the farm of origin. The 63 day-old turkeys were divided into four groups (n = 20) with five replicates. For individual identification, the turkeys were marked with coloured plastic rings on their legs. The birds were randomly allocated to 20 pens (1.10 × 2.25 m^2) at the beginning of the experiment (day 63) in such a manner that average body weight (BW) was similar between treatments (BW 4.03 kg/bird). All feeding groups were consistently distributed in the experimental rearing unit so that the pen location had no influence on the experimental results. In each pen, 5 kg of wood shavings (GOLDSPAN®, Goldspan GmbH & Co. KG, Goldenstedt, Germany) served as litter. The pens were equipped with drinking lines (Easy Line, LUBING Maschinenfabrik Ludwig Bening GmbH und Co. KG, Barnstorf, Germany) with two pendulum drinking nipples and a hanging feeder (automatic feeder, Siepmann GmbH, Herdecke, Germany). Feed and water were made available to the birds *ad libitum* during the trial period. The lighting programme consisted of 16 h light and 8 h darkness. The environmental temperature was controlled daily and kept at an adequate range (20.0 °C and 23.5 °C) throughout the experimental period. At the end of the study, the turkeys were slaughtered to analyse the livers and to take footpad samples for histology.

2.2. Feed Composition and Analysis

In order to study the response of turkeys to the reduction in the dietary crude protein supplied with different Met sources, the diets of the four experimental groups were formulated considering these two protein contents (standard as in practice high protein = HP vs. low protein = LP) and two Met sources (DLM vs. MHA-FA) as the main factors.

Initially, the HP diets were formulated to meet or exceed the amino acid recommendation provided for female turkeys in the breeder's guidelines [29]. Therefore Lysine sulfate and L-Threonine were added to the HP diets. To achieve a significant reduction in protein in LP diets, crystalline amino acids (L-Arginine, L-Isoleucine, L-Valine, L-Tryptophan, Lysine sulfate, L-Threonine) were supplied in order of limitation as the protein was reduced. In the formulation of the diets, it was considered either DLM (Evonik Industries AG, Essen, Germany) or MHA-FA (NOVUS Int., St. Charles, MO, USA) as the Met source and a bioavailability of MHA-FA of 65% relative to DLM was assumed considering previous results [25]. The nutritional composition of the experimental diets is shown in Table 1.

Table 1. Composition of the experimental diets.

Ingredients (%)	LP + DLM	LP + MHA-FA	HP + DLM	HP + MHA-FA
Wheat	36.0	36.0	36.0	36.0
Corn	43.4	43.4	32.3	32.3
Soybean meal	12.5	12.5	23.3	23.3
Soybean oil	1.50	1.50	3.33	3.33
Monocalcium phosphate	1.68	1.68	1.65	1.65
Calcium carbonate	1.18	1.18	1.11	1.11
Vitamin/mineral premix [1]	1.00	1.00	1.00	1.00
Sodium bicarbonate	0.26	0.26	0.26	0.26
Sodium chloride	0.21	0.21	0.21	0.21
Lysine sulfate (54.6%)	0.89	0.89	0.42	0.42
MHA-FA	-	0.46	-	0.32
DL-Methionine	0.30	-	0.21	-
L-Threonine	0.23	0.23	0.08	0.08
L-Arginine	0.31	0.31	-	-
L-Isoleucine	0.18	0.18	-	-
L-Valine	0.17	0.17	-	-
L-Tryptophan	0.05	0.05	-	-

LP + DLM = diet with 15% crude protein and DLM; LP + MHA-FA = diet with 15% crude protein and MHA-FA; HP + DLM = diet with 18% crude protein and DLM; HP + MHA-FA = diet with 18% crude protein and MHA-FA.
[1] Content per kg of feed: Vitamin A: 10,000 IU; Vitamin D3: 3000 IU; Vitamin E: 60 mg; Vitamin K3: 4 mg; Vitamin B1: 3 mg; Vitamin B2: 6 mg; Vitamin B6: 5 mg; Vitamin B12: 30 µg; Folic acid: 1.5 mg; Biotin: 150 µg; calcium pantothenate: 11 mg; Niacin: 55 mg; Mn: 70 mg; Zn: 80 mg; I: 0.8 mg; Fe: 25 mg; Cu: 15 mg; Se: 0.25 mg; choline chloride 346.00 mg.

The chemical composition of the diets and Met sources (DLM) and (MHA-FA) were analysed by EVONIK Operations GmbH, Hanau, Germany and is presented in Table 2. Dietary amino acids concentrations were determined by an Ion exchange chromatography analyzer (Biochrom 30+, Biochrom Ltd., Cambridge, UK) with post column derivatisation with ninhydrin. Amino acids were oxidised with performic acid which was neutralised with sodium metabisulfite [30,31]. Amino acids were liberated from the protein by hydrolysis with 6 *N* HCl for 24 h at 110 °C and quantified with the internal standard method by measuring the absorption of reaction products with ninhydrin at 570 nm. Supplemented amino acids were determined after extraction with 0.1 *N* HCl [31]. Supplemented MHA-FA was analysed using the method described by VDLUFA [32]. Contents of ash, protein, fat, fibre, starch and sugar were analysed using near-infrared spectroscopy (Foss NIR 6500, Foss A/S, Hilleroed, Denmark).

Table 2. Calculated and analysed nutrient contents of the experimental diets.

Nutrient Composition (g/kg, as is Basis)	LP + DLM	LP + MHA-FA	HP + DLM	HP + MHA-FA
Crude ash	44.0	43.0	50.0	51.0
Crude protein	155.5	151.5	185.4	186.5
Crude fat	37.0	39.0	55.0	52.0
Crude fibre	23.0	22.0	25.0	26.0
Nitrogen free extract [1]	651.0	650.9	598.2	597.3
Starch	491.0	485.0	430.0	420.0
Sugar	26.0	27.0	38.0	38.0
SID Lysine	10.1	10.1	10.1	10.1
SID Methionine	4.4	4.4	4.4	4.4
SID Methionine + Cystine	7.0	7.0	7.0	7.0
SID Threonine	6.2	6.2	6.2	6.2
SID Valine	7.4	7.4	7.4	7.4
SID Arginine	10.3	10.3	10.3	10.3
Lysine	10.8 (10.7)	10.8 (10.8)	11.1 (11.3)	11.1 (11.0)
Methionine [2]	5.1 (4.8)	5.1 (5.1)	4.7 (4.6)	4.7 (4.7)
Methionine + Cysteine [2]	7.6 (7.3)	7.6 (7.5)	7.7 (7.6)	7.7 (7.8)
Cysteine	2.5 (2.5)	2.5 (2.5)	3.0 (3.0)	3.0 (3.0)
Threonine	6.9 (6.9)	6.9 (6.6)	7.2 (7.3)	7.2 (7.2)
Valine	7.9 (7.9)	7.9 (7.8)	8.3 (8.4)	8.3 (8.3)
Arginine	10.0 (10.9)	10.0 (10.7)	11.3 (11.5)	11.3 (11.3)
Tryptophan	2.1 (2. 1)	2.1 (2.1)	2.2 (2.3)	2.2 (2.3)
Isoleucine	7.1 (7.1)	7.1 (6.9)	7.4 (7.6)	7.4 (7.5)
Leucine	(11.3)	(11.2)	(14.3)	(14.3)
Histidine	(3.40)	(3.40)	(4.50)	(4.50)
Phenylalanine	(6.70)	(6.60)	(9.00)	(9.00)
Glycine	(5.60)	(5.50)	(7.40)	(7.40)
Serine	(6.60)	(6.50)	(8.80)	(8.90)
Metabolizable energy [3] (MJ/kg)	12.2	12.1	12.4	12.2

LP + DLM = diet with 15% crude protein and DLM; LP + MHA-FA = diet with 15% crude protein and MHA-FA; HP + DLM = diet with 18% crude protein and DLM; HP + MHA-FA = diet with 18% crude protein and MHA-FA. [1] Nitrogen-free extract = DM − (crude ash + crude protein + crude fat + crud fibre). [2] Total methionine and methionine + cysteine in the MHA-FA treatments assume 65% Met efficiency. [3] Metabolisable energy (per kg) = 0.01551 × crude protein + 0.03431 × ether extracts + 0.01669 × starch + 0.01301 × sugar (as sucrose); SID: standardised ileal digestible for total amino acids; amino acids: calculated levels (in brackets: analysed values).

2.3. Parameters and Sampling Time Plan

Individual BW, feed intake, water intake, litter DM and gross footpad scoring were measured weekly. Individual BW and weighing in and weighing out of water and feed was carried out with a calibrated scale (Sytemwaage PCE-TB30, PCE Instruments, Meschede, Germany). At the end of the trial, the birds in each group were necropsied and samples of the liver were taken to analyse crude fat, crude protein contents and total antioxidant capacity. During slaughtering, the blood samples were taken for malondialdehyde analysis and footpads were sampled for histological analysis.

2.4. Litter Sampling

Litter samples were taken weekly at three defined spots along a diagonal line through each box to measure the DM content. For this purpose, a cup with a diameter of 5 cm was used to punch out a sample from the full depth of the bedding as described by Ullrich et al. (2019) [33]. The three samples from each box were pooled into one pool sample and the DM content was determined on the same day.

2.5. Gross Footpad Scoring

The footpads of the birds were scored weekly by the same person on a scale from 0 to 7 in accordance with Mayne et al. [34]; score 0 = healthy skin, score 7 = more than 50% of footpad area is necrotic (Figure 1). The average scoring of both legs was performed for each bird.

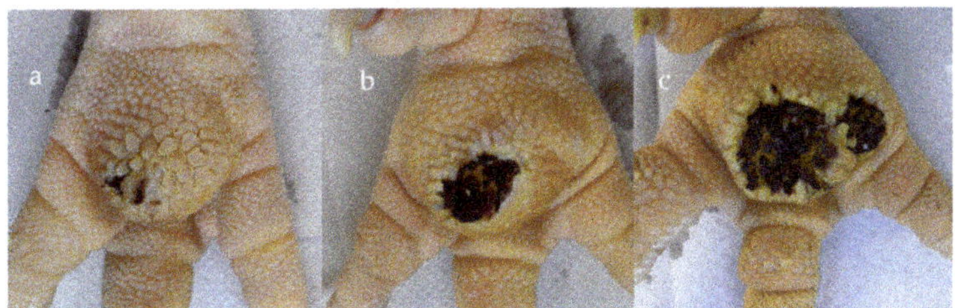

Figure 1. Scoring of foot pad: (**a**) Score 3 = small necrotic areas; (**b**) Score 5 = 25% of footpad is necrotic; (**c**) Score 7 = > 50% of footpad is necrotic (photo: ©Abd El-Wahab, A./University of Veterinary Medicine, Hannover, Germany).

2.6. Histological Footpad Scoring

In order to determine the histological footpad scoring in accordance with MAYNE et al. [34], approximately 2-cm tissue samples from footpads (middle part) were taken at dissection. The samples were first washed in phosphate-buffer saline, then fixed in 4% formaldehyde for 48 h. After fixation, the samples were embedded in paraffin and 4-μm sections of all samples were stained with haematoxylin and eosin using standard techniques. The sections were examined under a light microscope and categorised using the histopathological scoring system on a 7-point scale (0 = normal epidermis; 1 = hyperkeratosis; 2 = epidermal acanthosis; 3 = vacuoles in dermis and epidermis; 4 = presence of heterophils, macrophages and lymphocytes in dermis; 5 = increased density of heterophils, macrophages and lymphocytes; 6 = ulcer of the epidermis, only one lesion; 7 = more than one rupture or "ulcer" of the epidermis) in accordance with Mayne et al. [34].

2.7. Analysis of Liver Samples

The livers were first processed in accordance with the method described by Middendorf et al. (2019) for further chemical analysis [35]. The livers were first homogenised in a mixer (Grindomix GM 200, Retsch GmbH, Haan, Germany) and then freeze-dried (freeze dryer Gamma 1–20, Martin Christ Gefriertrocknungsanlagen GmbH, Osterode am Harz, Germany). The freeze-dried material was grinded in a mill (M20 universal mixer, IKA Werke GmbH und Co. KG, Staufen, Germany). The levels of DM and alpha-tocopherol determination were performed with the homogenised livers, whereas the crude fat and crude protein analyses were performed with the freeze-dried and grinded materials. The levels of liver DM, alpha-tocopherol, crude protein and crude fat were determined using the VDLUFA methods in accordance with Naumann and Bassler [36].

2.8. Oxidative Stress

Total antioxidant capacity (TAC) of liver parenchyma was determined in accordance with SHEN et al. [11]. Briefly, the TAC was determined by ELISA (ELISA Kit OxiSelect™ Total Antioxidant Capacity Assay Kit, Fa. Cell Biolabs, Inc., San Diego, CA, USA). Using the ELISA Kit, the TAC was calculated in relation to the liver protein. Malondialdehyde was determined in blood plasma of ten animals per group after dissection. This was performed by SYNLAB (Medizinisches Versorgungszentrum Labor München Zentrum GbR München, Germany) in accordance with the method by Chromsystems Instruments and Chemicals GmbH, Gräfelfing, Germany using the HPLC test kit "Malondialdehyde in Serum/Plasma HPLC" (Chromsystems Instruments and Chemicals GmbH, Gräfelfing, Germany).

2.9. Statistical Analysis

The statistical evaluation was carried out with the program SAS (Statistical Analysis System) Enterprise Guide 7.1 (SAS Institute Inc., Cary, NC, USA) in cooperation with the Institute for Biometry, Epidemiology and Information Processing of the University of Veterinary Medicine Hannover, Foundation, Germany. The sample size was calculated assuming an effect size of 0.25 for the main effect (footpad health), the alpha was set to 0.05 and the power to 0.60. With a numerator of the degrees of freedom (df) of 1 and 4 groups 81 animals were necessary. A Shapiro-Wilk test for normal distribution was performed and normally distributed data were checked for significant differences with the Ryan-Einot-Gabriel-Welsch-test (one-way ANOVA). For not normally distributed data, a Kruskal Wallis test was performed, followed by a Wilcoxon two-sample test. Histological and macroscopical foot pad scoring were performed using a Kruskal Wallis test. The significance level was set at $p < 0.05$.

3. Results

All birds were healthy and there were no mortalities during the experimental period. Additionally, there were no complications and no animals had to be medically treated during the experiment.

3.1. Performance

The results of the weekly BW recording are presented in Table 3. The BW was in accordance with the performance goals for B.U.T. Big 6 heavy lines [37].

Table 3. Average body weight (BW) of female turkeys fed experimental diets with different dietary protein content and using different methionine sources (DL-Methionine-DLM; liquid methionine hydroxyl analogue-free acid-MHA-FA) from day 63 to 98 (kg/bird, mean +/−SD).

Protein Content	Methionine Source	Experimental Period					
		Day 63	Day 70	Day 77	Day 84	Day 91	Day 98
Low [1]	DLM	4.05 ± 0.37	4.87 ± 0.50	5.73 ± 0.58	6.69 [ab] ± 0.63	7.69 [ab] ± 0.72	8.56 [bc] ± 0.74
Low [1]	MHA-FA	4.01 ± 0.32	4.76 ± 0.38	5.66 ± 0.42	6.50 [b] ± 0.47	7.43 [b] ± 0.57	8.27 [c] ± 0.64
High [2]	DLM	4.07 ± 0.30	5.08 ± 0.35	6.04 ± 0.40	7.06 [a] ± 0.48	8.15 [a] ± 0.49	9.06 [a] ± 0.57
High [2]	MHA-FA	3.99 ± 0.43	4.91 ± 0.45	5.88 ± 0.57	6.89 [ab] ± 0.60	7.95 [a] ± 0.64	8.85 [ab] ± 0.63
Factor		*p*-value					
Protein content (A)		0.9652	0.0628	0.0193	0.003	0.0006	0.0004
Methionine source (B)		0.4672	0.1461	0.2994	0.1419	0.1033	0.0889
A × B		0.8034	0.7658	0.7074	0.9563	0.8039	0.7956
Main effects							
Low protein content		4.03 ± 0.34	4.81 ± 0.45	5.69 [b] ± 0.50	6.60 [b] ± 0.56	7.56 [b] ± 0.65	8.41 [b] ± 0.70
High protein content		4.03 ± 0.37	4.99 ± 0.41	5.96 [a] ± 0.49	6.97 [a] ± 0.54	8.05 [a] ± 0.57	8.95 [a] ± 0.60
DLM source		4.06 ± 0.33	4.97 ± 0.44	5.88 ± 0.51	6.88 ± 0.58	7.92 ± 0.65	8.81 ± 0.70
MHA-FA source		4.00 ± 0.37	4.83 ± 0.42	5.77 ± 0.51	6.70 ± 0.57	7.69 ± 0.66	8.56 ± 0.69

[a,b,c] Means in a column within each main effects with different superscripts differ significantly ($p < 0.05$); [1] diet with 15% crude protein; [2] diet with 18% crude protein.

No interactions were observed between the crude protein level and methionine source ($p > 0.05$). Therefore, the responses of methionine sources and protein levels were independent and the main factors were analysed. There was no significant difference between the methionine sources ($p > 0.05$), but the groups fed DLM showed a tendency ($p = 0.0889$) for a higher final BW compared to MHA-FA. From the second week of the experiment onwards, protein reduction led to a significantly lower BW ($p < 0.05$). The performance data obtained in the 35-day period for feed intake, water intake, water to feed intake ratio (W:F), BW gain, protein efficiency ratio (PER) and feed conversation ratio (FCR) are presented in Table 4.

Table 4. Feed intake (FI), water intake (WI), water to feed ratio (W:F), body weight gain (BW), protein efficiency ratio (PER) and feed conversion ratio (FCR) of female turkeys fed experimental diets with different dietary protein content and using different methionine sources (DL-Methionine-DLM; liquid methionine hydroxyl analogue-free acid-MHA-FA) from day 63 to 98 (mean ± SD).

Protein Content	Methionine Source	Performance Parameters					
		FI (g)	WI (g)	W:F Intake (g/g)	BW Gain (g)	PER (g/g)	FCR (kg/kg)
Low [1]	DLM	12,532 [ab] ± 388	23,862 [bc] ± 1278	1.90 [b] ± 0.07	4505 [b] ± 460	2.30 [a] ± 0.05	2.78 [a] ± 0.05
Low [1]	MHA-FA	11,868 [b] ± 485	22,623 [c] ± 1025	1.91 [b] ± 0.06	4255 [b] ± 543	2.31 [a] ± 0.05	2.79 [a] ± 0.06
High [2]	DLM	13,006 [a] ± 411	26,464 [a] ± 735	2.04 [a] ± 0.03	4991 [a] ± 445	2.08 [b] ± 0.04	2.61 [b] ± 0.05
High [2]	MHA-FA	12,589 [a] ± 453	25,428 [ab] ± 999	2.02 [a] ± 0.05	4857 [a] ± 450	2.06 [b] ± 0.06	2.60 [b] ± 0.07
Factor				*p*-value			
Protein content (A)		0.0075	<0.0001	0.0001	<0.0001	<0.0001	<0.0001
Methionine source (B)		0.0137	0.0248	0.8061	0.0759	0.9002	0.966
A × B		0.5339	0.828	0.7063	0.5894	0.4755	0.7177
				Main effects			
Low protein content		12,200 [b] ± 543	23,243 [b] ± 1272	1.91 [b] ± 0.06	4380 [b] ± 513	2.30 [a] ± 0.05	2.79 [a] ± 0.06
High protein content		12,797 [a] ± 463	25,946 [a] ± 991	2.03 [a] ± 0.04	4924 [a] ± 447	2.07 [b] ± 0.05	2.60 [b] ± 0.06
DLM source		12,769 [a] ± 452	25,163 [a] ± 1687	1.97 ± 0.08	4748 ± 510	2.19 ± 0.12	2.69 ± 0.11
MHA-FA source		12,229 [b] ± 583	24,026 [b] ± 1759	1.96 ± 0.08	4556 ± 579	2.18 ± 0.14	2.69 ± 0.12

[a,b,c] Means in a column within each main effects with different superscripts differ significantly ($p < 0.05$); [1] diet with 15% crude protein; [2] diet with 18% crude protein.

No interactions were observed between the dietary protein level and Met source ($p > 0.05$) for the performance parameters in the entire period. The analysis of the main factors showed that birds fed diets with MHA-FA had a significantly lower feed intake and water intake compared to the birds fed DLM. The groups fed DLM showed a tendency ($p = 0.0759$) for a higher BW gain compared to MHA-FA. Independent of the evaluated Met source, the protein reduction led to a significantly lower feed intake, water intake, W:F intake ratio, BW gain and a higher PER and FCR ($p < 0.05$).

3.2. Litter Quality and Footpad Dermatitis

The results of the DM content of the litter material measured weekly are presented in Table 5.

As shown in Table 5, there were no interactions between the crude protein level and methionine source ($p > 0.05$) for dry matter content of the litter material during the trial, neither for the crude protein level effect ($p > 0.05$) nor methionine source effect ($p > 0.05$) when the main factors were taken into consideration for the statistical analysis. The results for FPD scores are presented in Table 6.

Table 5. Dry matter content (%) of the litter of female turkeys fed experimental diets with different dietary protein content and using different methionine sources (DL-Methionine-DLM; liquid methionine hydroxyl analogue-free acid-MHA-FA) from day 70 to 98 (mean ± SD).

Protein Content	Methionine Source	Experimental Period				
		Day 70	Day 77	Day 84	Day 91	Day 98
Low [1]	DLM	49.6 ± 8.46	58.1 ± 4.47	65.3 ± 6.60	65.5 ± 7.81	64.6 ± 7.16
Low [1]	MHA-FA	48.3 ± 6.48	52.3 ± 7.05	62.4 ± 5.05	62.3 ± 7.16	62.0 ± 3.98
High [2]	DLM	48.8 ± 6.20	57.1 ± 4.90	58.5 ± 4.71	58.6 ± 5.07	59.7 ± 5.10
High [2]	MHA-FA	54.1 ± 6.37	59.3 ± 9.23	66.0 ± 6.60	59.5 ± 5.58	66.5 ± 3.82
Factor				*p*-value		
Protein content (A)		0.4503	0.3899	0.5865	0.1519	0.9371
Methionine source (B)		0.5494	0.5956	0.4395	0.7312	0.4303
A × B		0.3207	0.2488	0.0911	0.5293	0.0903
				Main effects		
Low protein content		48.9 ± 7.97	55.2 ± 6.93	63.8 ± 6.38	63.9 ± 8.08	63.3 ± 6.26
High protein content		51.4 ± 5.99	58.2 ± 7.87	62.2 ± 7.23	59.0 ± 5.64	63.1 ± 5.94
DLM source		49.2 ± 6.76	57.6 ± 4.97	61.9 ± 7.03	62.0 ± 7.85	62.2 ± 7.04
MHA-FA source		51.2 ± 7.42	55.8 ± 9.40	64.2 ± 6.48	60.9 ± 6.92	64.3 ± 4.74

[1] diet with 15% crude protein; [2] diet with 18% crude protein.

Table 6. Gross lesions of FPD of female turkeys fed experimental diets with different dietary protein content and using different methionine sources (DL-Methionine-DLM; liquid methionine hydroxyl analogue-free acid-MHA-FA) from day 63 to 98 (mean ± SD).

Protein Content	Methionine Source	Experimental Period					
		Day 63	Day 70	Day 77	Day 84	Day 91	Day 98
Low [1]	DLM	3.90 ± 0.55	3.98 ± 0.60	2.90 ± 0.55	2.35 ± 0.67	1.98 ± 0.44	1.38 [c] ± 0.65
Low [1]	MHA-FA	4.00 ± 0.51	4.03 ± 0.62	2.83 ± 0.67	2.28 ± 0.55	1.83 ± 0.47	1.80 [bc] ± 0.71
High [2]	DLM	3.90 ± 0.58	3.88 ± 0.56	2.63 ± 0.46	2.33 ± 0.80	2.25 ± 0.79	2.18 [ab] ± 0.65
High [2]	MHA-FA	4.10 ± 0.55	3.95 ± 0.48	2.70 ± 0.57	2.43 ± 0.57	2.28 ± 0.55	2.38 [a] ± 0.51
Factor					*p*-value		
Protein content (A)		0.8767	0.5453	0.0885	0.8039	0.0099	0.0001
Methionine source (B)		0.3071	0.3306	0.9920	0.6195	0.7590	0.0580
					Main effects		
Low protein content		3.95 ± 0.53	4.00 ± 0.60	2.86 [a] ± 0.61	2.31 [b] ± 0.61	1.90 [b] ± 0.46	1.59 [b] ± 0.71
High protein content		4.00 ± 0.57	3.91 ± 0.52	2.66 [b] ± 0.51	2.38 [a] ± 0.69	2.26 [a] ± 0.67	2.28 [a] ± 0.59
DLM source		3.90 ± 0.56	3.93 ± 0.57	2.76 ± 0.52	2.34 ± 0.73	2.11 ± 0.65	1.78 ± 0.76
MHA-FA source		4.05 ± 0.53	3.99 ± 0.55	2.76 ± 0.62	2.35 ± 0.56	2.05 ± 0.55	2.09 ± 0.68

[a,b,c] Means in a column within each main effects with different superscripts differ significantly ($p < 0.05$). [1] diet with 15% crude protein; [2] diet with 18% crude protein.

As shown in Table 6, footpad health scores reduced during the course of the experiment from about 4.0 at day 63 to about 1.9 at day 98. An evaluation according to the factor protein level showed a significant effect on footpad health. While, there were no significant protein effects from day 63 to day 84, protein reduction resulted in a significant impairment of the footpad score at day 91 and day 98 ($p < 0.05$). The evaluation according to the factor Met source showed a tendency ($p < 0.10$) to lower footpad scores for the DLM diets compared to MHA-FA on the last day of the experiment.

As shown in Table 7, an evaluation according to the factor protein level showed a significant improvement in the histological footpad score for low protein diets, while the evaluation according to the factor Met source showed no effect of the Met source on the histological footpad score.

Table 7. Histological FPD scores of female turkeys fed experimental diets with different dietary protein content and using different methionine sources (DL-Methionine-DLM; liquid methionine hydroxyl analogue-free acid-MHA-FA) (mean ± SD).

Protein Content	Methionine Source	Day 98 of Life
Low [1]	DLM	2.25 [b] ± 1.02
Low [1]	MHA-FA	2.75 [ab] ± 1.07
High [2]	DLM	3.20 [a] ± 0.89
High [2]	MHA-FA	3.25 [a] ± 0.72
Factor		*p*-value
Protein content (A)		0.0016
Methionine source (B)		0.1733
		Main effects
Low protein content		2.50 [b] ± 1.06
High protein content		3.23 [a] ± 0.80
DLM source		2.73 ± 1.06
MHA-FA source		3.00 ± 0.93

[a,b] Means in a column within each main effects with different superscripts differ significantly ($p < 0.05$). [1] diet with 15% crude protein; [2] diet with 18% crude protein.

3.3. Liver Parameters

The absolute liver weight and relative liver weight, DM content, fat and protein contents of liver (mean ± SD) are presented in Table 8. With the exception of relative liver weight, there was no significant impact by the dietary protein level or Met source nor were there any interactions. The experiment showed a significantly increased relative liver weight due to the protein reduced diets ($p < 0.05$). DLM and MHA-FA nutrition had no significant effect on liver weight gain or fat and protein contents in livers.

Table 8. Absolute and relative liver weights as well as the contents of DM, fat and protein in the liver of female turkeys fed experimental diets with different dietary protein content and using different methionine sources (DL-Methionine-DLM; liquid methionine hydroxyl analogue-free acidMHA-FA) at day 98 (mean ± SD).

Protein Content	Methionine Source	Item				
		Absolute Weight (g)	Relative Weight (%)	DM (g/kg)	Fibre (g/kg DM)	Protein (g/kg DM)
Low [1]	DLM	147 ± 23.6	1.71 ± 0.23	304 ± 16.3	157 ± 49.7	620 ± 62.5
Low [1]	MHA-FA	140 ± 28.5	1.68 ± 0.29	300 ± 18.6	144 ± 49.9	621 ± 71.0
High [2]	DLM	142 ± 20.7	1.57 ± 0.18	303 ± 13.4	143 ± 40.2	633 ± 48.8
High [2]	MHA-FA	136 ± 22.9	1.54 ± 0.21	301 ± 12.2	132 ± 33.2	639 ± 44.4
Factor				*p*-value		
Protein content (A)		0.4793	0.0064	0.9074	0.1785	0.2239
Methionine source (B)		0.2461	0.5821	0.3611	0.2064	0.7842
A × B		0.9195	0.9952	0.7713	0.907	0.8309
				Main effects		
Low protein content		143 ± 26.1	1.70 [a] ± 0.26	302 ± 17.4	151 ± 49.7	620 ± 66.0
High protein content		139 ± 21.7	1.55 [b] ± 0.19	302 ± 12.7	137 ± 36.8	636 ± 46.2
DLM source		144 ± 22.0	1.64 ± 0.22	303 ± 14.7	150 ± 45.2	626 ± 55.7
MHA-FA source		138 ± 25.6	1.61 ± 0.26	300 ± 15.5	138 ± 42.3	630 ± 59.2

[a,b] Means in a column within each main effects with different superscripts differ significantly ($p < 0.05$). [1] diet with 15% crude protein; [2] diet with 18% crude protein.

3.4. Parameter of Oxidative Stress

The content of vitamin E and total antioxidant capacity (TAC) of the liver, as well as the malondialdehyde in blood plasma values are presented in Table 9. No interactions of main factors were found. Also, the dietary protein level did not affect either listed parameter.

The two-factorial ANOVA showed a significant increase in TAC of the liver parenchyma for DLM compared to MHA-FA ($p < 0.05$; 1.05 vs. 0.96 µmol UAE/g liver protein).

Table 9. Content of Vitamin E and total antioxidant capacity (TAC) of the liver, malondialdehyde in blood plasma used as an indicator of oxidative stress of female turkeys fed experimental diets with different dietary protein content and using different methionine sources (DL-Methionine-DLM; liquid methionine hydroxyl analogue-free acid-MHA-FA) at day 98 (mean ± SD).

Protein Content	Methionine Source	Oxidative Stress Parameter		
		Viamin E (mg/kg DM)	TAC (µmol UAE/g Protein)	Malondialdehyde (µmol/Liter)
Low [1]	DLM	16.2 ± 6.56	1.02 [ab] ± 0.20	0.38 ± 0.11
Low [1]	MHA-FA	15.8 ± 5.52	0.93 [b] ± 0.11	0.34 ± 0.10
High [2]	DLM	14.0 ± 4.57	1.08 [a] ± 0.17	0.36 ± 0.09
High [2]	MHA-FA	15.5 ± 4.65	1.00 [ab] ± 0.17	0.38 ± 0.10
Factor			*p*-value	
Protein content (A)		0.2950	0.1086	0.8367
Methionine source (B)		0.6471	0.0260	0.7633
A × B		0.4238	0.8563	0.3856
			Main effects	
Low protein content		16.0 ± 5.99	0.98 ± 0.17	0.36 ± 0.10
High protein content		14.8 ± 4.62	1.04 ± 0.17	0.37 ± 0.09
DLM source		15.1 ± 5.70	1.05 [a] ± 0.19	0.37 ± 0.10
MHA-FA source		15.7 ± 5.04	0.96 [b] ± 0.14	0.36 ± 0.10

[a,b] Means in a column within each main effects with different superscripts differ significantly ($p < 0.05$). [1] diet with 15% crude protein; [2] diet with 18% crude protein.

4. Discussion

In this study the effects of DLM and MHA-FA supplementation in protein reduced diets on the performance, footpad health and liver health of fattening turkeys were evaluated and the results obtained for the several parameters will be discussed in separate sections.

4.1. Performance

There were no complications or mortality observed during the experimental period. On arrival at the experimental facility, female turkeys had an average BW of 4.03 kg ± 0.355 and were therefore 15% below breeder recommendations of 4.73 kg for B.U.T. 6 females at day 63 [37]. However, the BW achieved at day 98 particularly regarding with standard protein levels exceeded 4640 g/bird which was suggested by the breeder as an efficient amount of feed. However, the average BW of turkeys at day 98 was 8.69 kg/bird and thus still 8% below 9.45 kg as suggested by the breeder. It should be noted that the FCR observed in the present trial ranged between 2.60 and 2.79 kg/kg and substantially exceeded the performance goals of the breeding company [37]. Therefore, the experimental feeds were basically well utilised. This is remarkable because specifications for the amino acid levels and applied profile are for heavy female turkeys 98 to 111 days-old, whereas recommendations for 70 to 97 days-old females would be somewhat higher [38].

To reduce nitrogen emissions, a protein reduction in the diets is necessary [16]. This is only possible by balancing in particular the essential amino acids in the feed using free amino acid supplements. In the diets of all experimental groups, the essential amino acids lysine, Met, threonine, tryptophan, arginine, valine and isoleucine were optimised on a digestible amino acid basis to achieve similar levels. This was achieved by individually adding crystalline amino acids (with the exception of MHA-FA which is liquid). Basically, analyses of diets confirmed successful feed production as targeted levels were recovered. However, protein reduction in diets containing 15% protein was strong (3% points) and showed lower contents of some amino acids than the diets containing 18% protein. In this context, the lower content of the essential amino acids leucine, histidine and phenylalanine and of the semi-essential amino acids serine and glycine in the protein reduced diets has to be noted. This is in line with observations by Lemme et al. [39] who reduced crude protein more moderately by only 1% point and who achieved a significant reduction in Gly-equivalents (Gly + 0.714 * Ser) in the protein-reduced feeds. Moreover, the researchers reported a significantly lower intake of Gly-equivalents at protein reduction.

Despite balancing for six essential amino acids using commercially available products in the experiment, all performance parameters (Tables 4 and 5) were negatively affected by the protein-reduced diets. In contrast to Lemme et al. [39,40], we formulated the experimental feed on a standardised ileal digestible amino acid basis using digestibility coefficient as proposed by Lemme et al. [39–41]. However, these digestibility coefficients were determined with broilers. It has been reported that digestibility coefficients differ between turkeys and broilers which might have interfered with protein reduction, as proportions of ingredients substantially changed [42–44]. Nonetheless, Kluth and Rodehutscord [44] did not report any substantial differences for amino acid digestibility with respect to soybean meal, but Kozslowski et al. [43] suggested differences. Probably more of importance are the amino acids which were not considered for balancing. As mentioned above, the degree of protein reduction was high in this trial, this having tremendous impact on levels of not considered essential amino acids as well as conditionally essential and non-essential amino acids. According to feed analysis, the total levels of lysine, methionine + cysteine, threonine, tryptophane, arginine, valine and isoleucine were 3–7% lower in low protein diets compared to standard protein, while on a digestible level, they were very similar. However, leucine, histidine and phenylalanine as well as glycine and serine were 20–25% lower in low protein feeds. While there is no information about requirements for turkeys for these amino acids, it is assumed that one or more of them were below the requirements limiting protein synthesis and, therefore, resulted in lower BW gain and higher PER and FCR. Lemme et al. [40] were not successful in protein reduction as performance dropped in any case. Nevertheless, their research indicated that balancing amino acids beyond lysine, methionine + cysteine, threonine and tryptophan were needed to achieve the same FCR and breast meat deposition as with protein reduction including the entire range of essential amino acids. Indeed, Lemme et al. [6] were able to at least maintain BW gain and FCR when reducing the protein level of commercial type diets by 1% point and balancing for lysine, methionine + cysteine, threonine, arginine, valine and isoleucine. However, meat deposition was slightly affected. They calculated the intake of essential amino acids as well as glycine-equivalents and concluded that a significantly lower glycine-equivalent intake was responsible for reduced meat deposition. The content of the semi-essential amino acids serine and glycine must be noted. The synthesis of glycine and serine can be limited at a high BW gain intensity, in this case either glycine or serine have to be supplemented to the diet [45]. Dean et al. [45] reported that glycine and serine first had a limiting effect on BW gain and FCR of broilers when protein content of diets was reduced from 22% to 19% and all essential amino acids were supplemented. Thus, there is a likelihood that the intake of certain amino acids in this trial was insufficient for maintaining the performance at the level of the standard protein treatments.

With special regard to methionine:cysteine ratio, literature data reporting the optimal methionine:cysteine ratio in poultry diets differ markedly [46]. One important influencing factor is that the metabolic degradation of the indispensable methionine may yield cysteine [47–49]. Consequently, an appropriate amount of cysteine leads to increases the availability of methionine [50]. However, the close metabolic link between methionine and cysteine makes it more difficult to assess both the animals' requirement for total sulfur containing amino acids and the optimal dietary methionine:cysteine [46]. However, Wheeler and Latshaw [51] also found an age-dependent effect on the optimal methionine:cysteine.

Dietary protein reduction significantly reduced water intake. This is an often described effect in the context of protein reduction and is assigned to decreased soybean meal use [4,16,52,53]. Soybean meal contributes relatively high levels of potassium to the feed, stimulating water intake [54]. In our trial, inclusion of soybean meal was reduced by 46%.

In addition, less nitrogen has to be excreted which reduces the water requirement [54]. Assuming 33 g nitrogen/kg weight gain and analysed protein levels of the experimental feeds, average nitrogen depositions of 149, 140, 165, and 160 g/bird and nitrogen intakes of 314, 295, 390 and 372 g/bird can be estimated for birds of treatments LP + DLM, LP + MHA-FA, HP + DLM and HP + MHA-FA, respectively, for the experimental period [6]. Accordingly, dietary protein reduction lowered nitrogen excretion by 27% and thus increased nitrogen utilisation from about 42.6% to 47.5%. Protein reduction by 1% point over the entire production cycle reduced N-excretion by 8%, which is a similar magnitude of response when related to 1% point protein reduction. It is remarkable that reduced water intake due to protein reduction was not reflected in dry matter content of litter in the current trial [6]. Also, Lemme et al. [6] were unable to observe a clear relationship between dietary protein level and litter DM content although they reported markedly improved foot pad health. In the current trial, water intake was not necessarily only influenced by dietary protein per se but also by feed intake. There was a high correlation between water intake and feed intake ($r = 0.94$) and, accordingly, also feed intake was reduced by protein reduction ($p < 0.05$). Still, W:F intake ratio would confirm the relationship with soybean meal and related potassium intake as well as with reduced N-excretion. However, the above discussed strong impact on the dietary amino acid profile and respective match with amino acid requirements negatively affected the feed intake. D'Mello [55] described in his review that amino acid imbalances result in reduced or even depressed feed intake in monogastrics including poultry, and the 20–25% lower levels of those amino acid not considered in feed optimisation may indicate an imbalance. Moreover, Schutte and Pack [56] reported that the increase in methionine levels (i.e., increase in availability and absorption) increased feed intake in broilers. It is well known that The DLM are actively absorbed (transported against a concentration gradient), while MHA-FA is absorbed by the H+ dependent system which is slower than the Na+ system in DLM [57]. According to the literature, the average relative bioavailability of MHA-FA products compared to DLM is approximately 75–80% on an equimolar basis [23,58]. Thus, MHA-FA may be less efficiently absorbed and utilised by birds than DLM and may consequently led to reduce feed intake.

Supplementation of MHA-FA followed the recommendation of Hoehler and Hooge [24] and Hoehler et al. [25]. Accordingly, a relative bioavailability of 65% is assigned to MHA-FA compared to DLM. This suggests that by replacing MHA-FA with DLM in a weight-to-weight ratio of 100:65 in feed, animal performance is not affected. Indeed, the results of the current study basically confirm no detrimental effect of the relatively low inclusion levels of DLM on any performance parameter. Rather the contrary is the case as the BW at the end of the experiments tended to be higher with DLM reflected in the BW gain. Moreover, the apparently lower weight gain can be explained by significantly lower feed intake of MHA-FA fed turkeys, which due to the strong correlation between feed and water intake, was also mirrored in a significantly lower water intake. It is emphasised that neither W:F intake ratio nor PER and FCR were affected by Met sources. Höhler et al. [25] also tested DLM and MHA-FA in a 65:100 ratio at three different supplementation levels in one of the reported trials and described the same growth performance and FCR. Hoehler and Hooge [24] confirmed this observation, as four commercial type trials with >5000 turkeys per treatment replacement of MHA-FA with DLM in a 100:65 ratio resulted in the same performance. Agostini et al. [59] fed turkeys over an entire production cycle and while final BW and FCR were not affected by the Met sources supplemented in a 65:100 ratio, they reported a significantly lower breast meat yield in the case of MHA-FA feeding. Lemme and Meyer [38] applying the same experimental design as Agostini et al. [59], also found significantly lower breast meat yield in MHA-FA fed turkeys compared to DLM fed turkeys. When correcting for mortality, FCR tended to be higher with MHA-FA ($p = 0.54$) in the latter study. When summarising literature findings and the current study, it can be concluded that replacing MHA-FA with DLM in turkey feed at a weight-to-weight ratio of 100:65 results in the same performance. Few results even indicated a slight advantage for DLM. There is quite a lot research on these Met sources with respect to physiology and metabolism providing evidence for the lower bioavailability of MHA-FA compared to DLM. Research performed by Maenz and Engele-Schaan [57], Drew et al. [60] and Malik et al. [61] suggested that a substantial portion of ingested MHA-FA is degraded by intestinal microbes and thus, not available to the host. DLM is absorbed by sodium-driven transport with high affinity of the transporters to DLM and a relatively high velocity, whereas MHA-FA is absorbed like short chain fatty acids with proton-driven transporters with relatively low affinity to MHA-FA and relatively low velocity [57]. Moreover, the latter study suggests that diffusion through tight junctions of the intestinal wall does not play a role. Finally, absorption studies by Mitchell and Lemme [62] suggest that especially the short chains of liquid MHA-FA are poorly absorbed. This finding is confirmed by Rhone Poulenc [63] and Van Weerden et al. [64] who conducted simultaneous dose-response studies with DLM, MHA-FA and only short MHA-FA chains, consistently reporting a lower relative bioavailability of MHA-FA oligomers compared to the commercial MHA-FA product which contains about 65% monomers, 23% monomers and 12% water.

4.2. Footpad Health

Impaired footpad health is often related to increased water intake, increased water excretion and, as a consequence, increased litter moisture. Indeed, we could not find a clear effect of protein reduction on litter moisture nor could Lemme et al. [6], but Lemme et al. [6] found a clear improvement in footpad health, as the average lesion score could be reduced especially by increased proportions of grade 0 and grade 1 scores. The same was observed in the current study, as footpad scoring numbers were significantly reduced with protein reduction. This effect was measurable after four weeks of feeding the diets. In their review on low protein nutrition of broilers, Greenhalgh et al. [54] clearly indicated the relationship with improved footpad health. At day 98, the evaluation of the influence of the Met source showed tendentiously lower FPD scores for DLM compared to MHA-FA ($p < 0.10$). This is in line with the positive effect of Met supplementation on footpad health which has been described many times [10,12,13,65,66]. Deficiency in Met intake has a significant negative impact on animals such as growth inhibition via interruption in protein synthesis (i.e., affect skin structure), the induction of metabolic disorder, and the reduction of disease defensive potential [67]. Abd El-Wahab et al. [10] found that at almost identical litter DM contents (measured weekly), the young turkeys fed high level of Met in diet had lower FPD scores than those fed diet with low Met content. It means that level of dietary Met plays an important role for health of skin rather than moisture content in the litter. Thus, it seems that Met has a specific function regards foot pad health (as known for skin and feathers) as mentioned by Abd El-Wahab et al. [14] via protein synthesis and continuous production of keratin. Consequently, low availability of Met could affect protein synthesis negatively and affect skin of foot pad.

Chavez and Kratzer [13] compared the influence of DLM and MHA-FA on the footpad health of white fattening turkeys. They found that supplementing a basal diet with 4.00 g DLM/kg diet resulted in an improvement in footpad health that was not achieved to the same extent by supplementing 4.03 g of MHA-FA/kg diet. When interpreting the studies of Chavez and Kratzer [13], it must be noted that the comparing 4 g DLM with 4.03 g MHA-FA, an equivalence of both substances was assumed regarding the biological activity, whereas the authors of other studies assumed a lower biological activity of MHA-FA [10]. The tendentiously lower FPD scores of the animals whose diets contained DLM compared to those with equivalent supplementation of MHA-FA (based on 65% biological activity) corresponds to the observations of Chavez and Kratzer [13] and Abd El Wahab et al. [68].

4.3. Liver

The significantly higher relative liver weight of the protein-reduced groups has to be seen in context to the lower BW of the animals which were fed low protein diets. Nonetheless, the livers of the animals in low protein groups still had numerically higher crude fat content of the livers than those in groups with 18% protein in the diet. Here the

7% higher starch content of protein-reduced diets must be noted, since a high carbohydrate content of a diet increases the de novo lipogenesis in the liver [69]. As a consequence of enhanced de novo lipo-genesis in the liver of birds fed the low protein diets [70,71], birds are expected to have increased liver weights and deposit more abdominal fat due to increased energy:protein ratio. Hidalgo et al. [72] also reported no differences in carcass yield, breast meat yield, and abdominal fat pad in broilers fed low-protein diets with a constant energy:protein ratio. Chrystal et al. [5] summarised 19 broiler feeding studies and reported that body fat deposition was increased by 36% on average with an average protein reduction of 5.5% points. They as well as Liu et al. [53] related this effect to the shift towards a higher starch content with protein reduction and further suggested related changing digestion dynamics of starch and protein in broilers. An increase in the fat content of the liver when "protein compounds" are replaced by high-starch "energy compounds" also corresponds to the observations of Zhang et al. [73]. They compared the use of soybeans, corn starch and lard in isoenergetic and isoproteic diets in laying hens and found a significant increase in fat content of livers when corn starch (fat content liver: 11.8%) and lard (fat content liver: 10.2%) were used compared to soybeans (fat content liver: 8.2%).

4.4. Oxidative Stress

Liver is a major organ attacked by reactive oxygen species and parenchymal cells are primary cells subjected to oxidative stress-induced injury in the liver. DLM may be more effectively used in glutathione synthesis compared to MHA-FA, which can potentially explain the higher TAC observed in the present study. The influence of various Met sources on parameters of oxidative capacity has previously been discussed in literature. Zeitz et al. [74] and Zhang et al. [75] studied the effects of the two Met sources on various parameters related to oxidative stress in broilers. Compared to deficient methionine + cysteine supply, supplementation of either Met source improved antioxidant status but no differences between products were observed. Liu et al. [76] tested DLM and MHA-FA under heat stress conditions and reported some effects and interactions of Met sources. Accordingly, DLM fed broilers in the grower phase had greater muscle GSH and hepatic unsaturated fatty acid concentrations than MHA-FA fed birds. Nonetheless, across all measured tissues, parameters and ages, only few data sets were affected by Met sources. Wang et al. [77] concluded from their research on broilers that DLM could be effectively used in glutathione synthesis to exert antioxidant functions, whereas MHA-FA favoured S-adenosylmethionine synthesis. The reduced availability of glutathione in the MHA-FA fed broilers may have led to an increased sense of oxidative stress, which is supported by the observation that antioxidant-related gene expression was increased in this treatment.

5. Conclusions

In light of the conducted experiment, it is possible to conclude that supplementing six essential amino acids proved insufficient to compensate the strong protein reduction in the diet, which led to a significantly lower performance despite the increase in footpad health. Additionally, the response observed regarding the crude protein reduction is independent of the Met sources. Regarding the Met sources, it was concluded that replacing MHA-FA with 65 parts of DLM in feed does not compromise performance and additionally has advantages for the antioxidant capacity of the liver parenchyma.

Author Contributions: Conceptualisation, J.B.L. and C.V.; Data curation, A.A.E.-W., J.B.L. and C.V.; Formal analysis, J.B.L. and A.A.E.-W.; Funding acquisition, J.C.d.P.D., A.L. and C.V.; Investigation, J.B.L., R.B. and C.V.; Methodology, J.B.L., A.A.E.-W., M.L. and R.B.; Project administration, C.V.; Resources, C.V.; Supervision, C.V.; Validation, C.V.; Visualisation, J.B.L., A.A.E.-W., R.B. and C.V.; Writing—Original Draft Preparation, J.B.L., A.A.E.-W. and C.V.; Writing—Review and Editing, J.B.L., A.A.E.-W., J.C.d.P.D., A.L., M.L., R.B. and C.V. All authors have read and agreed to the published version of the manuscript.

Funding: This study was funded by Evonik Operations GmbH, Essen, Germany. The role of the funding body was merely financial for purchasing the materials and analyzing the data for the trials at the Institute for Animal Nutrition, University of Veterinary Medicine Hannover, Foundation, Germany. It did not have any role in the collection, interpretation, analysis or writing of the data. This publication was supported by Deutsche Forschungsgemeinschaft and University of Veterinary Medicine Hannover, Foundation within the funding programme Open Access Publishing.

Institutional Review Board Statement: The experiment was carried out in accordance with German regulations. The Ethics Committee of Lower Saxony for Care and Use of Laboratory Animals (LAVES) approved the experiment in accordance with § 4, paragraph 3 of the Animal Protection Act (reference number: 33.8-42502-05-18A313).

Informed Consent Statement: Not applicable.

Data Availability Statement: The original contributions generated for the study are included in the article, further inquiries of which can be directed to the corresponding author.

Acknowledgments: We would like to thank Frances Sherwood-Brock for proof-reading the manuscript to ensure correct English.

Conflicts of Interest: J.C.P.D. and A.L. are employees of Evonik Operations GmbH. They had no role in the collection, analyses or interpretation of data nor in the writing of the manuscript or in the decision to publish the results. The manuscript has not been previously published.

References

1. Dawkins, M.S. Animal welfare and efficient farming: Is conflict inevitable? *Anim. Prod. Sci.* **2016**, *57*, 201–208. [CrossRef]
2. Kapell, D.; Hocking, P.; Glover, P.; Kremer, V.; Avendaño, S. Genetic basis of leg health and its relationship with body weight in purebred turkey lines. *Poult. Sci.* **2017**, *96*, 1553–1562. [CrossRef]
3. Vieira, S.L.; Stefanello, C.; Cemin, H.S. Lowering the dietary protein levels by the use of synthetic amino acids and the use of a mono component protease. *Anim. Feed. Sci. Technol.* **2016**, *221*, 262–266. [CrossRef]
4. Selle, P.H.; De Paula Dorigam, J.C.; Lemme, A.; Chrystal, P.V.; Liu, S.Y. Synthetic and Crystalline Amino Acids: Alternatives to Soybean Meal in Chicken-Meat Production. *Animals* **2020**, *10*, 729. [CrossRef]
5. Chrystal, P.V.; Greenhalgh, S.; Selle, P.H.; Liu, S.Y. Facilitating the acceptance of tangibly reduced-crude protein diets for chicken-meat production. *Anim. Nutr.* **2020**, *6*, 247–257. [CrossRef] [PubMed]
6. Lemme, A.; Naranjo, V.; De Paula Dorigam, J.C. Utilization of Methionine Sources for Growth and Met+Cys Deposition in Broilers. *Animals* **2020**, *10*, 2240. [CrossRef]
7. Rychen, G.; Aquilina, G.; Azimonti, G.; Bampidis, V.; De Lourdes Bastos, M.; Bories, G.; Chesson, A.; Cocconcelli, P.S.; Flachowsky, G.; Gropp, J.; et al. Safety and efficacy of hydroxy analogue of methionine and its calcium salt (ADRY+®) for all animal species. *EFSA J.* **2018**, *16*, e05198. [CrossRef]
8. Gazdzinski, P.; Squires, E.J.; Julian, R.J. Hepatic lipidosis in turkeys. *Avian Dis.* **1994**, *38*, 379–384. [CrossRef]
9. Popp, C.; Hauck, R.; Vahlenkamp, T.W.; Lüschow, D.; Kershaw, O.; Hoferer, M.; Hafez, H.M. Liver pathology associated with increased mortality in turkey breeder and meat turkey flocks. *Avian Dis.* **2014**, *58*, 474–481. [CrossRef]
10. Abd El-Wahab, A.; Kölln, M.; Kamphues, J. Is there a specific role of dietary methionine or its source regarding foot pad health in fattening turkeys? In Proceedings of the 10th Hafez International Symposium on Turkey Diseases, Berlin, Germany, 5–7 June 2014; Mensch und Buch Verlag: Berlin, Germany, 2014; pp. 22–32.
11. Shen, Y.B.; Ferket, P.; Park, I.; Malheiros, R.D.; Kim, S.W. Effects of feed grade L-methionine on intestinal redox status, intestinal development, and growth performance of young chickens compared with conventional dl-methionine. *J. Anim. Sci.* **2015**, *93*, 2977–2986. [CrossRef] [PubMed]
12. Chavez, E.; Kratzer, F.H. Prevention of foot pad dermatitis in poults with Methionine. *Poult. Sci.* **1972**, *51*, 1545–1548. [CrossRef]
13. Chavez, E.; Kratzer, F.H. Effect of diet on foot pad dermatitis in poults. *Poult. Sci.* **1974**, *53*, 755–760. [CrossRef]
14. Abd El-Wahab, A.; Ahmed, M.; Ibrahim, T. Impact of dietary methionine levels and sources on performance and health of foot pad in broilers. *Asian J. Anim. Vet. Adv.* **2016**, *11*, 357–362. [CrossRef]
15. Li, C.; Lesuisse, J.; Schallier, S.; Climaco, W.; Wang, Y.; Bautil, A.; Everaert, N.; Buyse, J. The effects of a reduced balanced protein diet on litter moisture, pododermatitis and feather condition of female broiler breeders over three generations. *Anim. Int. J. Anim. Biosci.* **2018**, *12*, 1493–1500. [CrossRef]
16. Lemme, A.; Hiller, P.; Klahsen, M.; Taube, V.; Stegemann, J.; Simon, I. Reduction of dietary protein in broiler diets not only reduces n-emissions but is also accompanied by several further benefits. *J. Appl. Poult. Res.* **2019**, *28*, 867–880. [CrossRef]
17. Kidd, M.T.; Maynard, C.W.; Mullenix, G.J. Progress of amino acid nutrition for diet protein reduction in poultry. *J. Anim. Sci. Biotechnol.* **2021**, *12*, 45. [CrossRef] [PubMed]

18. Visscher, C.; Middendorf, L.; Gunther, R.; Engels, A.; Leibfacher, C.; Mohle, H.; Dungelhoef, K.; Weier, S.; Haider, W.; Radko, D. Fat content, fatty acid pattern and iron content in livers of turkeys with hepatic lipidosis. *Lipids Health Dis.* **2017**, *16*, 98. [CrossRef] [PubMed]

19. Azis, R. Hepatic lipidosis in turkeys. *World Poult.* **2008**, *24*, 28–29.

20. Jariyahatthakij, P.; Chomtee, B.; Poeikhampha, T.; Loongyai, W.; Bunchasak, C. Effects of adding methionine in low-protein diet and subsequently fed low-energy diet on productive performance, blood chemical profile, and lipid metabolism-related gene expression of broiler chickens. *Poult. Sci.* **2018**, *97*, 2021–2033. [CrossRef]

21. Jansman, A.J.M.; Kan, C.A.; Wiebenga, J. *Comparison of the Biological Efficacy of* DL-*Methionine and Hydroxy4-Methylthiobutanoic Acid (HMB) in Pigs and Poultry*; Animal Sciences Group: London, UK, 2003.

22. Vazquez-Anon, M.; Kratzer, D.; Yi, I.G.; Knight, C.D.; Gonzalez-Esquerra, R. A multiple regression model approach to contrast the performance of 2-hydroxy-4-methylthio butanoic acid and DL-methionine supplementation tested in broiler experiments and reported in the literature. *Poult. Sci.* **2006**, *85*, 693–705. [CrossRef] [PubMed]

23. Sauer, N.; Emrich, K.; Piepho, H.P.; Lemme, A.; Redshaw, M.S.; Mosenthin, R. Meta-analysis of the relative efficiency of methionine-hydroxy-analogue-free-acid compared with DL-methionine in broilers using nonlinear mixed models. *Poult. Sci.* **2008**, *87*, 2023–2031. [CrossRef] [PubMed]

24. Hoehler, D.; Hooge, D.M. Relative Effectiveness of Methionine Sources in Turkeys—Scientific and New Commercial Data. *Int. J. Poult. Sci.* **2003**, *2*, 361–366.

25. Hoehler, D.; Lemme, A.; Jensen, S.K.; Vieira, S.L. Relative Effectiveness of Methionine Sources in Diets for Broiler Chickens. *J. Appl. Poult. Res.* **2005**, *14*, 679–693. [CrossRef]

26. Lemme, A.; Hoehler, D.; Brennan, J.J.; Mannion§, P.F. Relative effectiveness of methionine hydroxy analog compared to DL-methionine in broiler chickens. *Poult. Sci.* **2003**, *81*, 838–845. [CrossRef] [PubMed]

27. Agostini, P.S.; Dalibard, P.; Mercier, Y.; Van Der Aar, P.; Van der Klis, J.D. Comparison of methionine sources around requirement levels using a methionine efficacy method in 0 to 28 day old broilers. *Poult. Sci.* **2016**, *95*, 560–569. [CrossRef]

28. Jankowski, J.; Ognik, K.; Kubi'nska, M.; Czech, A.; Ju´skiewicz, J.; Zdu´nczyk, Z. The effect of DL-, L-isomers and DL-hydroxy analog administered at 2 levels as dietary sources of methionine on the metabolic and antioxidant parameters and growth performance of turkeys. *Poult. Sci.* **2017**, *96*, 3229–3238. [CrossRef] [PubMed]

29. Aviagen Turkeys. *Feeding Guidelines for Nicholas and B.U.T. Heavy Lines*; Aviagen Turkeys: Tattenhall, UK, 2018.

30. Llames, C.R.; Fontaine, J. Determination of Amino Acids in Feeds: Collaborative Study. *J. AOAC Int.* **2020**, *77*, 1362–1402. [CrossRef]

31. Directive, C. Establishing community methods for the determination of amino acids, crude oils and fats, and olaquindox in feeding stuff and amending directive 71/393/EEC, annex part a. Determination of amino acids. *Off. J. Eur. Communities L* **1998**, *257*, 14–23.

32. VDLUFA. Bestimmung von DL-2-Hydroxy-4-methyl-mercapto-buttersäure nach Hydrolyse (Gesamt MHAR). In *Methode 4.11.4 Methodenbuch. Band III. Die Chemische Untersuchung von Futtermitteln, 4th Supplement*; VDLUFA-Verlag: Darmstadt, Germany, 1997.

33. Ullrich, C.; Langeheine, M.; Brehm, R.; Taube, V.; Rosillo Galera, M.; Rohn, K.; Popp, J.; Visscher, C. Influence of different methionine sources on performance and slaughter characteristics of broilers. *Animals* **2019**, *9*, 984. [CrossRef] [PubMed]

34. Mayne, R.; Else, R.; Hocking, P. High litter moisture alone is sufficient to cause footpad dermatitis in growing turkeys. *Br. Poult. Sci.* **2007**, *48*, 538–545. [CrossRef] [PubMed]

35. Middendorf, L.; Radko, D.; Dungelhoef, K.; Sieverding, E.; Windhaus, H.; Mischok, D.; Visscher, C. Amino acid pattern in the liver and blood of fattening turkeys suffering from hepatic lipidosis. *Poult. Sci.* **2019**, *98*, 3950–3962. [CrossRef]

36. Naumann, C.; Bassler, R. Methoden der landwirtschaftlichen Forschungs-und Untersuchungsanstalt, Biochemische Untersuchung von Futtermitteln. In *Methodenbuch, Band III. Die Chemische Untersuchng von Futtermitteln (Einschließlich der Achten Ergänzungen)*; VDLUFA Verlag: Darmstadt, Germany, 2012.

37. Aviagen Turkeys. B.U.T. 6: Mastleistungsziele. Zugriff. 5 September 2019. Available online: http://www.aviagenturkeys.com/de-de/?redirectedfrom=en_us (accessed on 1 August 2021).

38. Lemme, A. Turkey production: Toward better welfare and health. In Proceedings of the 5th Symposium Turkey Production, Berlin, Germany, 28–30 May 2009; Volume 5.

39. Lemme, A. Turkey Times. In Proceedings of the 14th Turkey Science and Production Conference, Chester, UK, 4–6 March 2020; pp. 79–86.

40. Lemme, A.; Frackenpohl, U.; Petri, A.; Meyer, H. Effects of Reduced Dietary Protein Concentrations with Amino Acid Supplementation on Performance and Carcass Quality in Turkey Toms 14 to 140 Days of Age. *Int. J. Poult. Sci.* **2004**, *3*, 391–399. [CrossRef]

41. Lemme, A.; Ravindran, V.; Bryden, W.L. Ileal digestibility of amino acids in feed ingredients for broilers. *World's Poult. Sci. J.* **2004**, *60*, 423–438. [CrossRef]

42. Kozłowski, K.; Helmbrecht, A.; Jankowski1, A.L.J.; Kluge, H.; Jeroch, H. Standardized ileal amino acid digestibility (SIAAD) of wheat DDGS in male growing turkeys. *Arch. Geflugelkd.* **2012**, *76*, 136–139.

43. Kozłowski, K.; Helmbrecht, A.; Lemme, A.; Jankowski, J.; Jeroch, H. Standardized ileal digestibility of amino acids from high-protein feedstuffs for growing turkeys—A preliminary study. *Arch. Geflügelk.* **2011**, *75*, 185–190.

44. Kluth, H.; Rodehutscord, M. Comparison of Amino Acid Digestibility in Broiler Chickens, Turkeys, and Pekin Ducks. *Poult. Sci.* **2006**, *85*, 1953–1960. [CrossRef]

45. Dean, D.W.; Bidner, T.D.; Southern, L.L. Glycine supplementation to low protein, amino acid-supplemented diets supports optimal performance of broiler chicks. *Poult. Sci.* **2006**, *85*, 288–296. [CrossRef] [PubMed]

46. Brede, A.; Wecke, C.; Liebert, F. Does the Optimal Dietary Methionine to Cysteine Ratio in Diets for Growing Chickens Respond to High Inclusion Rates of Insect Meal from Hermetia illucens? *Animals* **2018**, *8*, 187. [CrossRef]

47. Stipanuk, M.H. Metabolism of sulfur-containing amino acids. *Annu. Rev. Nutr.* **1986**, *6*, 179–209. [CrossRef]

48. Stipanuk, M.H. Sulfur amino acid metabolism: Pathways for production and removal of homocysteine and cysteine. *Annu. Rev. Nutr.* **2004**, *24*, 539–577. [CrossRef]

49. Brosnan, J.T.; Brosnan, M.E. The sulfur-containing amino acids: An overview. *J. Nutr.* **2006**, *136*, 1636–1640. [CrossRef] [PubMed]

50. Bunchasak, C.; Kimura, G.; Tanaka, K.; Ohtani, S.; Collado, C.M. The Effect of Supplementing Cystine on the Growth Performance and Liver Lipid and Phospholipid Contents of Broiler Chicks. *Jpn. Poult. Sci.* **1998**, *35*, 60–66. [CrossRef]

51. Wheeler, K.; Latshaw, J. Sulfur amino acid requirements and interactions in broilers during two growth periods. *Poult. Sci.* **1981**, *60*, 228–236. [CrossRef] [PubMed]

52. Francesch, M.; Brufau, J. Nutritional factors affecting excreta/litter moisture and quality. *World's Poult. Sci. J.* **2004**, *60*, 64–75. [CrossRef]

53. Liu, S.Y.; Macelline, S.P.; Chrystal, P.V.; Selle, P.H. Progress towards reduced-crude protein diets for broiler chickens and sustainable chicken-meat production. *J. Anim. Sci. Biotechnol.* **2021**, *12*, 20. [CrossRef] [PubMed]

54. Greenhalgh, S.; Chrystal, P.V.; Selle, P.H.; Liu, S.Y. Reduced-crude protein diets in chicken-meat production: Justification for an imperative. *World's Poult. Sci. J.* **2020**, *76*, 537–548. [CrossRef]

55. D'mello, J.P.F. Adverse Effects of Amino Acids. *Amino Acids Anim. Nutr.* **2003**, *2*, 125–142.

56. Schutte, J.B.; Pack, M. Effects of dietary sulphur-containing amino acids on performance and breast meat deposition of broiler chicks during the growing and finishing phases. *Br. Poult. Sci.* **1995**, *36*, 747–762. [CrossRef] [PubMed]

57. Maenz, D.D.; Engele-Schaan, M. Methionine and 2-Hydroxy-4-Methylthioobutanoic Acid are transported by distinct Na^+-dependent and H^+-dependent systems in the brush border membrane of the chick intestinal epithelium. *J. Nutr.* **1996**, *126*, 529–536. [CrossRef] [PubMed]

58. Baker, D.H. Comparative species utilization and toxicity of sulfur amino acids. *J. Nutr.* **2006**, *136*, 1670S–1675S. [CrossRef]

59. Agostini, P.S.; Aar, P.v.d.; Naranjo, V.D.; Lemme, A. Effect of methionine source at marginal and adequate methionine levels in turkeys. In Proceedings of the 21st European Symposium on Poultry Nutrition, Salou/Vila-seca, Spain, 8–11 May 2017.

60. Drew, M.D.; Syed, N.A.; Goldade, B.G.; Laarveld, B.; Kessel, A.G.V. Effects of Dietary Protein Source and Level on Intestinal Populations of Clostridium perfringens in Broiler Chickens. *Poult. Sci.* **2004**, *83*, 414–420. [CrossRef] [PubMed]

61. Malik, G.; Hoehler, D.; Rademacher, M.; Drew, M.D.; Kessel1, A.G.V. Apparent absorption of methionine and 2-hydroxy-4methylthiobutanoic acid from gastrointestinal tract of conventional and gnotobiotic pigs. *Animal* **2009**, *3*, 1378–1386. [CrossRef] [PubMed]

62. Mitchell, M.A.; Lemme, A. Examination of the Composition of the Luminal Fluid in the Small Intestine of Broilers and Absorption of Amino Acids under Various Ambient Temperatures Measured In Vivo. *Int. J. Poult. Sci.* **2008**, *7*, 223–233.

63. Uzu, G. Comparison of nutritional efficacy of DL-methionine and its hydroxy analog free acid in broilers. *Rhône-Poulenc Feed Addit. Dev.* **1983**, *125*, 33.

64. Van Weerden, E.J.; Schutte, J.B.; Bertram, H.L. Utilization of the polymers of methionine hydroxy analogue free acid (MHA-FA) in broiler chicks. *Archiv. Geflügelkunde* **1992**, *56*, 63–68.

65. Murillo, M.G.; Jensen, L.S. Sulfur amino acid requirement and foot pad dermatitis in turkey poults. *Poult. Sci.* **1976**, *55*, 554–562. [CrossRef] [PubMed]

66. Abd El-Wahab, A.; Aziza, A.; El-Adl, M. Impact of dietary excess methionine and lysine with or without addition of L-carnitine on performance, blood lipid profile and litter quality in broilers. *Asian J. Anim. Vet. Adv.* **2015**, *10*, 191–202. [CrossRef]

67. Bunchasak, C. Role of dietary methionine in poultry production. *J. Poult. Sci.* **2009**, *46*, 169–179. [CrossRef]

68. Abd El-Wahab, A.; Visscher, C.; Ratert, C.; Kolln, M.; Diephaus, D.; Beineke, A.; Kamphues, J. Outcome of an Experimental Study in Growing Turkeys Suspected of Having a Diet Related, Uncommon and Uncoordinated Gait. *Vet. Sci.* **2017**, *4*, 49. [CrossRef]

69. Parks, E.J.; Krauss, R.M.; Christiansen, M.P.; Neese, R.A.; Hellerstein, M.K. Effects of a low-fat, high-carbohydrate diet on VLDL-triglyceride assembly, production, and clearance. *J. Clin. Invest.* **1999**, *104*, 1087–1096. [CrossRef]

70. Rosebrough, R.; Steele, N. Energy and protein relationships in the broiler: 1. Effect of protein levels and feeding regimens on growth, body composition, and In Vitro lipogenesis of broiler chicks. *Poult. Sci.* **1985**, *64*, 119–126. [CrossRef] [PubMed]

71. Swennen, Q.; Janssens, G.; Collin, A.; Le Bihan-Duval, E.; Verbeke, K.; Decuypere, E.; Buyse, J. Diet-induced thermogenesis and glucose oxidation in broiler chickens: Influence of genotype and diet composition. *Poult. Sci.* **2006**, *85*, 731–742. [CrossRef]

72. Hidalgo, M.; Dozier, W., III; Davis, A.; Gordon, R. Live performance and meat yield responses of broilers to progressive concentrations of dietary energy maintained at a constant metabolizable energy-to-crude protein ratio. *J. Appl. Poult. Res.* **2004**, *13*, 319–327. [CrossRef]

73. Zhang, J.W.; Chen, D.W.; Yu, B.; Wang, Y.M. Effect of dietary energy source on deposition and fatty acid synthesis in the liver of the laying hen. *Br. Poult. Sci.* **2011**, *52*, 704–710. [CrossRef] [PubMed]

74. Zeitz, J.O.; Mohrmann, S.; Fehse, L.; Most, E.; Helmbrecht, A.; Saremi, B.; Eder, K. Tissue and plasma antioxidant status in response to dietary methionine concentration and source in broilers. *J. Anim. Physiol. Anim. Nutr.* **2018**, *102*, 999–1011. [CrossRef]
75. Zhang, S.; Gilbert, E.R.; Saremi, B.; Wong, E.A. Supplemental methionine sources have a neutral impact on oxidative status in broiler chickens. *J. Anim. Physiol. Anim. Nutr.* **2018**, *102*, 1274–1283. [CrossRef]
76. Liu, G.; Magnuson, A.D.; Sun, T.; Tolba, S.A.; Starkey, C.; Whelan, R.; Lei, X.G. Supplemental methionine exerted chemical form-dependent effects on antioxidant status, inflammation-related gene expression, and fatty acid profiles of broiler chicks raised at high ambient temperature. *J. Anim. Sci.* **2019**, *97*, 4883–4894. [CrossRef] [PubMed]
77. Wang, Y.; Yin, X.; Yin, D.; Lei, Z.; Mahmood, T.; Yuan, J. Antioxidant response and bioavailability of methionine hydroxy analog relative to DL-methionine in broiler chickens. *Anim. Nutr.* **2019**, *5*, 241–247. [CrossRef]

 *agriculture*

Article

Evolutionary Subdivision of Domestic Chickens: Implications for Local Breeds as Assessed by Phenotype and Genotype in Comparison to Commercial and Fancy Breeds

Tatyana A. Larkina [1], Olga Y. Barkova [1], Grigoriy K. Peglivanyan [1], Olga V. Mitrofanova [1], Natalia V. Dementieva [1], Olga I. Stanishevskaya [1], Anatoly B. Vakhrameev [1], Alexandra V. Makarova [1], Yuri S. Shcherbakov [1], Marina V. Pozovnikova [1], Evgeni A. Brazhnik [2], Darren K. Griffin [3] and Michael N. Romanov [3,*]

[1] Russian Research Institute of Farm Animal Genetics and Breeding—Branch of the L. K. Ernst Federal Science Center for Animal Husbandry, Pushkin, 196625 St. Petersburg, Russia; tanya.larkina2015@yandex.ru (T.A.L.); barkoffws@list.ru (O.Y.B.); peglivanian_grig@mail.ru (G.K.P.); mo1969@mail.ru (O.V.M.); dementevan@mail.ru (N.V.D.); olgastan@list.ru (O.I.S.); ab_poultry@mail.ru (A.B.V.); admiralmak@mail.ru (A.V.M.); yura.10.08.94.94@mail.ru (Y.S.S.); pozovnikova@gmail.com (M.V.P.)

[2] BIOTROF+ Ltd., Pushkin, 196602 St. Petersburg, Russia; bea@biotrof.ru

[3] School of Biosciences, University of Kent, Canterbury CT2 7NJ, UK; D.K.Griffin@kent.ac.uk

*   Correspondence: m.romanov@kent.ac.uk

Citation: Larkina, T.A.; Barkova, O.Y.; Peglivanyan, G.K.; Mitrofanova, O.V.; Dementieva, N.V.; Stanishevskaya, O.I.; Vakhrameev, A.B.; Makarova, A.V.; Shcherbakov, Y.S.; Pozovnikova, M.V.; et al. Evolutionary Subdivision of Domestic Chickens: Implications for Local Breeds as Assessed by Phenotype and Genotype in Comparison to Commercial and Fancy Breeds. *Agriculture* **2021**, *11*, 914. https://doi.org/10.3390/agriculture11100914

Academic Editor: István Komlósi

Received: 25 August 2021
Accepted: 19 September 2021
Published: 24 September 2021

Abstract: To adjust breeding programs for local, commercial, and fancy breeds, and to implement molecular (marker-assisted) breeding, a proper comprehension of phenotypic and genotypic variation is *a sine qua non* for breeding progress in animal production. Here, we investigated an evolutionary subdivision of domestic chickens based on their phenotypic and genotypic variability using a wide sample of 49 different breeds/populations. These represent a significant proportion of the global chicken gene pool and all major purposes of breed use (according to their traditional classification model), with many of them being characterized by a synthetic genetic structure and notable admixture. We assessed their phenotypic variability in terms of body weight, body measurements, and egg production. From this, we proposed a phenotypic clustering model (PCM) including six evolutionary lineages of breed formation: egg-type, meat-type, dual purpose (egg-meat and meat-egg), game, fancy, and Bantam. Estimation of genotypic variability was carried out using the analysis of five SNPs, i.e., at the level of genomic variation at the *NCAPG-LCORL* locus. Based on these data, two generally similar genotypic clustering models (GCM1 and GCM2) were inferred that also had several overlaps with PCM. Further research for SNPs associated with economically important traits can be instrumental in marker-assisted breeding programs.

Keywords: chicken breeds; evolutionary lineages; phenotypic traits; *NCAPG-LCORL*; synthetic genetic structure; admixture

1. Introduction

The poultry industry is currently the most highly efficient livestock production sector, and the domestic chicken, *Gallus gallus* (Linnaeus, 1758), is the most common farm animal and a notable model organism (e.g., [1,2]). During long-term domestication of poultry, various breeds have been formed that differ significantly, both at the phenotypic and genetic levels, in terms of body weight, plumage color, and many other traits [3–6]. According to Bennett et al. [7], breeding efforts in the period from the late Middle Ages to the present day resulted in a doubling of the body size and a change in the skeleton morphology in chickens. Significant changes in overall appearance, arising from the synthesis of breeds through crossbreeding and selection, reflect the acquired biological characteristics of these birds and can be closely related to the purpose of use (i.e., utility types), and admixture of breeds formed from the moment of domestication and during further breeding. Therefore, information on the phenotypic diversity of poultry and its genetic background, including candidate genes and quantitative trait loci (QTLs) associated with conformation, body weight and egg production, is very important and relevant [8,9]. This information may be of particular importance for local breeds and populations of chickens (e.g., [10–14]), since it can be used for significant acceleration of breeding progress in these populations [15].

One of the areas in the chicken genome presumably associated with growth traits is a region on chromosome 4 (GGA4) that embraces the candidate genes *LCORL* (ligand-dependent nuclear receptor corepressor-like) and *NCAPG* (non-SMC condensin I complex, subunit G). At the *NCAPG-LCORL* locus and in the area close to this region, SNPs were found that are associated with the weight of internal organs in chickens [16], egg weight [17], and oviduct size [18]. The *NCAPG-LCORL* region has also been studied in mammals and has been identified as a locus associated with growth and developmental traits. Furthermore, significant associations of this region with height in humans were shown among the European [19], Japanese [20], and African American populations [21], as well as in relation to body weight, growth, and skeletal size in cattle, horses, pigs, and sheep [22–27]. According to Lyu et al. [28], *LCORL* is one of the key genes that determine the characteristics of body weight in vertebrates and can be considered as potentially affecting the growth of chickens.

A unique bioresource collection farm at the Russian Research Institute of Farm Animal Genetics and Breeding (RRIFAGB) involves almost 50 chicken breeds and populations of various purposes, some of which are synthetic (Table S1). As an extensive sample of the world gene pool of chicken breeds [29,30], the bioresource collection is a good model for studying the phenotypic and genetic diversity of chickens [31]. It encompasses many commercial, local, and fancy breeds that conform to the traditional classification model (TCM) in terms of productivity and purpose of use, i.e., egg-type breeds (ETBs), meat-type breeds (MTBs), dual purpose breeds (DPBs), game breeds (GBs), and fancy breeds (FBs; also, ornamental or 'decorative' breeds) [32]. In particular, ETBs available in the collection are distinguished by high egg production, lower body weight, and light bones. MTBs are characterized by the development of pectoral and pelvic girdle muscles, and coarse skeleton, being also selected to reduce bone weight without reducing muscle weight. DPBs, which are often synthetic, include egg-meat breeds (EMBs) and meat-egg breeds (MEBs), and are intermediate between ETBs and MTBs. Depending on the preferred targets of selection, they belong either to MEBs (if meat productivity traits are more targeted), or to EMBs (if egg performance traits are predominantly selected for), while there is basically no essential difference between them in body weight and exterior. GB breeds are distinguished by unusually strong bones, have an often high and vertical body posture, are not very developed, but have very dense muscles. FB breeds are distinguished by the presence and strong development of one or several unusual (ornamental) traits, for example, dwarfism in Bantams. This has become the main selection factor in these breeds; therefore, the traits of meat and egg productivity fade into the background [32].

As a result of the assessment of chicken breeds and populations by phenotype and genotype (i.e., at certain genetic loci), it is possible to clarify their belonging to the main evolutionary branches of domestic chickens and types of their breed formation. As postulated by Moiseyeva et al. [3], there are four main evolutionary types of chicken breeds: egg, meat, game, and Bantam. The first of these, egg type, is suggested to be inherent in the birds of Mediterranean roots, and the other three have Asian roots. At the same time, there is a significant number of breeds that are difficult to relate with complete certainty to one of these four evolutionary lineages and which are mainly created synthetically, i.e., due to the crossing of different breeds and random or deliberate introgression during further breeding. As a rule, these breeds have a wide variety of geographic origins and pronounced genetic admixture and belong to DPB and FB breeds.

With this in mind, we contend that it is essential to conduct a broad survey of a wide range of various chicken breeds and populations to perform an advanced analysis of their phenotypic and genotypic diversity, admixture, and phylogeny. Along with examining evolutionary aspects of the subdivision of domestic chickens and their breed formation, it would be interesting to identify the respective variants at the *NCAPG-LCORL* locus, which may be selection markers associated with egg production, growth, and development in chickens. A better understanding of genomic variation at this locus can help to adjust breeding programs for commercial, local, and fancy breeds and include marker-assisted selection (MAS) using *NCAPG-LCORL* SNPs. In this regard, the main goal of this study was to study the phenotypic and genomic variability at the *NCAPG-LCORL* region on GGA4 to characterize genetic differences between 49 purebred and crossbred chicken populations of various purposes of use (i.e., utility types) that belong to different evolutionary branches and TCM categories of domestic chicken. Our extensive survey of various chicken breeds and populations enabled to reveal variation at the *NCAPG-LCORL* locus in connection with the evolutionary lineage affiliation and purpose of use of these breeds. We also discussed use of this locus as a possible candidate for MAS for egg and meat productivity traits.

2. Materials and Methods

2.1. Experimental Design

All chicken experiments were conducted with an ethical approval of the RRIFAGB— Branch of the L. K. Ernst Federal Research Center for Animal Husbandry (Protocol No. 2020-4 dated 3 March 2020).

The objects of the present experiments were chickens from the RRIFAGB bioresource collection farm, officially referred to as the "Genetic Collection of Rare and Endangered Chicken Breeds" (Pushkin, St. Petersburg, Russia), based on which this investigation was carried out.

In total, the study included 954 individuals from the following 49 purebred and crossbred populations that, according to TCM, are characterized by different purpose of use, with the alphabetic codes of each population being given in the parentheses: ETB, Leghorn Light Brown (LLB), Minorca Black (MB), Russian White (RW, of the most recent RRIFAGB collection strain sampled in 2016; RWG), RW (of a historical RRIFAGB collection strain sampled in 2001; RWS), RW (of an All-Russian Poultry Research and Technological Institute collection strain sampled in 2001; RWP); MTB, White Cornish (of commercial cross Smena-6, Line 1; WC1), White Cornish (of commercial cross Smena-6, Line 2; WC2), *inter se* crossbreds White Cornish × (Brahma Light × Sussex Light) (WC × (BL × SL)) and White Cornish × (Sussex Light × Amrock) (WC × (SL × Ar)), Red White-tailed Dwarf (RWD); DPB/EMB, Zagorsk Salmon (ZS), Pushkin (Pu), Rhode Island Red (RIR), Leningrad Mille Fleur (LMF), New Hampshire (NH), Leningrad Golden-and-gray (LGG), Pantsirevka Black (PB); DPB/MEB, Australorp Black Speckled (ABS), Aurora Blue (AB), Australorp Black (AoB), Amrock (Ar), Naked Neck (NN), Pervomai (Pm), Plymouth Rock Barred (PRB), Poltava Clay (PC), Sussex Light (SL), Faverolles Salmon (FS), Tsarskoye Selo (Ts), Yurlov Crower (YC), *inter se* crossbreds Sussex Light × Amrock (SL × Ar), Tsarskoye Selo × Sussex Light (Ts × SL), Tsarskoye Selo × (Sussex Light × Amrock) (Ts × (SL × Ar)), Brahma Light × Sussex Light (BL × SL); GB, Orloff Mahogany (OM), Moscow Game (MG), Uzbek Game (UG), *inter se* crossbreds Uzbek Game × Amrock (UG × Ar); FB, Russian Crested (RC), Ukrainian Muffed (UM), Bantam Mille Fleur (BMF), Brahma Buff (BB), Brahma Light (BL), Hamburg Silver Spangled Dwarf (HSSD), Poland White-crested Black (PWB), Silkie White (SW), Cochin Bantam (CB), Frizzle (F), Pavlov Spangled (PS), Pavlov White (PW) (see also Table S1 for further details). The basic diet and maintenance conditions were similar for all the experimental birds and comply with zootechnic/zoohygienic standards.

2.2. Phenotypic Characteristics of the Studied Chicken Populations

To assess the phenotypic diversity, a core of the RRIFAGB bioresource collection was examined including the 39 breeds and populations of chickens that represent, according to TCM, various purposes of breed use (Tables S2 and S3). We collected a series of phenotypic traits including body weight of sexually mature birds at the age of 52 weeks and 13 main body measurements at the age of 330 days were obtained. The body measurements were produced using a compass and a tape measure and included: body length (cm), body slanting length (cm), body and neck length (cm), keel length (cm), chest girth (cm), chest depth (cm), pectoral angle (°), distance between shoulder joints (cm), distance between hip joints (cm), femur length (cm), tibia length (cm), shank length (cm), and shank girth (cm). Morphometric parameters for hens and cocks specific to each breed/population are presented in Tables S2 and S3, respectively. Egg production was assessed over a 52-week life period and egg weight at 35 weeks of age (Table S2).

A variety of chicken exterior parameters was further examined relative to TCM and other clustering models (evolutionary affiliation), and genotypes at the *NCAPG-LCORL* locus.

2.3. Genotyping of Chickens and SNP Data Processing

DNA used in this study was isolated from whole blood cells by the conventional phenol-chloroform method. SNP scanning was performed using an Illumina Chicken 60K SNP iSelect BeadChip (Illumina, San Diego, CA, USA). Ten SNPs were selected in the *NCAPG-LCORL* region on GGA4 using the PLINK 1.9 program [33] and the extract function. Quality control of genotyped SNP loci was proceeded using PLINK 1.9. Additionally, DNA samples with a SNP genotyping quality of more than 90% were included in the further analysis as assessed using the GenomeStudio 2.0 software (Illumina Inc., San Diego, CA, USA). A threshold was set for Hardy–Weinberg equilibrium (HWE) errors (at $p < 0.0001$), and only SNPs with a minor allele frequency (MAF) of more than 5% were taken into account. After all filtering steps, five SNPs were selected for subsequent computations.

2.4. Mathematical and Statistical Analyzes

To assess variability of phenotypic traits, a heatmap for the distribution of the 39 phenotyped chicken breeds and populations was built using the heatmap function in the stats package (R v. 4.2; [34]). Fuzzy analysis clustering (FAC), principal component analysis (PCA), and average linkage clustering (ALC) were implemented using R and libraries for the R environment, along with Euclidean distance metric. For the FAC method, we applied the fanny function from the cluster package [35,36]. In the case of ALC, the unweighted pair group method with arithmetic mean (UPGMA) method was employed using the Euclidean distance matrix between objects with a bootstrapping validation using 1000 bootstrap samples. The pvclust software package for R [37] was used to calculate approximately unbiased (AU) p-values and bootstrap probability (BP) values. Clusters with AU values over 95% were considered significant. The optimal number of clusters was selected for interpretation using the Elbow method [38] and the factoextra package for R [39].

To estimate genotypic variation, biometric data processing was performed using PLINK 1.9 and Microsoft Excel programs. Generation of admixture models for distinguishing clusters based on genotype data for the five selected SNPs at the *NCAPG-LCORL* locus, including cross-validation (CV) error plots for determining the number of ancestral populations (K), was carried out using the ADMIXTURE program [40].

Degree of genetic diversity of the analyzed populations was evaluated by the indicators of observed (H_o) and expected (H_e) heterozygosity as calculated based on data for the genotype and allele occurrence frequencies at each polymorphic locus. Assessment of genetic diversity between populations was also carried out using the Hudson F_{ST} statistics implemented in the EIGENSOFT 6.1.4 software package [41].

The correspondence of the allele and genotype frequency distribution to HWE in each population was verified using the χ^2 test. In particular, we analyzed the $H_o - H_e$ deviations for each locus in accordance with the HWE principle that also states that genotype frequencies are related to allele frequencies by simple (quadratic) relationships [42]. Estimation of the reliability of the data obtained was carried out using the Pearson χ^2 test. If the obtained value of the χ^2 statistic is greater than the critical one (i.e., 3.84, the number of degrees of freedom being 1), it was concluded that there was a shift in the genetic equilibrium in an analyzed population. For the studied populations, the inbreeding coefficient F_{is} was also calculated [43].

Using the STATISTICA 10.0 package (Statsoft, Inc./TIBCO, Palo Alto, CA, USA), a comparison of differences in allele frequencies between population groups (clusters) according to TCM and other clustering models was conducted by employing the non-parametric Kruskal–Wallis test for one-way analysis of variance (ANOVA) by ranks because the data did not pass the normality test. Boxplots were generated using STATISTICA 10.0, as well.

Pairwise values of the F_{ST} statistic and the average within-group and between-group divergence D were calculated using the SMARTPCA package (part of EIGENSOFT, [41]), as described elsewhere [44]. Unrooted neighbor-joining dendrograms were constructed based on pairwise F_{ST} statistics and average between-group divergence D using the online T-REX tool with the tree inference, also known as the NJ option [45].

The Chicken QTLdb database [46] was explored to test if any of the five significant SNPs identified in the present study would overlap with the previously established QTLs at the *NCAPG-LCORL* locus.

Linkage disequilibrium (LD) between pairs of SNPs in the 49 chicken populations was assessed using the PLINK 1.9 software, the D' coefficient proposed by Lewontin (see for review [47]), and the squared Pearson correlation coefficient r^2 [48]. LD block structure was determined according to the solid spine algorithm as implemented in the Haploview 4.2 software [48] wherein the specified parameters corresponded to the values $D' \geq 0.75$.

3. Results

3.1. Analysis of Phenotypic Traits and Breed Clustering by Phenotype

As a result of the survey across the 39 breeds and populations of chickens, values of phenotypic traits were obtained for adult birds of both sexes (Tables S2 and S3). Variation was observed in terms of body weight, 13 main body measurements, egg number over a 52-week period of life, and egg weight. Herewith, the average body weight of females was 2.23 ± 0.80 kg, and that of males 2.49 ± 0.90 kg. The average egg production was 149.55 ± 27.67 eggs, and the average egg weight was 57.14 ± 4.60.

The clustering patterns by phenotypic traits in the breeds and populations observed were produced using more sophisticated mathematical approaches and are discussed in detail below.

Before the onset of the experiments, all breeds and populations in the RRIFAGB bioresource collection were a priori divided into groups in accordance with TCM, i.e., with five generally accepted categories (purposes of breed use, or utility types), namely ETBs, DPBs (with two subcategories, EMBs and MEBs), MTBs, GBs, and FBs (Table S1).

For all phenotypic traits (Tables S2 and S3) measured in hens and cocks, FAC and PCA clustering of breeds and populations (Figure 1a,b) revealed configurations (patterns) of breed grouping that differed from that in TCM. In these analyses, the first component corresponded to 75.5% of genetic variability, and the second one was responsible for 9.1% of genetic variability. In general, using both FAC (Figure 1a) and PCA (Figure 1b), six clusters were identified that we attributed, with a certain degree of conventionality, to ETBs (cluster I), DPBs (II) (including subclusters EMB/IIa, and MEB/IIb), MTBs (III), GBs (IV), FBs (V), and BTBs (VI; i.e., carriers of the dwarf mutation). We designated this breed grouping pattern as the phenotypic clustering model (PCM).

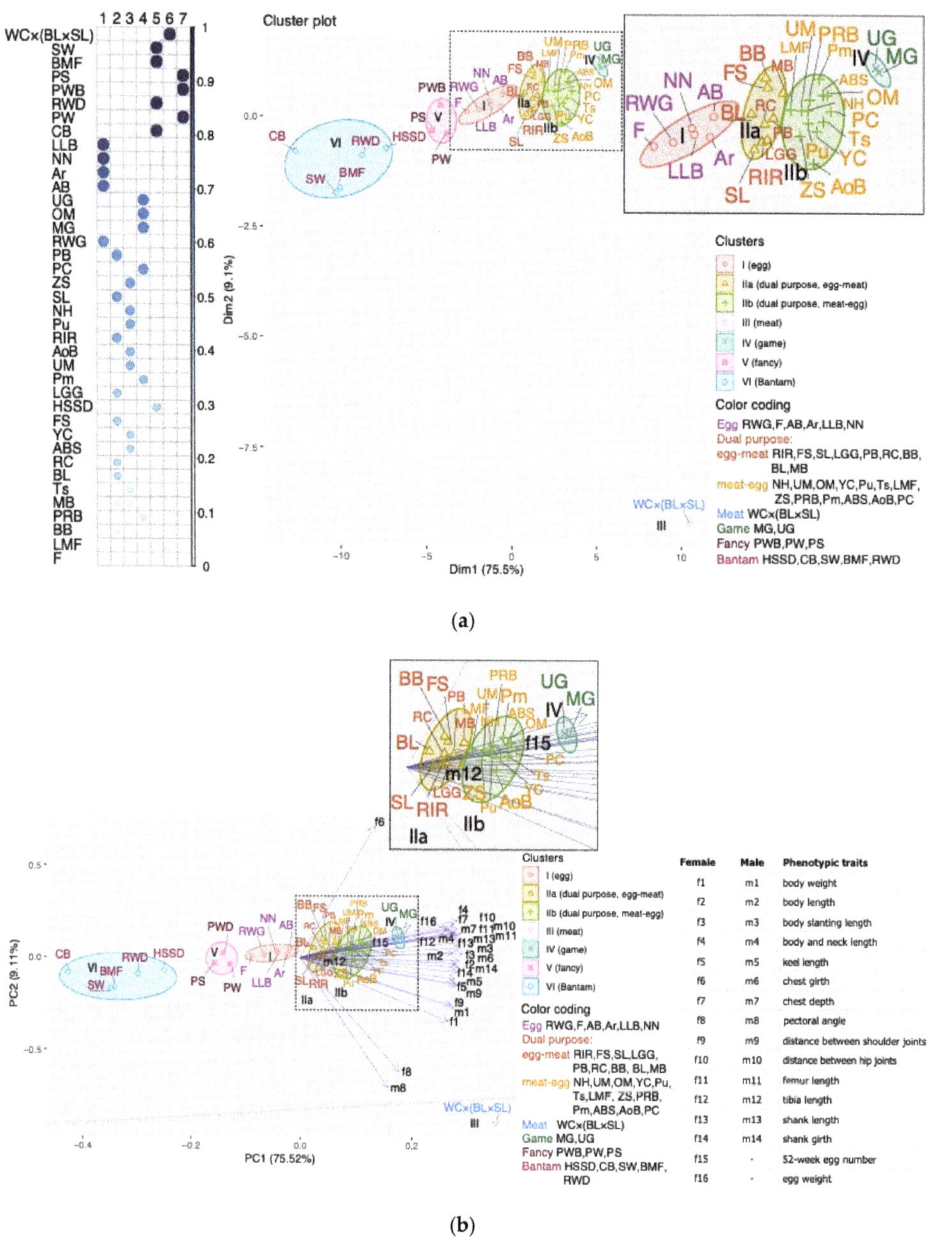

(a)

(b)

Figure 1. Clustering patterns for the 39 chicken breeds/populations based on their phenotypic traits and using fuzzy analysis clustering (**a**) and PCA (**b**). The observed clustering pattern of six clusters formed the basis of PCM. Population codes are given according to Supplementary Table S1.

Within clusters I, IIa and IIb, there was a core of breeds that more or less adequately matched the basic characteristics of each breed category, as well as several other breeds that joined the core due to the similarity of the main phenotypic traits, regardless of their affiliation according to TCM and often in contrast to TCM. For example, the RWD breed was assigned to BTBs instead of MTBs by its phenotypic characteristics (Tables S2–S4).

A clear differentiation and a significantly isolated position on the FAC and PCA plots (Figure 1a,b) were observed for the only representative of the MTB category, i.e., the WC × (BL × SL) population (single cluster III). A separate position was also occupied, on the one hand, by clusters of two true GBs (IV), i.e., MGs and UGs, and, on the other hand, by BTBs and similar dwarf chickens (VI). Interestingly, a separate FB cluster (V) included the following three closely spaced and similar breeds with the most pronounced ornamental characters: PWB, and two varieties of the Pavlov breed (PS and PW). The suggested PCM and the respective subdivision of breeds into six phenotypic clusters were further used for comparison with other clustering patterns (models) obtained on the basis of both phenotypic traits and SNP genotypes at the *NCAPG-LCORL* locus.

Similarly to PCM (as in Figure 1a,b), the populations were located on the FAC and PCA plots that were produced for morphometric characters only (Figures S1a and S2a). At the same time, a different pattern can be seen in the FAC and PCA plots for egg production characteristics (Figures S1b and S2b).

A similar (as in Figure 1a,b) distribution of breeds and populations by phenotypic traits was discovered when using other mathematical methods of comparative analysis, i.e., building a heatmap based on indicators of body weight, egg weight, and egg number (Figure 2), and phylogenetic UPGMA trees using Euclidean distances for different groups of characters (all, Figure S3a; morphometric, Figure S3b; and egg characters, Figure S3c). Some incongruent clustering patterns observed in comparison with the FAC and PCA plots (Figure 1a,b) were explained in each case by peculiarities of the applied mathematical algorithms and a sample (set) of the phenotypic indicators used.

3.2. Analysis of Genetic Variation at the Locus NCAPG-LCORL

3.2.1. Population Genetic Parameters

As a result of SNP scanning, five significant SNPs were identified in all the 49 breeds and crossbreds in a region on GGA4 that harbors *NCAPG-LCORL*, as well as in the flanking regions as follows: GGaluGA265966, GGaluGA265969, rs15619223, rs14491017, and rs14491028 (Table 1). Genotype frequency distributions for the five SNP markers in the 49 chicken populations are presented in Table S4.

Table 1. Summary for the five SNPs at the *NCAPG-LCORL* locus on GGA4.

SNP	Position in GRCg6a Build [49]	Nearest Gene	Nucleotide Change	Location
GGaluGA265966	75,796,627	*NCAPG*	A/G	Intergenic region
GGaluGA265969	75,827,200	*NCAPG* *LCORL*	T/C	Intron
rs15619223	75,850,294	*NCAPG* *LCORL*	A/C	Intron
rs14491017	75,885,777	*NCAPG*	C/T	Intron
rs14491028	75,903,919	*NCAPG*	C/T	Intron

Analysis of the actual and theoretical distribution of genotypes (data not shown) revealed a significant shift ($p < 0.05$) of the genetic equilibrium for the rs14491017 substitution in the Ts × SL crossbred chickens ($\chi^2 = 4.25$), that for the rs14491028 substitution in the chicken breeds Ts ($\chi^2 = 5.71$), NH ($\chi^2 = 5.00$), and Pu ($\chi^2 = 3.95$), and for GGaluGA265969 substitution in the chicken populations F ($\chi^2 = 4.36$) and PRB ($\chi^2 = 5.14$).

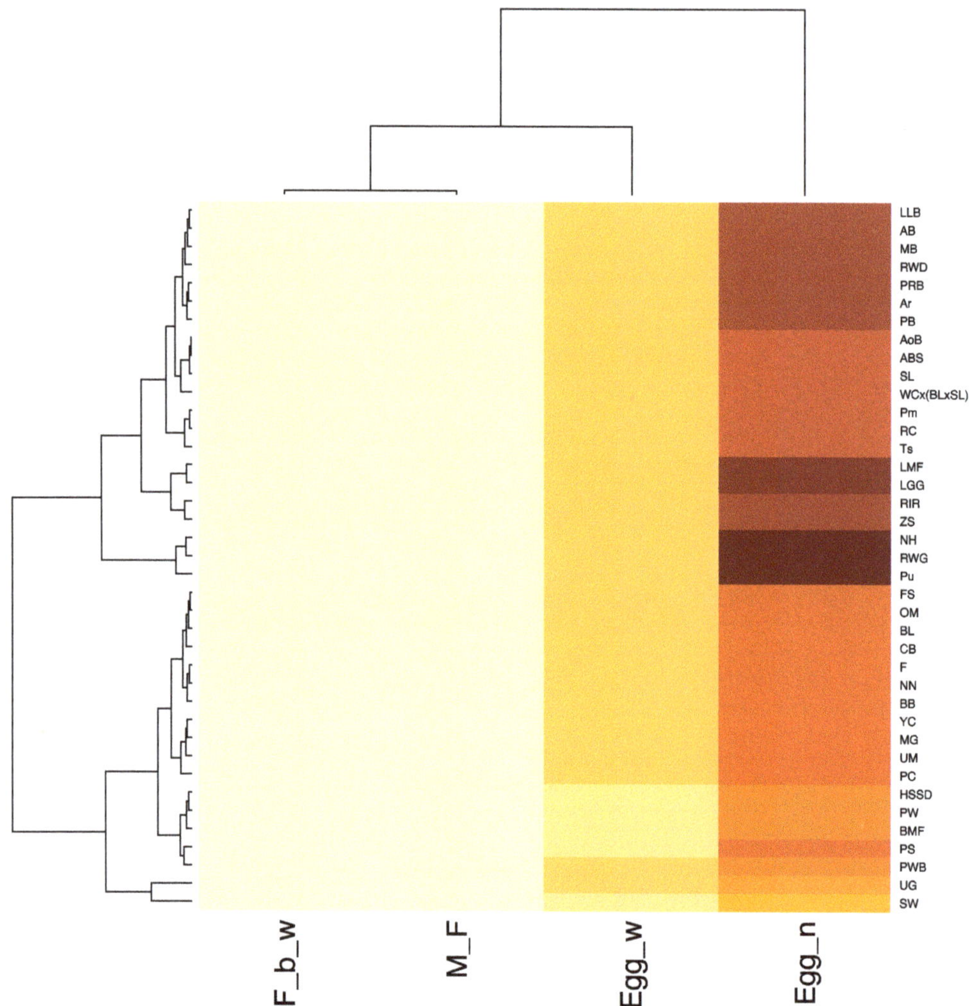

Figure 2. A heatmap showing distribution of the 39 phenotyped chicken breeds/populations and built using the heatmap function in stats package (R v. 4.2) on the base of the following phenotypic traits: female body weight (F_b_w), a sum of male and female body weights (M_F), egg weight (Egg_w), and 52-week egg number (Egg_N). The rows (populations) and columns (traits) are ordered in accordance with hierarchical clustering as displayed through the respective dendrograms. Lighter colors correspond to lower correlation, and darker ones to higher correlation. Population codes are given according to Supplementary Table S1.

For the substitutions GGaluGA265966 and rs15619223 in all analyzed populations of the bioresource collection, regardless of breed, the χ^2 values did not significantly exceed the critical value, that is, there was no significant difference between the observed and expected heterozygosity. Lower occurrence frequency of individual genotypes did not correlate with a shift in the genetic equilibrium since it was a consequence of lower frequency of individual alleles. In general, it can be noted that the populations were in the genetic equilibrium.

Inbreeding coefficient, F_{is}, in the studied populations of the bioresource collection varied in the range $-1 \le F_{is} \le 1$ (data not shown). The maximum negative value was found in the RWP chicken population, and the maximum positive one in the CB population. With increasing positive value of the coefficient, a rise in inbreeding occurred in the studied populations. At the GGaluGA265969 locus, two breeds, Pu (F_{is} = 0.44) and F (F_{is} = 0.46), had higher positive values. FB chickens of the BMF, SW, PS, and PW breeds differed from all other analyzed groups in that the C allele was found in all their individuals, which could indicate a strong selection pressure on this locus. As for GGaluGA265966, the UM population had F_{is} = 0, with H_{obs} = H_{exp}, that suggested a panmixia in this population at this marker locus.

In almost each category of chicken breeds, except GBs, some separate populations were monomorphic and had only one certain genotype for the rs15619223 marker. For instance, in three MTB populations (WC1, WC2, and WC × (BL × SL)), only the AA genotype was found. Interestingly, for this same SNP, associations with body weight and egg weight were previously found in RWG [50,51]. Thus, this locus can be considered as a potentially effective candidate for selection, since it could be selected in the process of targeted selection based on the desired trait of body weight. It can also be noted that, in the ETB populations of RWS and RWP, only the CC genotype was found at rs14491017 in all individuals.

3.2.2. Clustering of the Analyzed Chicken Breeds by Genotypes

Using the results of the genotype frequency distributions for the five SNPs in the 49 genotyped breeds and chicken populations (Table S4), we performed the appropriate clustering procedures that showed population grouping patterns that was somewhat different from those for TCM and PCM. On the respective FAC and PCA plots (Figure S4a,b), a quite clearly distinguishable, isolated localization of ETB and MTB clouds was observed, with a cloud of DPB populations between them. In this case, the first component contributed to ~30% of the total genetic variability, i.e., significantly less than the analogous first component when using the phenotypic traits, and the second component accounted for 17% of genetic variability. Hereby, the observed clouds (clusters) were more diffuse along the contours. Besides the true representatives of each cluster (ETBs, MTBs, or DPBs), breeds and populations classified in other TCM and PCM categories were joining them. The ETB cluster was subdivided into two distinct subclusters (Ia and Ib), and the DPB cluster could also be subdivided into two subclusters (IIa and IIb). In addition, each cluster (subcluster) had outliers, e.g., LLB and HSSD in the ETB subcluster Ia, AoB in the DPB subcluster IIb, and CB and Pu in the MTB cluster, which also did not always correspond to the definition of either cluster. We designated the resulting distribution pattern of populations relative to each other (Figure S4a,b) as the genotypic clustering model 1 (GCM1).

Construction of the UPGMA tree using Euclidean metrics led to a slightly different clustering of the analyzed groups of chickens (Figure S4c). We conditionally designated this pattern as genotypic clustering model 2 (GCM2), which consists, as it were, of four population groups (clusters). In the first GCM2 group, we conditionally combined two ETB and two BTB populations from the GCM1 subcluster Ia. The second GCM2 cluster involved one DPB/MEB and one BTB population from the GCM1 cluster III; and the third GCM2 cluster involved three FB populations (those that formed a separate FB cluster in PCM), one ETB, and one DPB/EMB population from the GCM1 subcluster Ia. The largest was the fourth GCM2 cluster that consisted of the following two large subclusters: IVa (mainly MTBs with adjoining purebred populations and crossbreds from the GCM1 cluster III) and IVb (breeds and crossbreds from the GCM1 subclusters Ib, IIa, and IIb).

3.2.3. Assessment of Breed Admixture by Genotypes

In general, there was a significant admixture of the 49 chicken genotyped breeds and populations, which made it somewhat difficult to identify clearly pronounced admixture models. Comparative analysis of admixture was carried out between groups of breeds subdivided both in relation to their purpose of use (TCM) and according to other clustering models (PCM, GCM1, and GCM2). The plots for determining the number of K groups in the admixture models that could best fit the obtained data showed that the optimal number of ancestral populations was close to the values of K = 6 (Figure S5).

Analysis of admixture by genotype frequencies revealed some genetic characteristics of individual populations and their groups at the *NCAPG-LCORL* locus using TCM, PCM, GCM1, and GCM2 (Figure S6). Particularly, clustering analysis of admixture models revealed to a certain extent the subdivision of individuals according to their origin. At the same time, ETB and MTB chickens, as a rule, tended to form two separate patterns (clusters), which is consistent with the results of genotype frequencies and variants of the compared clustering (classification) models.

3.2.4. Genetic Differentiation of Populations by Alleles

At the first stage of determining the genetic differentiation of the 49 breeds and populations by allele frequencies in the five SNPs (as presented in Supporting Information (SI) S1), they were combined into classes and subclasses according to TCM. When comparing allele frequencies for the GGaluGA265969 substitution using the Kruskal–Wallis test (SI S2), we found out that ETB chickens significantly differed from MTBs and FBs, MTBs from DPBs, DPBs from FBs (at $p < 0.05$; SI S2a and SI S3a), and the MEB subgroup from FBs (at $p < 0.05$; SI S2f and SI S3a). For the GGaluGA265966 substitution, differences were found for the ETB group in comparison with the MTB and DPB groups ($p < 0.05$; SI S2b and SI S3b), as well as with MEB ($p < 0.05$; SI S2g and SI S3b). For the SNPs rs15619223 and rs14491028, ETB chickens differed from all other groups and subgroups at $p < 0.05$ (SI S2c,h and SI S3c; and SI S2e,j and SI S3e, respectively). MTB chickens differed in frequency from the DPB group in the SNPs rs15619223 and rs14491017 ($p < 0.05$; SI S2c and SI S3c; and SI S2d and SI S3d, respectively). In addition, the MTB group had significant differences with the MEB subgroup in allele frequencies at the rs15619223 locus ($p < 0.05$; SI S2h and SI S3c), as well as with the EMB and MEB subgroups at the rs14491017 locus ($p < 0.05$; SI S2i and SI S3d). FB chickens significantly differed from the ETB, DPB, and MEB subgroups at GGaluGA265969 ($p < 0.05$; SI S2a and SI S3a), and also from ETB, MTB, and DPB at rs15619223 ($p < 0.05$; SI S2c,h and SI S3c).

Next, we compared clusters and subclusters if combined according to PCM for the significance of differences in allele frequencies in the five SNPs (SI S3f–j). All PCM clusters/subclusters differed in pairs at all five marker loci, except for FBs which had significant pairwise differences at four loci. At the same time, FBs had the greatest number of significant pairwise differences, i.e., 17. For other clusters/subclusters, the following number of significant pairwise differences was found: ETBs and MTBs, 12 each; MEBs, 11; EMBs, 8; and GBs, 6.

Finally, when comparing GCM1 clusters/subclusters, numbers of their significant pairwise differences were revealed in the following descending order: ETB/Ia, 11 (in four SNPs); ETB Ib, 10 (in five SNPs); DPB/IIa, 11 (in five SNPs); DPB/IIb, 8 (in four SNPs); and MTB/III, 14 (in five SNPs) (SI S3k–o).

If we estimate the number of significant pairwise differences for individual SNP loci for a total of all three models (TCM, PCM, and GCM1), the following data were obtained: GGaluGA265966, 14; GGaluGA265969, 16; rs15619223, 28; rs14491017, 23; and rs14491028, 14 differences (SI S3).

Genetic differentiation between the analyzed groups of chickens by allele frequencies was also compared based on paired F_{ST}. The differences for the calculated pairwise F_{ST} values were significant at $p < 0.05$. In the case of TCM (data not shown), the maximum F_{ST} distances relative to other groups were obtained for ETBs when paired with MTBs (0.330), EMBs (0.237), GBs (0.225), and MEBs (0.178), as well as for the pair MTB–FB (0.153). If PCM was applied (Table 2), FBs demonstrated the greatest F_{ST} distances when comparing to MTBs (0.393), GBs (0.321), DPBs (0.219), BTBs (0.142), and ETBs (0.100).

Table 2. Pairwise D and F_{ST} values [1] based on allele frequencies in the five SNPs at the *NCAPG-LCORL* locus among the studied chicken breeds/populations grouped by differences in phenotypic traits (according to PCM that involves six clusters as in Figure 1a,b).

No. of Animals	Chicken Groups	Egg-Type and Related	Dual Purpose and Related	Meat-Type and Related	Game	Fancy	Bantam and Related
132	Egg-type and related	(1.20)	1.22	1.30	1.21	1.21	1.27
470	Dual purpose and related	0.0290 ± 0040	(1.11)	1.08	1.01	1.46	1.27
65	Meat-type and related	0.1490 ± 0.0522	0.0690 ± 0.0322	(0.79)	0.86	1.68	1.26
48	Game	0.0970 ± 0.0328	0.0290 ± 0.0130	0.0420 ± 0.0242	(0.79)	1.56	1.27
53	Fancy	0.1000 ± 0.0125	0.2190 ± 0.0258	0.3930 ± 0.0548	0.3210 ± 0.0325	(0.84)	1.30
92	Bantam and related	0.0160 ± 0.0124	0.0310 ± 0.0279	0.1200 ± 0.0526	0.1020 ± 0.0457	0.1420 ± 0.0493	(1.27)

[1] Genetic differentiation values as calculated in EIGENSOFT. Above the diagonal: D (average divergence) values between population groups; on the diagonal: D (average divergence) values within population groups (in parentheses); below the diagonal: pairwise F_{ST} values between population groups $\pm$ standard deviations (in italics). All the pair-wise differences as assessed by F_{ST} were significant (shown by analysis of variance, $p < 0.05$).

An unrooted neighbor-joining dendrogram built on the basis of pairwise F_{ST} values for six PCM groups (Figure 3) seemed to adequately reflect both TCM and PCM, and the contemporary concept of the main evolutionary branches of domestic chickens [3]. Herewith, FBs and BTBs were closer to ETBs (on the right side of the tree), GBs to MTBs (on the left side of the tree), and DPBs occupied an intermediate position between these two major parts of the dendrogram.

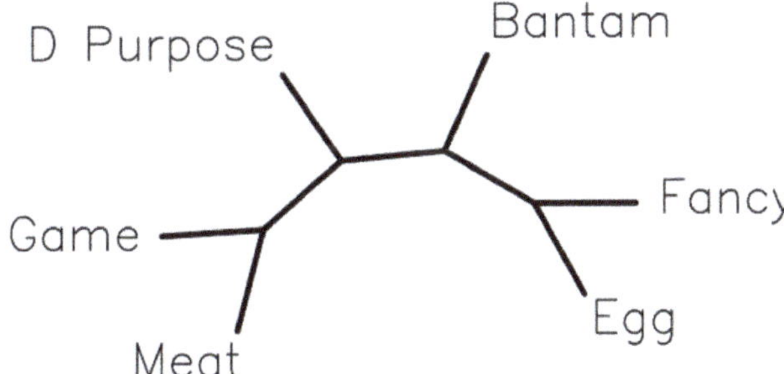

Figure 3. An unrooted phylogenetic tree constructed using the neighbor-joining method and based on the F_{ST} values (Table 2) inferred from allele frequencies in the five SNPs for six chicken breed/population groups as clustered according to PCM (Figure 1a,b).

Similar relationships between the six PCM groups were also found for paired average between-group divergence values D based on allele frequencies (Table 2). Using the D metrics, a phylogenetic neighbor-joining tree was built (data not shown), which completely coincides in its topology with the F_{ST} tree (Figure 3). Additionally, an UPGMA tree was generated based on the analysis of genotype frequencies (Figure S7). Although its topology had a slightly different spatial configuration, mutual positions of individual branches relative to each other generally coincided with the topology of the F_{ST} tree.

In addition, we note that, judging from values of the average within-group divergence D, a clear interrelation was also observed within each of the six PCM groups (Table 2). BTBs and ETBs had the highest D values (1.20 and 1.27, respectively); MTBs, GBs, and FBs had the lowest D values (0.79–0.84); and DPBs showed a more intermediate value (1.11).

In general, the results of examining the populations and their groups by allele frequencies in the five SNPs showed that the compared groups of chickens differ genetically from each other, which has also been confirmed by the data on genotype frequencies.

3.2.5. Overlapping of SNPs with QTLs

Using the Chicken QTLdb database [46], we found that the *NCAPG-LCORL* locus (75,897,761 ... 75,920,718; GRCg6a build [49]) harbors 15 known SNPs linked to the previously established QTLs (SI S4).

Out of the five significant SNPs discovered in the present study (Table 1), two, rs15619223 and rs14491017, were directly listed in the Chicken QTLdb database. For one more, rs14491028, there was a very close interval flanked by two known SNPs. Anyway, all the five SNPs were located near and within the *NCAPG-LCORL* region and overlapped with known QTLs for egg weight, egg production and growth traits (SI S4).

3.2.6. LD Structure of Gene Pool Breeds Based on Haplotypes

LD analysis in the 49 chicken populations resulted in identifying the following 11 populations with similar haploblocks at the *NCAPG-LCORL* locus: PW, BMF, PS, two historical populations of the RW breed (RWS, RWP), FS, CB, BB, NH, PC, and ZS (Figure 4). Further, a comparative assessment of these populations was carried out. As a result, a population specificity of the LD structure at the *NCAPG-LCORL* locus was found in these 11 chicken populations. In Figure 4, SNPs GGaluGA265966 in PW, BMF, and PS, and rs14491017 in RWS were removed from the LD analysis due to their complete heterozygosity. Two populations of the RW breed (RWP and RWS), FS, and CB showed a strong linkage between all the five SNPs, making up a common 102-kb block.

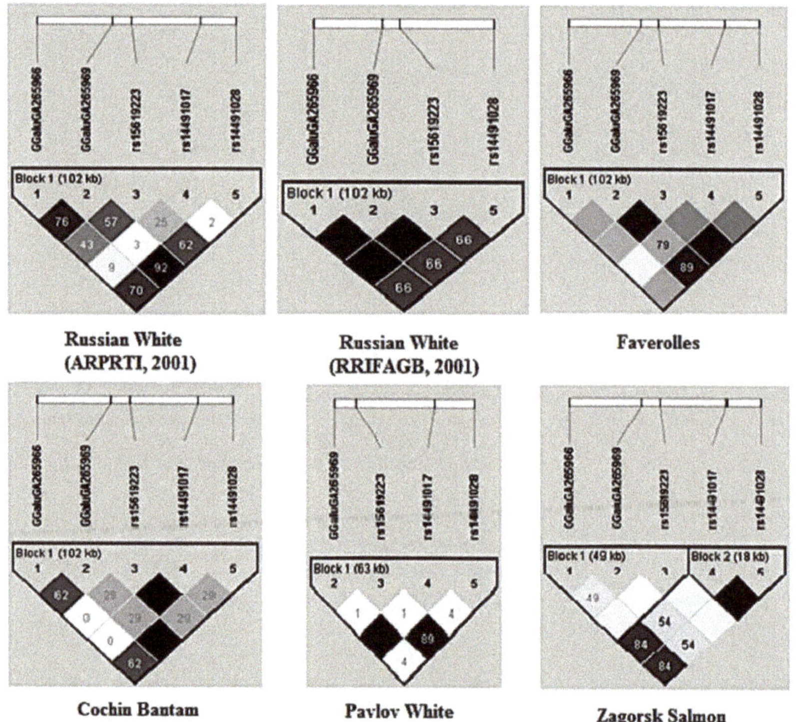

Figure 4. *Cont.*

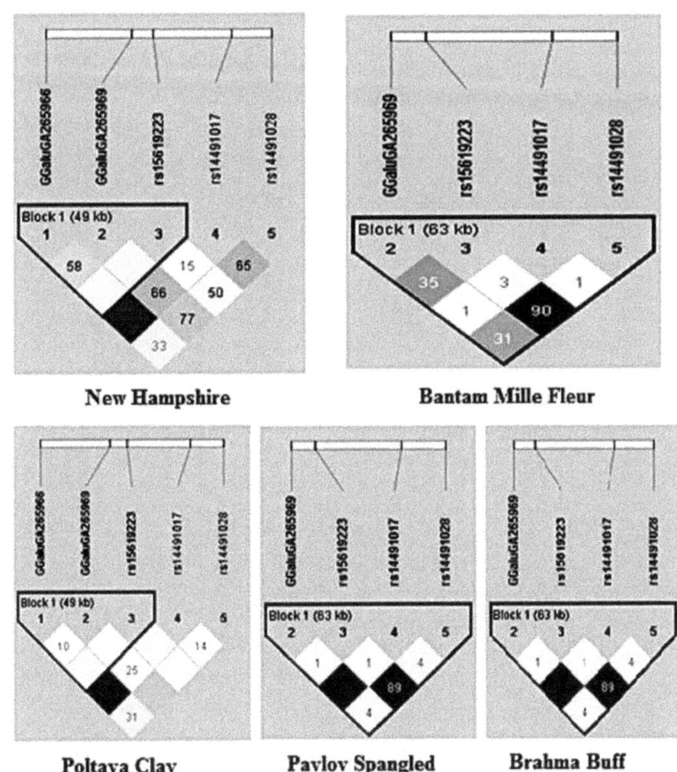

Figure 4. Structure of linkage disequilibrium at the *LCORL* gene in the 11 studied chicken populations using Haploview 4.2. The color scheme represents a linkage rate between the SNPs: white, no linkage ($r^2 = 0$); light gray, slight linkage ($0 < r^2 \leq 0.5$); dark gray, strong linkage ($0.5 < r^2 < 1$); black, full linkage ($r^2 = 1$).

There were an analogous block structure and almost complete LD between the substitutions GGaluGA265969 and rs15619223 observed in the FS and CB breeds ($r^2 = 1$ and $r^2 = 0.89$, respectively), as well as a similar LD between the substitutions GGaluGA265969 and rs15619223 in BB and FS ($r^2 = 1$ and $r^2 = 0.73$, respectively).

In the breeds ZS, PC, and NH, a 49-kb block of similar structure can be distinguished, which included the following three polymorphisms: GGaluGA265969, GGaluGA265966, and rs15619223. In the same three breeds, linkage was revealed between the SNPs GGaluGA265966 and rs14491017: at $r^2 = 1$ for NH and PC, and at $r^2 = 0.84$ for ZS.

The PC and BMF breeds had the same 63-kb block structure between the substitutions GGaluGA265969, rs15619223, rs14491017, and rs14491028. Moreover, high r^2 values (0.90 and 0.80) were found in these two breeds between the rs15619223 and rs14491028 nucleotide substitutions.

Figure 5 illustrates the distribution of haplotypes in the 11 studied populations. In the populations studied, a different degree of haplotype diversity was found, while only three breeds contained the same haplotypes. Of the hypothetically possible 45 haplotypes (when a single diploid individual has nine possible haplotypes) and based on the five SNPs at the *NCAPG-LCORL* locus, we identified 36 haplotypes. The same haplotypes (CAA, CGA, TGA, and TAA) were found in the breeds NH, PC, and ZS, with haplotype CGC being found in NH and PC. It should also be noted that there were the formation of a common

SNP haploblock and the presence of complete LD between nucleotide substitutions in the above breeds.

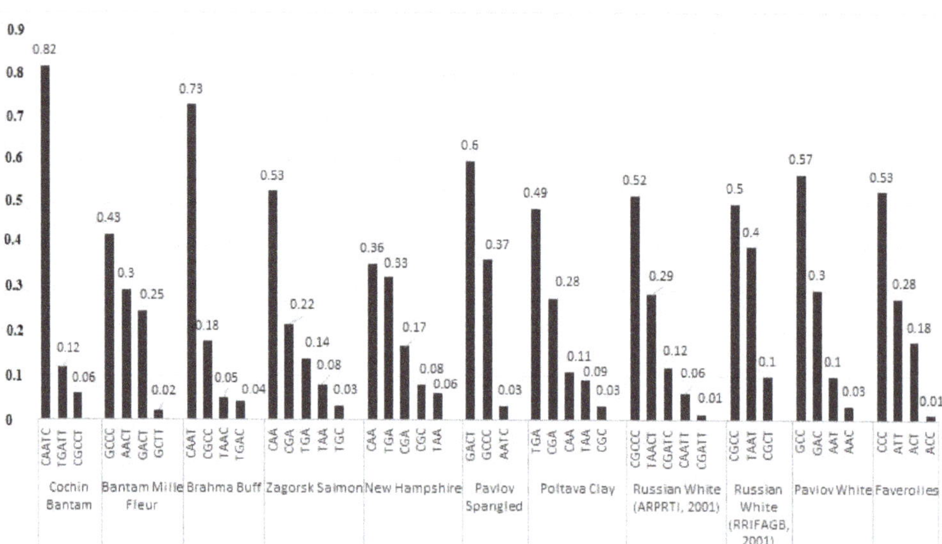

Figure 5. Distribution of haplotype frequencies in the 11 gene pool chicken populations.

4. Discussion

4.1. Analysis of Phenotypic Traits and a Model for Clustering Breeds by Phenotypes (PCM)

To preserve and use the available poultry genetic resources effectively, priority should be given to phenotypic characterization [52,53]. Phenotypic traits, primarily body weight and linear body measurements, are not only of economic importance, but also significant, alongside genetic parameters, in the classification of domestic animals, their assessment, and the search for ways to improve their productivity (e.g., [3,54–56]). Phenotypic and morphometric evaluation is often used in chicken breeding because it is simple, quick, and cost-effective (e.g., [3,53,57–59]). Body measurements are not uncommon in assessing the diversity of indigenous chicken breeds (e.g., [53,56,59,60]).

Based on the phenotypic diversity analysis of a wide range of global chicken gene pool breeds, we constructed PCM and compared it with TCM. For this purpose, we took phenotypic characteristics reflecting the morphometric parameters of the breeds including body weight and 13 body measurements of hens and cocks, as well as two main traits of egg productivity: egg number for 52 weeks of life and egg weight (Figure 1). We have shown that morphometric characters provided the greatest contribution to the formation of the breed clustering pattern in accordance with the PCM. This was clearly seen on the FAC and PCA plots (Figures S1a and S2a), in which the morphometric characters were used alone (i.e., without egg production traits), while fully resembling the PCM pattern. If clustering was achieved only based on egg production traits, the PCM pattern was disrupted (Figures S1b and S2b). At the same time, if we compare the characteristics of egg number and body weight, we can see that the contribution of the former to the clustering of breeds significantly exceeds the contribution of the latter ((Figures 2 and S1a,b). Thus, we can assume that among the morphometric and all phenotypic traits in general, the main contribution to PCM seems to be made by the traits of body measurements in females and males. Moiseyeva et al. [3] stated that body measurements, being quantitative traits, have a high heritability coefficient ($h^2 \approx 0.5$) and a lower intrapopulation variability. These features determine their high correlation with the genetic structure of breeds that was created over a relatively long evolutionary process of their development.

To the best of our knowledge, the proposed assessment of phenotypic diversity using phenotypic clustering of chicken breeds that we designated PCM, as well as the assessment of the contributions of certain phenotypic traits to this PCM, have not been specifically and purposefully examined in other previous studies. It is noteworthy that we carried out such assessments using quite a vast part of the world gene pool of chicken breeds that have different origins (often mixed), purpose of use, and the degree of admixture. Moiseyeva et al. [3] were also able to demonstrate good distinguishability of phenotypic (morphometric) traits to reproduce plausible breed differentiations and topologies of appropriate phylogenetic trees, even using smaller samples of 8–10 different breeds.

Our research not only has confirmed the main insights of the work by Moiseyeva et al. [3], but also significantly expanded them. Due to a larger sample of the global gene pool, including a great number of synthetic breeds and populations, and in the light of the clustering data obtained, we were able to revise the concept of four main evolutionary lineages of breed formation in domestic chickens as postulated by Moiseyeva et al. [3], which embraced ETBs, MTBs, GBs, and BTBs. According to our updated concept based on PCM, it is also necessary to distinguish two more evolutionary branches of chicken breeds, namely DPBs and FBs (Figure 1a,b). The PCM postulated by us and the appropriate concept of six major evolutionary lineages of chicken breed formation would, in our opinion, most fully reflect the entire spectrum of phenotypic diversity of the world gene pool of domestic chickens.

It should also be noted that our concept of evolutionary subdivision of domestic chickens and breed formation based on PCM also has certain discrepancies with TCM adopted in conventional and specialized literature on poultry (e.g., [32]). TCM is grounded on a rather simplified, speculative, and, to a certain extent, very artificial scheme for dividing chicken breeds into conditional utility types, i.e., purposes of use, when only one criterion for selecting poultry for egg or meat productivity can determine if a breed belongs to a certain TCM class. This traditional approach does not consider history of the origin (often mixed) and development of this breed, the degree of its synthetic nature and introgression, as well as the whole complex of its phenotypic features. This also results in significant artificiality, fuzziness, and uncertainty in terms of a defined phenotypic characterization and classification among numerous synthetic breeds from the EMB and MEB subcategories, which are commonly referred to as DPBs. The FB class looks no less artificial and speculative in TCM because it involves any breeds (including BTB) kept by fanciers for their phenotypic diversity and not related to explicit ETBs, MTBs, and GBs. Our proposed concept and PCM produced through an appropriate analysis of the phenotypic diversity of chicken breeds would highlight and correct the said shortcomings of TCM.

4.2. Genotypic Models of Clustering and Admixture

Concerning the development of two genotypic clustering models (GCM1 and GCM2), in a number of instances we noted both their congruence with each other and some similarities with PCM. For example, breeds and populations from the GCM2 cluster I were fully included in the GCM1 subcluster Ia (ETB), and members of the GCM2 subcluster IVa in the GCM1 cluster III (MTBs). Breeds of the PCM cluster V (FB) were stably clustered together in the GCM1 subcluster Ia and in the GCM2 cluster III, etc. The existing differences between the models can be attributed to the various algorithms used to construct them, and to the differences in the nature of the compared traits. It should also be noted that, for PCM, we used a fairly wide set of phenotypic traits describing both morphology and performance characteristics of poultry, while for genotypic models, one region of the chicken genome corresponding to the *NCAPG-LCORL* locus was employed, which apparently narrowed the resolving power of analyzing the breeds and populations by genotypes. Many other investigations including Moiseyeva et al. [3] also observed an unequal contribution of certain sets of phenotypic/genetic factors to the resulting patterns of phylogeny/clustering among examined breeds.

Nevertheless, combining data on genotyping (i.e., allele frequencies in the five SNPs) and on their grouping according to PCM showed significant differences between groups (clusters) of breeds and an expected topology of the breed phylogenetic tree (Figure 3) in full accordance with our proposed updated concept of the evolutionary subdivision of domestic chickens into six evolutionary lineages of breed formation. In addition, the observed differences at the *NCAPG-LCORL locus* between the identified groups (clusters) of breeds did not contradict the previous information on the relationship of variability at this locus with certain phenotypic traits (e.g., [50,51,61,62]). Considering the above, we would suggest the importance of our findings at the *NCAPG-LCORL* locus that demonstrated a genetic variability among a wide sample of commercial, local, and fancy chicken breeds, including synthetic ones, which largely overlaps with the patterns of their phenotypic variability and does not contradict the general ideas about history of formation and development of a particular breed.

An important aspect of our study was the detection of significant admixture among the 49 genotyped chicken breeds and populations. This was expressed in complex patterns of admixture models (Figure S6), which, as a rule, did not allow us to clearly distinguish among themselves the classes (clusters) that we proposed for four models (TCM, PCM, GCM1, and GCM2). This significant admixture can be explained by the history of mixed origin of the studied breeds, i.e., a synthetic nature of their formation, when genetic makeup of many breeds was composed by mixing the genomes of several original, distinctive breeds and/or due to individual crossbreeding events with other breeds and introgression during their breeding (e.g., [63]). Nevertheless, by exploring the obtained patterns of admixture models, one could generally notice certain differences between ETB and MTB chickens, which is consistent with the results of the genotype frequency analysis. We would also suggest that the admixture models themselves can be considered as an auxiliary tool in clarifying evolutionary signatures of subdivision and inference of demographic history among chicken breeds and populations.

4.3. Analysis of Genetic Variation at the Locus NCAPG-LCORL

Characterization of genetic variation and population genetic structure based on SNPs at key chicken performance-related genes not only helps to determine the characteristics of various commercial, fancy, and local breeds and populations, including synthetic and highly admixed ones, but also to assess whether this information can be useful in MAS [50,64]. We genotyped chickens of various purposes of use (TCM) and phenotypes (PCM) using the Illumina Chicken 60K SNP iSelect BeadChip, which revealed the presence of the five significant SNPs on GGA4 in the area of the *NCAPG-LCORL* locus and flanking regions.

The differences found at the genetic level, i.e., in the SNPs within the locus covering the *NCAPG-LCORL* genes may be of great interest, since these genes are one of the key regulators of RNA polymerase II transcription and have pleiotropic effects in terms of body weight/size and egg weight/size [65,66]. Guo et al. [67] identified genomic variants associated with the size and mass of chicken bones at the *NCAPG-LCORL* locus. SNPs at the *NCAPG* gene (rs14491030) and the *LCORL* gene (rs14699480) were associated with egg weight [68]. According to Yi et al. [17], rs14491030 at the *NCAPG* gene can simultaneously affect both egg weight and body weight. In an investigation by Barkova and Smaragdov [69], significant associations of rs14491030 at the *NCAPG* gene with egg weight and shell elastic deformation were found. Sun et al. [70] showed that *NCAPG* had a definite effect on the eggshell quality in pullets. The pleiotropic effect of the *NCAPG-LCORL* locus may be related to the fact that egg weight affects the body weight of chickens at birth, their physical shape and further performance [71]. Eggshell, providing gas exchange between an embryo and environment and supplying calcium to embryo bones, is of biological importance for the development of avian embryos [72]. In a study by Liu et al. [61], the *LCORL* locus was significantly associated with the body weight of chickens. It was also shown that the *LCORL* gene has different levels of expression in slow-growing and fast-growing broiler chickens.

In the present investigation, we found that following even TCM, ETB chickens in general were significantly different from chickens of other purposes of use (utility types). So, for the GGaluGA265969 substitution, significant differences were shown when comparing the ETB–MTB and ETB–FB group pairs. Presumably, such differences could be due to the fact that this SNP might be associated with the body weight of hens.

SNP rs15619223 showed the largest number (28) of significant pairwise differences between chicken groups of various use/productivity types (TCM) or various clusters/subclusters (PCM and GCM1) (SI S3). According to our previous studies [50,51], this SNP is probably associated with body weight and egg weight in RW chickens. Although for SNPs GGaluGA265966 (14 significant pairwise differences), rs14491028 (14 differences), and rs14491017 (23 differences) we did not find any other reports on their relationship with productive traits in chickens, the presence of polymorphism and a significant difference in allele frequencies between groups of contrasting chicken breeds (SI S3) would suggest considering these SNPs as potentially momentous.

In general, for three models, TCM, PCM, and GCM1, we demonstrated a reliable resolution of the obtained data on allele frequencies in the SNPs of the NCAPG-LCORL locus to discriminate chicken breeds and populations related to ETBs, MTBs, DPBs (EMBs, MEBs), GBs, or FBs (SI S3). Thus, based on our own data available for the region on GGA4 that involves *NCAPG-LCORL*, we can confirm that there is a definite relationship between genetic variation at this locus and phenotypic diversity in chicken breeds, and this is an important QTL for body weight and egg weight [61].

The maximum heterozygosity for a particular locus is achieved when the frequencies of its alleles are equal and depends on the number of alleles. This is relevant for populations under selection for a desired trait. In the conditions of artificial breeding of populations, parental pairs are assorted, while relationship of individuals participating in mating increases. As a result of mating closely related individuals (i.e., inbreeding), proportion of homozygous genotypes is growing in a population. An important feature of inbreeding is the constancy of allele frequencies in all inbred generations observed while the number of heterozygous genotypes declines [73]. In our study, we used closed gene pool populations with a small population size and identified isolated cases of inbreeding (see Section 3.2.1). However, due to paucity of the studied populations, assessment of the inbreeding level in them would be possible with full confidence in further detailed genome-wide studies.

Among the identified SNPs, a shift in genetic equilibrium was found in crossbred chickens Ts × SL and in purebred populations Ts, NH, Pu, F and PRB ($\chi^2 > 3.84$), which may be indicative of strong selection pressure and a consequence of the selection of offspring from the best sire.

In the SNP-assisted analysis of genetic differentiation between chicken groups using TCM and paired F_{ST}, significant differences were detected for ETB chickens in comparison with MTBs, EMBs, GBs, and MEBs, as well as MTBs and FBs. With applying PCM (Table 2), greatest F_{ST} distances were shown for FBs relative to other clusters. These results of the F_{ST}-based analysis suggest a certain "predictive power" of the five SNPs for discriminating chickens of different other groups (clusters) when employing TCM and PCM.

The five significant SNPs identified in this study overlapped with the previously established QTLs for egg weight, egg production and growth traits at the *NCAPG-LCORL* locus (SI S4), confirming that these SNPs are relevant to and important for evaluating genomic variation among the studied local, commercial and fancy chicken breeds.

In the present study, we also investigated LD structure at the *LCORL* gene among various chicken breeds. Close LD was observed in the FS, CB, and BB breeds, suggesting their common origin as the Cochin, Brama, Dorking, and Houdan were FS ancestors. Chickens of the ZS breed were developed by crossing the RW, RIR, NH, and YC breeds. PC, such as NH, was created using the RIR breed [31]. The common origin of the listed breeds could explain their almost complete identity in LD. A similar conformation and diminutive sizes of the BMF and PS chicken breeds probably determined the same structure of their LD patterns. This may be due to the fact that polymorphism in the *LCORL* gene in animals can have a significant effect on the height, skeletal size, bone formation, and muscle development during embryogenesis [62].

When examining LD haploblocks between the SNPs in the studied chicken populations, important information was obtained regarding a population specificity of the haploblock structure at the *NCAPG-LCORL* locus for the 11 breeds/populations. These included two ETB populations (RWS and RWP), two EMB populations (FS and BB), three MEB populations (NH, PC and ZS), two FB populations (PS and PW), and two BTB populations (BMF and CB).

LD analysis revealed 36 haplotypes for these breeds. Common haplotypes are confirmed by the origin of breeds and similar phenotypes, suggesting a common mechanism for the formation of these LD patterns. Four similar haplotypes in the three MEB (NH, PC, and ZS) breeds, as well as one more common haplotype in NH and PC, can be associated with their common descent from the same ancestral breeds.

Due to the intensive selection in the breeding program and a significant reduction in the population size, changes in LD were observed in the chicken gene pool populations over several generations. Characterization of LD is of fundamental importance in carrying out genome-wide association analysis and genomic selection, as well as in identifying recent genomic rearrangements. LD can make a certain positive contribution when used in MAS in the future poultry farming [74]. At the same time, the costs of genotyping SNPs can be significantly reduced since markers are usually linked with each other in the area of influence on a trait [75]. Comparative assessment of populations with different demographic histories is an important source of information on changes in the genome when breeding small groups and assessing the results of crossing [64,76].

5. Conclusions

According to the FAO recommendations [52], to characterize the genetic resources of agricultural animals, it is necessary to rely on three types of information: phenotypic, genetic, and historical, which we attempted to do in the framework of this study. Herein, we proposed and studied four models of clustering (classification), i.e., TCM, PCM, GCM1, and GCM2, using a large spectrum of the global gene pool of chicken breeds that often have a synthetic genetic structure, significant admixture, and possible introgression. Based on these models, we expanded the earlier concept of four evolutionary lineages in domestic chicken breeding postulated by Moiseyeva et al. [3]. Our updated concept of evolutionary subdivision and breed formation includes six evolutionary branches of domestic chickens: ETBs, FBs, BTBs, DPBs, MTBs, and GBs as represented on the appropriate phylogenetic tree (Figure 3). We also found a complex and indistinct character of admixture models for many phenotypically and genotypically diverse breeds, many of which have a specific demographic history and a synthetic genetic blueprint.

Additionally, we discovered significant differences in allele frequencies for the analyzed SNPs at the *NCAPG-LCORL* locus in the small chicken populations of different phenotypes and purposes of use. This may point out that the importance of this locus in understanding the genetic basis for the formation of productive traits in poultry. Chickens of ETB and MTB are genetically far apart and formed two distinct clusters. Analysis of LD patterns revealed a close linkage between the SNPs in some DPB populations, which may prove a history of their origin from common ancestral breeds. Based on the data obtained, it can be assumed that the presence of LD can be an auxiliary tool in carrying out MAS for a small number of markers associated with each other in the area of influence on a trait, which can significantly reduce the costs of genotyping SNPs in general.

In addition, the current authentic allele composition in FB chickens at the *NCAPG-LCORL* locus was characterized, suggesting their gene pool as unique "inexhaustible" genetic resources. Since the *NCAPG-LCORL* locus is one of the key regulators of RNA polymerase II transcription and has a noticeable pleiotropic effect in relation to body weight and size, as well as egg size in chickens, we characterized genetic variants at this locus in various commercial, fancy, and local breeds and populations. Our research findings seem to be relevant for the purpose of identifying individuals that are carriers of SNP variants potentially associated with economically important QTLs. This information can also be in demand in MAS for meat and egg performance in chickens.

Genome-wide association studies can identify SNPs related to formation of phenotypic traits. This provides a unique opportunity to improve the methods for choosing markers of choice for MAS using information about many SNP markers across the entire genome, and thus increasing the selection accuracy. The data obtained can also be used in determining parental pairs and correcting selection targets in local chicken populations. The tested approaches in analyzing SNPs show their high efficiency and will be put into practical use in breeding and preservation programs for small chicken breeds and populations. Importantly, the current thorough phenotypic and genotypic survey involved several remarkable native breeds of Russia and the former USSR including OM [77], YC [78,79], RW [50,51,64], PC [80–83], PS/PW, RC, MG, UM, and UG. The obtained information will be helpful in their future conservation and commercial use.

Overall, our investigation has further contributed to solving the problem of evolutionary subdivision of domestic chickens including implications for synthetic breeds and admixture through a comprehensive assessment of phenotypic variation and genotypic structure at the *NCAPG-LCORL* locus, an important QTL for body meat and egg production traits.

Supplementary Materials: The following are available online at https://www.mdpi.com/article/10.3390/agriculture11100914/s1, Figure S1: Fuzzy Analysis Clustering-based plots showing the distribution of the 39 phenotyped breeds/populations and built using (**a**) morphometric traits and (**b**) egg production traits; Figure S2: PCA plots showing the distribution of the 39 phenotyped breeds/populations and built using (**a**) morphometric traits and (**b**) egg production traits; Figure S3: Plots of phylogenetic UPGMA-based trees constructed in comparison with PCM (shown via color coding) and using Euclidean distances for different sets of phenotypic traits; Figure S4: Clustering plots for the 49 chicken breeds/populations based on their genotypes in the five SNPs at the NCAPG-LCORL locus and produced using (**a**) Fuzzy Analysis Clustering, (**b**) PCA, and (**c**) an UPGMA-assisted tree using Euclidean metrics; Figure S5: Number of ancestral populations using genotypes in the five SNPs at the *NCAPG-LCORL* locus for the 49 genotyped chicken breeds/populations and applying different clustering models; Figure S6: Population structure based on the genetic variability in the 49 genotyped chicken breeds/populations for five SNP markers at the *NCAPG-LCORL* locus and produced by Bayesian clustering using the ADMIXTURE program (as visualized in color in R v.4.1.0); Figure S7: An UPGMA-based tree using Euclidean distances for six chicken breed clusters according to PCM and based on genotype frequencies in the five SNPs at the *NCAPG-LCORL* locus among the 49 genotyped chicken breeds/populations; Table S1: Forty-nine chicken breeds, strains and crosses used in the study and listed in accordance with TCM; Table S2: Phenotypic traits in females among the 39 studied chicken gene pool breeds/populations listed in accordance with PCM; Table S3: Phenotypic traits in males among the 39 studied chicken gene pool breeds/populations listed in accordance with PCM; Table S4: Frequency distribution of genotypes for five SNPs at the *NCAPG-LCORL* locus on GGA4 among the 49 studied chicken gene pool breeds/populations listed in accordance with PCM; Supporting Information (SI) S1: Allele frequencies in the five studied SNPs at the *NCAPG-LCORL* locus; SI S2: Significance levels of differences in allele frequencies for the five SNPs; SI S3: Boxplots and significance of differences in allele frequencies for the studied five SNPs at the locus *NCAPG-LCORL* as assessed for chicken breed groups; SI S4: Known QTLs at the *NCAPG-LCORL* locus and relevant SNPs in the Chicken QTLdb database.

Author Contributions: Conceptualization, T.A.L., M.N.R. and M.V.P.; methodology, M.N.R., M.V.P., Y.S.S. and E.A.B.; software, M.V.P., Y.S.S. and E.A.B.; validation, O.Y.B., T.A.L. and O.V.M.; formal analysis, M.V.P., N.V.D., A.B.V., O.V.M. and M.N.R.; investigation, G.K.P., A.V.M., A.B.V., T.A.L. and O.Y.B.; resources, O.I.S., A.B.V., A.V.M. and G.K.P.; data curation, O.I.S., A.B.V., N.V.D., M.V.P., O.V.M., G.K.P. and T.A.L.; writing—original draft preparation, T.A.L. and M.N.R.; writing—review and editing, T.A.L., M.N.R. and D.K.G.; visualization, M.V.P., Y.S.S., E.A.B.; supervision, T.A.L., M.V.P.;

project administration, T.A.L.; funding acquisition, T.A.L. All authors have read and agreed to the published version of the manuscript.

Funding: This research was funded by the Ministry of Science and Higher Education of the Russian Federation (State Assignment Program 0445-2021-0010).

Institutional Review Board Statement: The study was conducted according to the guidelines of the Declaration of Helsinki and approved by the Institutional Review Board of the Russian Research Institute of Farm Animal Genetics and Breeding—Branch of the L. K. Ernst Federal Research Center for Animal Husbandry (Protocol No. 2020-4 dated 3 March 2020).

Informed Consent Statement: Not applicable.

Data Availability Statement: The data presented in this study are available in this article and Supplementary Material.

Conflicts of Interest: The authors declare no conflict of interest. The funders had no role in the design of the study; in the collection, analyses, or interpretation of data; in the writing of the manuscript, or in the decision to publish the results.

References

1. Romanov, M.N. Genetics of Broodiness in Poultry—A Review. *Asian-Australas. J. Anim. Sci.* **2001**, *14*, 1647–1654. [CrossRef]
2. Dodgson, J.B.; Romanov, M.N. Use of Chicken Models for the Analysis of Human Disease. *Curr. Protoc. Hum. Genet.* **2004**, *40*, 15.5.1–15.5.15. [CrossRef] [PubMed]
3. Moiseyeva, I.G.; Romanov, M.N.; Nikiforov, A.A.; Sevastyanova, A.A.; Semyenova, S.K. Evolutionary Relationships of Red Jungle Fowl and Chicken Breeds. *Genet. Sel. Evol.* **2003**, *35*, 403–423. [CrossRef]
4. Kholofelo Malomane, D.; Simianer, H.; Reimer, C.; Weigend, A.; Weigend, S. Different Evolutionary Dynamics Revealed by Functional SNP Classes in Global Chicken Groups. *Worlds Poult. Sci. J.* **2018**. Suppl.: The XVth European Poultry Conference. Conference Information and Proceedings, Dubrovnik, Croatia, 17–21 September 2018, p. 87, Abstract ID 207.
5. Makarova, A.V.; Mitrofanova, O.V.; Vakhrameev, A.B.; Dementeva, N.V. Molecular-Genetic Bases of Plumage Coloring in Chicken. *Vavilovskii Zh. Genet. Sel. (Vavilov J. Genet. Breed.)* **2019**, *23*, 343–354. [CrossRef]
6. Wang, M.-S.; Thakur, M.; Peng, M.-S.; Jiang, Y.; Frantz, L.A.F.; Li, M.; Zhang, J.-J.; Wang, S.; Peters, J.; Otecko, N.O.; et al. 863 Genomes Reveal the Origin and Domestication of Chicken. *Cell Res.* **2020**, *30*, 693–701. [CrossRef]
7. Bennett, C.E.; Thomas, R.; Williams, M.; Zalasiewicz, J.; Edgeworth, M.; Miller, H.; Coles, B.; Foster, A.; Burton, E.J.; Marume, U. The Broiler Chicken as a Signal of a Human Reconfigured Biosphere. *R. Soc. Open Sci.* **2018**, *5*, 180325. [CrossRef] [PubMed]
8. Wang, M.-S.; Huo, Y.-X.; Li, Y.; Otecko, N.O.; Su, L.-Y.; Xu, H.-B.; Wu, S.-F.; Peng, M.-S.; Liu, H.-Q.; Zeng, L.; et al. Comparative Population Genomics Reveals Genetic Basis Underlying Body Size of Domestic Chickens. *J. Mol. Cell Biol.* **2016**, *8*, 542–552. [CrossRef] [PubMed]
9. Li, J.-J.; Zhang, L.; Ren, P.; Wang, Y.; Yin, L.-Q.; Ran, J.-S.; Zhang, X.-X.; Liu, Y.-P. Genotype Frequency Distributions of 28 SNP Markers in Two Commercial Lines and Five Chinese Native Chicken Populations. *BMC Genet.* **2020**, *21*, 12. [CrossRef] [PubMed]
10. Wężyk, S.; Cywa-Benko, K.; Romanov, M.N. Ochrona Przed Zagładą Rodzimych Ras Drobiu w Krajach Wschodniej Europy/Protection against the Extinction of the Native Breeds of Poultry in the Countries of Eastern Europe. In Proceedings of the International Symposium Conservation Measures for Rare Farm Animal Breeds, Balice, Poland, 17–19 May 1994; National Research Institute of Animal Production: Balice, Poland; Polish Society of Animal Production: Balice, Poland, 1994; pp. 25–26.
11. Romanov, M.N.; Sakhatsky, N.I. Ochrona Zasobów Genetycznych Drobiu na Ukrainie/Conservation of Poultry Genetic Resources in Ukraine. In *Summaries of the 9th International Symposium of Young Poultry Scientists, Siedlce, Poland, 8–10 June 1995*; World's Poultry Science Association, Polish Branch: Siedlce, Poland, 1995; pp. 87–88.
12. Romanov, M.N.; Sakhatsky, N.I. Inventory of Poultry Genetic Resources in Ukraine. In *Naučno-Tehničeskij Bûlleten-Institut Pticevodstva*; Poultry Research Institute, Ukrainian Academy of Agrarian Sciences: Kharkiv, Ukraine, 1995; No. 34; pp. 3–14.
13. Wezyk, S.; Cywa-Benko, K.; Romanov, M.N. Protection against the Extinction of the Native Breeds of Poultry in Eastern European Countries. In Proceedings of the International Symposium "Conservation Measures for Rare Farm Animal Breeds, Balice, Poland, 17–19 May 1994; pp. 89–101.
14. Romanov, M.N.; Weigend, S.; Bondarenko, Y.V.; Podstreshny, A.P.; Kutnyuk, P.I.; Sakhatsky, N.I. Studies on poultry germplasm diversity and conservation in Ukraine. In Proceedings of the Poultry Genetics Symposium, Mariensee, Germany, 6–8 October 1999; Working Group 3 of WPSA: Mariensee, Germany, 1999; p. 140.
15. Dementeva, N.V.; Vakhrameev, A.B.; Larkina, T.A.; Mitrofanova, O.V. Efficiency of Using SNP Markers in the MSTN Gene in the Selection of the Pushkin Breed Chickens. *Vavilovskii Zh. Genet. Sel. (Vavilov J. Genet. Breed.)* **2019**, *23*, 993–998. [CrossRef]
16. Dou, T.; Shen, M.; Ma, M.; Qu, L.; Li, Y.; Hu, Y.; Lu, J.; Guo, J.; Wang, X.; Wang, K. Genetic Architecture and Candidate Genes Detected for Chicken Internal Organ Weight with a 600 K Single Nucleotide Polymorphism Array. *Asian-Australas. J. Anim. Sci.* **2019**, *32*, 341–349. [CrossRef] [PubMed]

17. Yi, G.; Shen, M.; Yuan, J.; Sun, C.; Duan, Z.; Qu, L.; Dou, T.; Ma, M.; Lu, J.; Guo, J.; et al. Genome-Wide Association Study Dissects Genetic Architecture Underlying Longitudinal Egg Weights in Chickens. *BMC Genom.* **2015**, *16*, 746. [CrossRef] [PubMed]

18. Shen, M.; Qu, L.; Ma, M.; Dou, T.; Lu, J.; Guo, J.; Hu, Y.; Wang, X.; Li, Y.; Wang, K.; et al. A Genome-Wide Study to Identify Genes Responsible for Oviduct Development in Chickens. *PLoS ONE* **2017**, *12*, e0189955. [CrossRef] [PubMed]

19. Wood, A.R.; Esko, T.; Yang, J.; Vedantam, S.; Pers, T.H.; Gustafsson, S.; Chu, A.Y.; Estrada, K.; Luan, J.; Kutalik, Z.; et al. Defining the Role of Common Variation in the Genomic and Biological Architecture of Adult Human Height. *Nat. Genet.* **2014**, *46*, 1173–1186. [CrossRef]

20. Okada, Y.; Kamatani, Y.; Takahashi, A.; Matsuda, K.; Hosono, N.; Ohmiya, H.; Daigo, Y.; Yamamoto, K.; Kubo, M.; Nakamura, Y.; et al. A Genome-Wide Association Study in 19 633 Japanese Subjects Identified *LHX3-QSOX2* and *IGF1* as Adult Height Loci. *Hum. Mol. Genet.* **2010**, *19*, 2303–2312. [CrossRef] [PubMed]

21. Carty, C.L.; Johnson, N.A.; Hutter, C.M.; Reiner, A.P.; Peters, U.; Tang, H.; Kooperberg, C. Genome-Wide Association Study of Body Height in African Americans: The Women's Health Initiative SNP Health Association Resource (SHARe). *Hum. Mol. Genet.* **2012**, *21*, 711–720. [CrossRef]

22. Soranzo, N.; Rivadeneira, F.; Chinappen-Horsley, U.; Malkina, I.; Richards, J.B.; Hammond, N.; Stolk, L.; Nica, A.; Inouye, M.; Hofman, A.; et al. Meta-Analysis of Genome-Wide Scans for Human Adult Stature Identifies Novel Loci and Associations with Measures of Skeletal Frame Size. *PLoS Genet.* **2009**, *5*, e1000445. [CrossRef]

23. Lindholm-Perry, A.K.; Sexten, A.K.; Kuehn, L.A.; Smith, T.P.; King, D.A.; Shackelford, S.D.; Wheeler, T.L.; Ferrell, C.L.; Jenkins, T.G.; Snelling, W.M.; et al. Association, Effects and Validation of Polymorphisms within the *NCAPG-LCORL* Locus Located on BTA6 with Feed Intake, Gain, Meat and Carcass Traits in Beef Cattle. *BMC Genet.* **2011**, *12*, 103. [CrossRef]

24. Rubin, C.-J.; Megens, H.-J.; Barrio, A.M.; Maqbool, K.; Sayyab, S.; Schwochow, D.; Wang, C.; Carlborg, Ö.; Jern, P.; Jørgensen, C.; et al. Strong Signatures of Selection in the Domestic Pig Genome. *Proc. Natl. Acad. Sci. USA* **2012**, *109*, 19529–19536. [CrossRef]

25. Metzger, J.; Schrimpf, R.; Philipp, U.; Distl, O. Expression Levels of *LCORL* are Associated with Body Size in Horses. *PLoS ONE* **2013**, *8*, e56497. [CrossRef]

26. Al-Mamun, H.A.; Kwan, P.S.; Clark, A.; Ferdosi, M.H.; Tellam, R.; Gondro, C. Genome-Wide Association Study of Body Weight in Australian Merino Sheep Reveals an Orthologous Region on OAR6 to Human and Bovine Genomic Regions Affecting Height and Weight. *Genet. Sel. Evol.* **2015**, *47*, 66. [CrossRef]

27. Takasuga, A. PLAG1 and NCAPG-LCORL in Livestock. *Anim. Sci. J.* **2016**, *87*, 159–167. [CrossRef] [PubMed]

28. Lyu, S.; Arends, D.; Nassar, M.K.; Brockmann, G.A. Fine Mapping of a Distal Chromosome 4 QTL Affecting Growth and Muscle Mass in a Chicken Advanced Intercross Line. *Anim. Genet.* **2017**, *48*, 295–302. [CrossRef]

29. Dementieva, N.V.; Kudinov, A.A.; Larkina, T.A.; Mitrofanova, O.V.; Dysin, A.P.; Terletsky, V.P.; Tyshchenko, V.I.; Griffin, D.K.; Romanov, M.N. Genetic Variability in Local and Imported Germplasm Chicken Populations as Revealed by Analyzing Runs of Homozygosity. *Animals* **2020**, *10*, 1887. [CrossRef] [PubMed]

30. Dementieva, N.V.; Mitrofanova, O.V.; Dysin, A.P.; Kudinov, A.A.; Stanishevskaya, O.I.; Larkina, T.A.; Plemyashov, K.V.; Griffin, D.K.; Romanov, M.N.; Smaragdov, M.G. Assessing the Effects of Rare Alleles and Linkage Disequilibrium on Estimates of Genetic Diversity in the Chicken Populations. *Animal* **2021**, *15*, 100171. [CrossRef] [PubMed]

31. Paronyan, I.A.; Plemyashov, K.V.; Segal, E.L.; Yurchenko, O.P.; Shabanova, S.A.; Vakhrameev, A.B.; Karpukhina, I.V.; Makarova, A.V.; Pervushina, A.T. *Breeds and Populations of Chickens Bred at the Germplasm Farm of the State Scientific Institution VNIIGRZh of the Russian Agricultural Academy: Album*; GNU VNIIGRZh: St. Petersburg, Russia, 2014. Available online: http://vniigen.ru/wp-content/uploads/2017/04/Katalog-Kur-1.pdf (accessed on 9 July 2021).

32. Bogolyubsky, S.I. *Poultry Breeding*; Agropromizdat: Moscow, Russia, 1991.

33. Chang, C.C.; Chow, C.C.; Tellier, L.C.; Vattikuti, S.; Purcell, S.M.; Lee, J.J. Second-Generation PLINK: Rising to the Challenge of Larger and Richer Datasets. *Gigascience* **2015**, *4*, 7. [CrossRef] [PubMed]

34. RDocumentation. Heatmap: Draw a Heat Map. Available online: https://www.rdocumentation.org/packages/stats/versions/3.6.2/topics/heatmap (accessed on 9 July 2021).

35. Shitikov, V.K.; Mastitsky, S.E. 10.4.2 Fuzzy k-Means Method (Fuzzy Analysis Clustering). In *Classification, Regression, Data Mining Algorithms Using R*. Electronic Book. 2017. Available online: https://ranalytics.github.io/data-mining/104-Other-Clustering-Methods.html#sec_10_4_2 (accessed on 9 July 2021).

36. Fanny: Fuzzy Analysis Clustering. Documentation for Package 'Cluster' Version 2.1.0. "Finding Groups in Data": Cluster Analysis Extended Rousseeuw et al. 2019. Available online: https://rdrr.io/cran/cluster/man/fanny.html (accessed on 9 July 2021).

37. Suzuki, R.; Shimodaira, H. Pvclust: An R Package for Assessing the Uncertainty in Hierarchical Clustering. *Bioinformatics* **2006**, *22*, 1540–1542. [CrossRef]

38. Zhao, Q.; Hautamaki, V.; Fränti, P. Knee Point Detection in BIC for Detecting the Number of Clusters. In *Lecture Notes in Computer Science*; Blanc-Talon, J., Bourennane, S., Philips, W., Popescu, D., Scheunders, P., Eds.; Springer: Berlin/Heidelberg, Germany, 2008; Volume 5259, pp. 664–673.

39. Kassambara, A.; Mundt, F. Factoextra: Extract and Visualize the Results of Multivariate Data Analyses. Version 1.0.5; 22 August 2017. Available online: https://cran.r-project.org/web/packages/factoextra/index.html (accessed on 9 July 2021).

40. Alexander, D.H.; Novembre, J.; Lange, K. Fast Model-Based Estimation of Ancestry in Unrelated Individuals. *Genome Res.* **2009**, *19*, 1655–1664. [CrossRef] [PubMed]

41. Price, A.L.; Patterson, N.J.; Plenge, R.M.; Weinblatt, M.E.; Shadick, N.A.; Reich, D. Principal Components Analysis Corrects for Stratification in Genome-Wide Association Studies. *Nat. Genet.* **2006**, *38*, 904–909. [CrossRef]

42. Ayala, F.J.; Kiger, J.A., Jr. *Modern Genetics*, 2nd ed.; Benjamin/Cummings: Menlo Park, CA, USA; London, UK, 1984.

43. Wright, S. The Genetical Structure of Populations. *Ann. Eugen.* **1951**, *15*, 323–354. [CrossRef] [PubMed]

44. Pilot, M.; Greco, C.; vonHoldt, B.M.; Jędrzejewska, B.; Randi, E.; Jędrzejewski, W.; Sidorovich, V.E.; Ostrander, E.A.; Wayne, R.K. Genome-Wide Signatures of Population Bottlenecks and Diversifying Selection in European Wolves. *Heredity* **2014**, *112*, 428–442. [CrossRef] [PubMed]

45. Boc, A.; Diallo, A.B.; Makarenkov, V. T-REX: A Web Server for Inferring, Validating and Visualizing Phylogenetic Trees and Networks. *Nucleic Acids Res.* **2012**, *40*, 573–579. [CrossRef]

46. The Chicken Quantitative Trait Locus (QTL) Database (Chicken QTLdb). Available online: https://www.animalgenome.org/cgi-bin/QTLdb/GG/index (accessed on 4 September 2021).

47. Wellek, S.; Ziegler, A. A Genotype-Based Approach to Assessing the Association between Single Nucleotide Polymorphisms. *Hum. Hered.* **2009**, *67*, 128–139. [CrossRef] [PubMed]

48. Barrett, J.C.; Fry, B.; Maller, J.; Daly, M.J. Haploview: Analysis and Visualization of LD and Haplotype Maps. *Bioinformatics* **2005**, *21*, 263–265. [CrossRef]

49. Genome Reference Consortium Chicken Build 6a. Available online: https://www.ncbi.nlm.nih.gov/assembly/GCF_000002315.6 (accessed on 4 September 2021).

50. Kudinov, A.A.; Dementieva, N.V.; Mitrofanova, O.V.; Stanishevskaya, O.I.; Fedorova, E.S.; Larkina, T.A.; Mishina, A.I.; Plemyashov, K.V.; Griffin, D.K.; Romanov, M.N. Genome-Wide Association Studies Targeting the Yield of Extraembryonic Fluid and Production Traits in Russian White Chickens. *BMC Genom.* **2019**, *20*, 270. [CrossRef] [PubMed]

51. Dementeva, N.V.; Larkina, T.A.; Mitrofanova, O.V.; Fedorova, E.S.; Pozdnyakova, T.E. Association of Single Nucleotide Polymorphism in LCORL Gene with the Productive Traits in Russian White Chicken Breed. *Ptitsevodstvo* **2019**, *5*, 14–17. [CrossRef]

52. FAO. *Phenotypic Characterization of Animal Genetic Resources*; FAO Animal Production and Health Guidelines No. 11; FAO: Rome, Italy, 2012.

53. Dorji, N.; Sunar, S.K. Short Communication: Morphometric Variations among Five Bhutanese Indigenous Chickens (*Gallus domesticus*). *J. Anim. Poult. Sci.* **2014**, *3*, 76–85.

54. Mel'nyk, Y.F.; Mykytyuk, D.M.; Bilous, O.V.; Kudryavs'ka, N.V.; Burkat, V.P.; Huzyev, I.V.; Podoba, B.Y.; Sharan, P.I.; Kovtun, S.I.; Platonova, N.P.; et al. *Program of Preservation of the Gene Pool of Main Types of Farm Animals in Ukraine for the Period Till 2015*; Aristey: Kyiv, Ukraine, 2009.

55. Negassa, D.; Melesse, A.; Banerjee, S. Phenotypic Characterization of Indigenous Chicken Populations in Southeastern Oromia Regional State of Ethiopia. *Anim. Genet. Resour. Inf.* **2014**, *55*, 101–113. [CrossRef]

56. Assefa, S.; Melesse, A.; Banerjee, S. Egg Production and Linear Body Measurement Traits of Local and Three Exotic Chicken Genotypes Reared under Two Agroecological Zones. *Int. J. Ecol. Ecosolut.* **2018**, *5*, 18–23.

57. Romanov, M.N. Study of Feathering Phenotypes in Hisex Brown Parent Lines and Their Hybrids—A Population-Genetic Analysis. In Proceedings of the 9th European Poultry Conference, Glasgow, UK, 7–12 August 1994; World's Poultry Science Association, UK Branch: Glasgow, UK; Volume 1, pp. 352–354.

58. Weigend, S.; Romanov, M.N. Current Strategies for the Assessment and Evaluation of Genetic Diversity in Chicken Resources. In *Abstracts and Proceedings of the XXI World's Poultry Congress, Montreal, QC, Canada, 20–24 August 2000*; CD-ROM; World's Poultry Science Association: Montreal, QC, Canada, 2000.

59. Otecko, N.O.; Ogali, I.; Ng'ang'a, S.I.; Mauki, D.H.; Ogada, S.; Moraa, G.K.; Lichoti, J.; Agwanda, B.; Peng, M.S.; Ommeh, S.C.; et al. Phenotypic and Morphometric Differentiation of Indigenous Chickens from Kenya and Other Tropical Countries Augments Perspectives for Genetic Resource Improvement and Conservation. *Poult. Sci.* **2019**, *98*, 2747–2755. [CrossRef]

60. Tyasi, T.L.; Mashiloane, K.; Mokoena, K. Comparison of Some Linear Body Measurement Traits of Local and Commercial Chicken Breeds of South Africa. *Sib. J. Life Sci. Agric.* **2021**, *13*, 134–143. [CrossRef]

61. Liu, R.; Sun, Y.; Zhao, G.; Wang, H.; Zheng, M.; Li, P.; Liu, L.; Wen, J. Identification of Loci and Genes for Growth Related Traits from a Genome-Wide Association Study in a Slow- × Fast-Growing Broiler Chicken Cross. *Genes Genom.* **2015**, *37*, 829–836. [CrossRef]

62. Han, Y.J.; Chen, Y.; Liu, Y.; Liu, X.L. Sequence Variants of the *LCORL* Gene and Its Association with Growth and Carcass Traits in Qinchuan Cattle in China. *J. Genet.* **2017**, *96*, 9–17. [CrossRef] [PubMed]

63. Tiley, G.P.; Pandey, A.; Kimball, R.T.; Braun, E.L.; Burleigh, J.G. Whole Genome Phylogeny of *Gallus*: Introgression and Data-Type Effects. *Avian Res.* **2020**, *11*, 1–15. [CrossRef]

64. Dementeva, N.V.; Romanov, M.N.; Kudinov, A.A.; Mitrofanova, O.V.; Stanishevskaya, O.I.; Terletsky, V.P.; Fedorova, E.S.; Nikitkina, E.V.; Plemyashov, K.V. The Study of the Structure of the Gene Pool of the Russian White Breed of Chickens by the Method of Genomic SNP-Scanning. *Selskokhoziaĭstvennaia Biol. (Agric. Biol.)* **2017**, *52*, 1166–1174. [CrossRef]

65. Sasaki, O.; Odawara, S.; Takahashi, H.; Nirasawa, K.; Oyamada, Y.; Yamamoto, R.; Ishii, K.; Nagamine, Y.; Takeda, H.; Kobayashi, E.; et al. Genetic Mapping of Quantitative Trait Loci Affecting Body Weight, Egg Character and Egg Production in F2 Intercross Chickens. *Anim. Genet.* **2004**, *35*, 188–194. [CrossRef]

66. Schreiweis, M.A.; Hester, P.Y.; Settar, P.; Moody, D.E. Identification of Quantitative Trait Loci Associated with Egg Quality, Egg Production, and Body Weight in an F2 Resource Population of Chickens. *Anim. Genet.* **2006**, *37*, 106–112. [CrossRef]

67. Guo, J.; Qu, L.; Dou, T.-C.; Shen, M.-M.; Hu, Y.-P.; Ma, M.; Wang, K.-H. Genome-Wide Association Study Provides Insights into the Genetic Architecture of Bone Size and Mass in Chickens. *Genome* **2020**, *63*, 133–143. [CrossRef]

68. Wolc, A.; Arango, J.; Settar, P.; Fulton, J.E.; O'Sullivan, N.P.; Preisinger, R.; Habier, D.; Fernando, R.; Garrick, D.J.; Hill, W.G.; et al. Genome-Wide Association Analysis and Genetic Architecture of Egg Weight and Egg Uniformity in Layer Chickens. *Anim. Genet.* **2012**, *43*, 87–96. [CrossRef]

69. Barkova, O.Y.; Smaragdov, M.G. Association of a Nonsynonymous Substitution in the Condensin NCAPG Gene with Traits of Eggs in Laying Hens. *Russ. J. Genet. Appl. Res.* **2016**, *6*, 804–808. [CrossRef]

70. Sun, C.; Qu, L.; Yi, G.; Yuan, J.; Duan, Z.; Shen, M.; Qu, L.; Xu, G.; Wang, K.; Yang, N. Genome-Wide Association Study Revealed a Promising Region and Candidate Genes for Eggshell Quality in an F2 Resource Population. *BMC Genom.* **2015**, *16*, 565. [CrossRef]

71. Nangsuay, A.; Ruangpanit, Y.; Meijerhof, R.; Attamangkune, S. Yolk Absorption and Embryo Development of Small and Large Eggs Originating from Young and Old Breeder Hens. *Poult. Sci.* **2011**, *90*, 2648–2655. [CrossRef]

72. Lourens, A.; van den Brand, H.; Meijerhof, R.; Kemp, B. Effect of Eggshell Temperature during Incubation on Embryo Development, Hatchability, and Posthatch Development. *Poult. Sci.* **2005**, *84*, 914–920. [CrossRef]

73. Gaevsky, N.A. *Acquaintance with Evolutionary Genetics*; Krasnoyarsk State University: Krasnoyarsk, Russia, 2002.

74. Dekkers, J. Commercial Application of Marker- and Gene-Assisted Selection in Livestock: Strategies and Lessons. *J. Anim. Sci.* **2004**, *82*, 313–328. [CrossRef]

75. Boichard, D.; Chung, H.; Dassonneville, R.; David, X.; Eggen, A.; Fritz, S.; Gietzen, K.; Hayes, B.; Lawley, C.; Sonstegard, T.; et al. Design of a Bovine Low-Density SNP Array Optimized for Imputation. *PLoS ONE* **2012**, *7*, e34130. [CrossRef] [PubMed]

76. Dementiva, N.V.; Kudinov, A.A.; Mitrofanova, O.V.; Mishina, A.I.; Smaragdov, M.G.; Yakovlev, A.F. WPSI-6 Chicken Resource Population as the Source of Study Genetic Improvement of Indigenous Breeds. *J. Anim. Sci.* **2018**, *96*, 513. [CrossRef]

77. Moiseyeva, I.G.; Sevastyanova, A.A.; Aleksandrov, A.V.; Vakhrameev, A.B.; Romanov, M.N.; Dmitriev, Y.I.; Semenova, S.K.; Sulimova, G.E. Orloff Chicken Breed: History, Current Status and Studies. *Izv. Timiryazev. S-Kh. Akad. (Proc. Timiryazev Agric. Acad.)* **2016**, (Issue 1), 78–96.

78. Moiseyeva, I.G.; Nikiforov, A.A.; Romanov, M.N.; Aleksandrov, A.V.; Moysyak, Y.V.; Semyenova, S.K. Origin, History, Genetics and Economic Traits of the Yurlov Crower Chicken Breed. In *Elektronnyi zhurnal (Electronic Journal)*; Laboratory of Animal Comparative Genetics, N.I. Vavilov Institute of General Genetics: Moscow, Russia, 2007. Available online: https://web.archive.org/web/20120210170800/http://www.lab-cga.ru/articles/Yurlovskaya/Yurlovskaya.htm (accessed on 9 July 2021).

79. Moiseyeva, I.G.; Romanov, M.N.; Aleksandrov, A.V.; Nikiforov, A.A.; Sevastyanova, A.A. Evolution and Genetic Diversity of Old Domestic Hen's Breed—Yurlovskaya Golosistaya: System Analysis of Variability Forms. *Izv. Timiryazev. S-Kh. Akad. (Proc. Timiryazev Agric. Acad.)* **2009**, (Issue 3), 132–147.

80. Romanov, M.N.; Bondarenko, Y.V. Introducing the Ukrainian Indigenous Poultry—The Poltava Chickens. *Fancy Fowl* **1994**, *14*(2), 8–9.

81. Moiseyeva, I.G.; Kovalenko, A.T.; Mosyakina, T.V.; Romanov, M.N.; Bondarenko, Y.V.; Kutnyuk, P.I.; Podstreshny, A.P.; Nikiforov, A.A.; Tkachik, T.E. Origin, History, Genetics and Economic Traits of the Poltava Chicken Breed. In *Elektronnyi zhurnal (Electronic Journal)*; Laboratory of Animal Comparative Genetics, N.I. Vavilov Institute of General Genetics: Moscow, Russia, 2006; Issue 4, Available online: https://web.archive.org/web/20120205195904/http://www.lab-cga.ru/articles/Jornal04/Statia2.htm (accessed on 9 July 2021).

82. Moiseyeva, I.G.; Romanov, M.N.; Kovalenko, A.T.; Mosyakina, T.V.; Bondarenko, Y.V.; Kutnyuk, P.I.; Podstreshny, A.P.; Nikiforov, A.A. The Poltava Chicken Breed of Ukraine: Its History, Characterization and Conservation. *Anim. Genet. Resour. Inf.* **2007**, *40*, 71–78. [CrossRef]

83. Kulibaba, R.; Tereshchenko, A. Transforming Growth Factor β1, Pituitary-Specific Transcriptional Factor 1 and Insulin-Like Growth Factor I Gene Polymorphisms in the Population of the Poltava Clay Chicken Breed: Association with Productive Traits. *Agric. Sci. Pract.* **2015**, *2*, 67–72. [CrossRef]

 *agriculture*

Article

Examination of the Usage of a New Beak-Abrasive Material in Different Laying Hen Genotypes (Preliminary Results)

Tamás Péter Farkas [1,*], Attila Orbán [2], Sándor Szász [1], András Rapai [1], Erik Garamvölgyi [1] and Zoltán Sütő [1]

[1] Institute of Animal Science, Hungarian University of Agriculture and Life Sciences, Guba Sándor Str. 40, H-7400 Kaposvár, Hungary; szasz.sandor@uni-mate.hu (S.S.); rapaiandras123@gmail.com (A.R.); garamvolgyi.erik@uni-mate.hu (E.G.); suto.zoltan@uni-mate.hu (Z.S.)

[2] Bábolna TETRA Ltd., Radnóti Miklós Str. 16, H-2943 Bábolna, Hungary; orbanattila@babolnatetra.com

* Correspondence: farkas.tamas.peter@uni-mate.hu; Tel.: +36-7-0775-4633

Abstract: The aim of the experiment was to investigate the use and effect of a new beak-abrasive material not yet examined on mortality of non-beak trimmed laying hens of different genotypes housed in an alternative pen. The study was performed on 636 females belonging to three genotypes of Bábolna TETRA Ltd. (a1 = commercial brown layer hybrid (C); a2 = purebred male line offspring group (maternal); a3 = purebfigure ed female line offspring group (paternal)). A total of 318 hens, i.e., 106 hens/genotype distributed in six pens (53 hens/pen), were evaluated. Cylindrical beak-abrasive blocks of 5.3–5.6 kg were suspended (0.1–0.4 mm diameter gravel, limestone grit, lime hydrate, and cement mixture) in six alternative pens. In six control pens without abrasive material, 318 hens, i.e., 106 hens/genotype (2 pens control group/genotype, i.e., C1 = commercial brown layer hybrid, C2 = purebred male line offspring group, C3 = purebred female line offspring group; 53 hens/pen;) were placed where there were no beak-abrasive materials. The rate of change in the weight of the beak-abrasive materials and the mortality rate were recorded daily. In the six pens equipped with beak-abrasive materials, infrared cameras were installed, and 24 h recordings were made. The number of individuals pecking the beak-abrasive material, the time and duration of dealing with the material were recorded. Data coming from one observation day are given. During the 13 experimental weeks of observation, the weight loss of beak-abrasives differed significantly in the different genotypes (a1 = 27.4%; a2 = 29.6%; a3 = 56.6%). During the only day analyzed, the hens from all the genotypes mostly stayed between 17:00 and 21:00 h in the littered scratching area where the beak-abrasive material was placed (a1 = 48.4%; a2 = 49.2%; a3 = 54.4%). In the case of each genotype, the rate of the hens dealing with beak-abrasives in the first two periods of the day was relatively low (0.2%–0.7%). Peaks of the activity were between 17:00 and 21:00 (a1 = 0.8%; a2 = 1.3%; a3 = 1.8%). The a3 dealt with the beak-abrasive materials to a significantly greater extent in the period from 13:00 to 17:00 (0.8%) and from 17:00 to 21:00 (1.8%) than the a1 (0.2% and 0.8%, respectively). Due to the use of the beak-abrasive materials, the mortality rate decreased the most in the genotypes that used them (a1 with beak-abrasive material 0.0% vs. C1 9.4%; a2 with beak-abrasive material 2.9% vs. C2 12.4%; a3 with beak-abrasive material) 15.4% vs. C3 5.7%). It can be concluded that the insertion of beak-abrasive materials increased the behavioral repertoire of hens, which is particularly beneficial from an animal welfare point of view. Further and longer-term research is needed to determine whether the insertion of the beak-abrasive material has a beneficial effect on the mortality data of the experimental groups through enrichment, either through physical abrasion of the beak or both.

Keywords: laying hen; non-caged; non-trimmed; beak abrasion; behavior; aggressiveness; mortality

Citation: Farkas, T.P.; Orbán, A.; Szász, S.; Rapai, A.; Garamvölgyi, E.; Sütő, Z. Examination of the Usage of a New Beak-Abrasive Material in Different Laying Hen Genotypes (Preliminary Results). *Agriculture* **2021**, *11*, 947. https://doi.org/10.3390/agriculture11100947

Academic Editor: Vincenzo Tufarelli

Received: 20 July 2021
Accepted: 24 September 2021
Published: 29 September 2021

Publisher's Note: MDPI stays neutral with regard to jurisdictional claims in published maps and institutional affiliations.

1. Introduction

The examination of the abrasive material is actual nowadays; that is why the alternative husbandry technologies are increasingly gaining prominence in European laying hen farming, in which the abandonment of beak trimming is encouraged [1]. In addition, a significant proportion of consumers want to eat eggs from a husbandry technology where the hens live in a larger area, in more stimulus-rich conditions, without beak trimming.

In general, those who advocate the preamble of alternative housing systems and the abandonment of beak trimming are referred to the behavior of wild hen species (junglefowl) in their natural environment. In its natural environment, one of the ancestors of domestic hens (*Gallus gallus domesticus*) is the red junglefowl (*Gallus gallus* or *Gallus ferrugineus*) searching for food, which is also assisted by the beak [2]. Their beaks constantly make contact with materials of different hardness in the environment while they search for food. This activity is carried out almost throughout the whole day, from which it follows that their beaks are subject to constant friction and wear.

This natural process does not apply in this form to layer hybrids kept in cages. They do not have to search for food, as it is at their disposal within a few meters, and they do not have the same substances in the cages as in nature for the wear of their beaks [3].

Laying hens in alternative housing systems usually live in large groups but in limited places and environments, and their lack of stimuli can lead to abnormal behaviors and cannibalism due to abandonment of the beak trimming.

Unfortunately, aggression and cannibalism may be higher in non-beak trimmed flocks, which may result in higher mortality rates, as well as higher rates of egg breakage and feed waste [4,5]. For this reason, beak trimming of hens is widely used [6,7], which solved these problems somewhat [8], both in alternative systems [9] and cage housing [10]. However, it should be mentioned that there are several negative effects of beak trimming. Beak shortening can cause tissue and nerve damage [11], which may result in abnormalities that may inhibit the hen from feathering [12–15] and feeding [16], furthermore may have a negative effect on body weight [17] and egg production. Moreover, the shortened beak is not fully capable of performing its function against ectoparasites [18–20]. These changes not only jeopardize the welfare of the hens but can also be a serious economic loss for the producers [21]. Partly because of these, beak trimming is now banned in Norway (1974), in Finland (1986), in Sweden (1988), and Denmark (2013) due to the pressure of animal welfare organizations, commercial interest groups, and certain sections of the consumers. Other countries, such as Germany (from 2017) and England (scheduled for 2025), have initiated voluntary agreements to phase out the practice of beak trimming. Belgium, Luxembourg, Switzerland have urged for stricter animal welfare laws. This practice was banned in Austria as early as 2013 and has been banned in The Netherlands since 1 September 2018 [1].

Based on these facts, it would be desirable to try other alternative solutions that facilitate beak wear and enrich the environment for laying hens, which can reduce the adverse effects of long sharp beaks on production. In addition, adequate environmental enrichment would also be important for laying hens, which can diversify the environment of the animals, thus broadening the behavioral repertoire and reducing the incidence of abnormal behaviors.

Without beak trimming, an excellent way to promote natural beak wear is inserting various beak-abrasive materials into the place where the animals are kept, which can simultaneously be an enriching element, an opportunity to reduce aggression and wear the beak.

Therefore, in our research, we aimed to investigate the use and effect of a new, hitherto unexamined beak-abrasive material on mortality and aimed to investigate the behavior and location during the different parts of the day for non-beak trimmed laying hens of different genotypes housed in an alternative pen.

2. Materials and Methods

The study was conducted at the Poultry Testing Station (Kaposvár, HU, Somogy) of the Kaposvár Campus Training and Experimental Plant of Hungarian University of Agriculture and Life Sciences, with a herd of three different genotypes of laying hens provided by Bábolna TETRA Ltd (Bábolna, Hungary). (a1 = commercial brown layer hybrid (K);

a2 = purebred male line offspring group (maternal); a3 = purebred female line offspring group (paternal)) (Figure 1). The set numbers are: N = 636; n = 212 hen/genotype; 53 hen/pen. The laying hens were not beak trimmed. The barn typically had a temperature of 15–18 °C and humidity between 65% and 70%. Lighting of 16 h (continuously between 5:00 and 21:00; 30 LUX, warm white) per day was used during the experimental period. Laying hens were allowed to consume ad libitum commercially available hen feed (175.1 g/kg of CP, 39.8 g/kg of CF, 11.50 MJ of ME, 34.2 g/kg of Ca, 5.4 g/kg of P, 8 mg/kg of Cu, 80 mg/kg of Zn, 50 mg/kg of Fe, 100 mg/kg of Mn, 1 mg/kg of I, 0.3 mg/kg of Se) from 2 suspended hand-filled feeders (trough length: 120 cm) and drinking water from a suspended open water drinker (trough length: 120 cm) in each pen (Figure 2).

Figure 1. Experimental stocks with three different genetic backgrounds (commercial hybrid (**left**), purebred offspring of maternal (in the **middle**) and paternal lines (**right**)).

Figure 2. Infrared camera installed in the pen; HD-quality recording; pen interior layout with suspended beak wear material.

A total of 53 nineteen-week-old pullets (1041 cm2/hen) were housed in the 12 alternative pens, each with a floor area of 5.52 m2. In each pen, 1/3 of the floor space consisted of a scratching area littered with wood shavings, while the remaining 2/3 of the floor space consisted of an elevated plastic grid floor (Figure 2).

There were provided 14 laying nests with artificial grass placed on two levels per pen for the hens (3.8 hens/nest). In front of the laying nest row, 2 perches per level helped approach the nests. Cylindrical beak-abrasive blocks of 5.3–5.6 kg (0.1–0.4 mm diameter gravel, a mixture of limestone grit, lime hydrate, and cement) were hung above the scratching area (Figure 1) in six pens (2 pens/genotype) identified as a1, a2 and a3 for genotype. The beak-abrasive blocks were provided to us by Bábolna TETRA Ltd(Bábolna, Hungary). and were manufactured by the company itself. We adjusted the height of the beak-abrasive blocks so that their center fell into the back height of the laying hens. For each observed pen, we recorded daily the rate of block weight change (loss) and hens mortality rate of the abrasive. As a control group, no beak-abrasives were placed in 6 pens (2 pens control group/genotype, i.e., C1, C2, C3).

Infrared cameras (GeoVision Target(Budapest, Hungary) H.265 4.0 Mpixel outdoor IP Eyeball dome camera) were installed over the 6 pens equipped with abrasive materials, and 24 h recordings were made on the test week using special software (GeoVision GV-NVR System) (Budapest, Hungary).

Flock monitoring and data collection began at week 24 of life after 5 weeks of adaptation and production. To date, a total of 13 study weeks have been evaluated. On each study day, the location and activity of the hens within the pen were recorded and observed every quarter hours (96 times per day). In our research presenting the preliminary results, we have so far evaluated one day in two repeats per genotype, which meant the analysis of a total of $(1 \times 3 \times 2 \times 96 =)$ 576 recordings.

Analyzing the camera footage, we calculated the time proportion of hens spent in the laying nests, perch, plastic grid floor, and scratching area. In addition, we also recorded the activity of the birds in the different compartments, i.e., differentiated between eating, drinking, scratching, resting-feathering behavior, and the use of abrasive material.

The weight change of all beak-abrasive blocks was measured daily. For the pen paternal), which produced the greatest weight loss, a complete lighting period, i.e., 16 h were observed without interruption. The number of individuals 'dealing' with the beak-abrasive materials, the date and duration of dealing with the substance were recorded. Dealing with the beak-abrasive material means: The hens touch only the blocks with their beak to feed or any other reason.

Based on this, three categories were distinguished:

1. The individual's attention is only drawn to one or two pecks, after which they continue to search for another activity;
2. The individual pecks more than two times, using the abrasive for 5–10 s;
3. The birds peck it several times and use it for at least 15 s.

The degree of abrasion of the beak-abrasive blocks, the location of the hens, and the distribution of different behaviors, as well as the mortality rate, were evaluated by the likelihood ratio test using the SPSS 10.0 software package.

3. Results and Discussion

3.1. The Degree of Weight Change of the Abrasive Materials

A statistically verifiable difference was observed in the extent of weight loss of the measured beak-abrasive materials between the examined genotypes (Table 1). The slightest weight loss was observed in the commercial hybrid; little more was measured in the maternal genotype. In contrast, for the paternal genotype, more than 56% of the abrasive material was depleted during the study period. In this period, the paternal genotype dealt the most with the beak-abrasive material and wore the largest proportion of it. This observation is strongly correlated with the results in Table 4, which show that this group had the lowest mortality.

Table 1. Extent of weight loss of beak-abrasive materials in the case of different genotypes over a 13-week study period (%).

	Commercial Hybrid a_1	Maternal a_2	Paternal a_3	Prob.
Extent of weight loss	27.42 [a]	29.60 [b]	56.61 [c]	<0.001

[a, b, c] indicate significant differences among the different genotypes ($p < 0.05$).

3.2. Location and Behavior of Laying Hens, Dealing with Abrasive Material

In order to explore in more detail the background of the laying hen's handling of beak-abrasive materials, we need to know the location of the animals in the pen at a given time of day because a higher degree of expression of the specific behavioral repertoires, characteristic for a given part of the pen, is expected when there is a greater presence of animals in it.

Table 2 shows the percentual distribution of laying hens between different parts of the laying hen house depending on genotype and time of day. For all the genotypes examined, it is clear that only in the last light period of the day, i.e., between 17:00 and 21:00, the laying hens were staying the most in the littered scratching area, where the beak abrasion material was also suspended. In the case of the paternal genotype, there were significantly more laying hens in the litter scratching area than on the grid floor, which is consistent with the data in Table 3 and Figures 3 and 4. The only interesting difference in terms of location data is that the maternal genotype was much more in the scratching area during the dark period than the other two genotypes.

Table 2. Percentual distribution of laying hens among different parts of the pen depending on genotype and time of day (%).

Periods of the Day	Distribution of Location Choice of Laying Hens, %				
	In the Nest	On the Perches	On the Plastic Grid Floor	Littered Scratching Area	Prob.
Commercial hybrid a_1					
5:00–9:00	5.4 [aD]	6.5 [aC2]	48.6 [cA]	39.5 [bB1]	<0.001
9:00–13:00	2.7 [aC2]	7.4 [bC3]	52.5 [dB1]	37.4 [cB]	<0.001
13:00–17:00	0.0 [aA1]	6.7 [bC3]	54.2 [dB2]	39.0 [cB1]	<0.001
17:00–21:00	0.7 [aB2]	3.9 [bB2]	47.0 [cA]	48.4 [cC1]	<0.001
21:00–5:00	1.9 [aC2]	2.8 [bA3]	87.7 [dC2]	7.5 [cA1]	<0.001
Prob.	<0.001	<0.001	<0.001	<0.001	-
Paternal a_2					
5:00–9:00	4.0 [bD]	2.4 [aC1]	47.8 [cB]	45.8 [cC2]	<0.001
9:00–13:00	1.5 [aC1]	1.9 [aBC1]	56.8 [cC2]	39.7 [bB]	<0.001
13:00–17:00	0.2 [aB2]	1.7 [bBC1]	50.4 [cB1]	47.7 [cC3]	<0.001
17:00–21:00	0.1 [aB1]	1.3 [bB1]	44.2 [cA]	54.4 [dD2]	<0.001
21:00–5:00	0.0 [aA1]	0.0 [aA1]	87.4 [cD2]	12.6 [bA2]	<0.001
Prob.	<0.001	<0.001	<0.001	<0.001	-
Maternal a_3					
5:00–9:00	5.1 [aC]	5.9 [aC2]	47.9 [cAB]	41.1 [bAB1]	<0.001
9:00–13:00	4.1 [aC3]	5.2 [aC2]	50.1 [cB1]	40.5 [bAB]	<0.001
13:00–17:00	1.6 [aAB3]	4.8 [bBC2]	50.2 [dB1]	43.4 [cB2]	<0.001
17:00–21:00	1.0 [aA2]	3.6 [bB2]	46.3 [cA]	49.2 [cC1]	<0.001
21:00–5:00	1.9 [aB2]	1.9 [aA2]	56.3 [cC1]	39.8 [bA3]	<0.001
Prob.	<0.001	<0.001	<0.001	<0.001	-

[a, b, c, d] indicate significant differences among the different parts of the pen ($p < 0.05$); [A, B, C, D] indicate significant differences among the different periods of the day ($p < 0.05$); [1, 2, 3] indicate significant differences among the different genotypes ($p < 0.05$).

Overall, the laying hens frequently used the littered scratching area and also scratched there because the hen's natural behavior and instinct include searching and scratching for food [22].

To determine the extent of dealing with beak-abrasive material, we gathered important information about the animals' total daily behavioral repertoire, and this way, we assessed the extent to which this activity occurs relative to other behaviors. Moreover, the daily change and rhythm of each form of behavior also have an effect on the development of the handling of the beak-abrasive materials, as this activity satisfies a "secondary" need for poultry, as eating, drinking, and laying eggs are basic needs and behavior.

For all the studied genotypes, a statistically verifiable difference was obtained between the different behaviors for all periods, and daily fluctuations were also observed between the periods (Table 3). Interestingly, compared to other behaviors, dealing with the beak-abrasive material represents a significantly lower rate at most times of the day. For all genotypes, the usage of beak-abrasives in the first two periods of the day showed relatively low levels.

Table 3. Distribution of different behaviors (%) in several parts of the pen depending on genotype and period of the day.

Periods of the Day	Staying at Nests	Staying on Perches	On the Plastic Grid Floor				On the Littered Scratching Area					Prob.
			Eating	Drinking	Resting, Feathering	Total	Eating	Scratching	Resting, Feathering	Usage of Beak-Abrasive Material	Total	
Commercial hybrid a₁												
5:00–9:00	5.4 [cdD]	6.5 [dC2]	4.3 [bcC1]	1.5 [bB]	42.7 [gB2]	48.6 [A]	3.5 [bB1]	15.4 [eBC2]	20.3 [fD1]	0.2 [aB]	39.5 [B1]	<0.001
9:00–13:00	2.7 [bC2]	7.4 [dC3]	10.6 [eD1]	4.0 [cC2]	37.9 [hA2]	52.5 [B1]	6.8 [dD2]	17.1 [gC2]	13.4 [fB1]	0.2 [aB]	37.4 [B]	<0.001
13:00–17:00	0.0 [aA1]	6.7 [dC2]	4.1 [cC2]	3.0 [cBC2]	47.1 [hC2]	54.2 [B1]	9.6 [eE2]	13.4 [fB]	15.9 [gC1]	0.2 [bB1]	39.0 [B1]	<0.001
17:00–21:00	0.7 [aB2]	3.9 [dB3]	1.4 [bcB1]	2.2 [cB]	43.3 [fBD]	47.0 [A]	5.0 [dC]	20.9 [eD2]	21.6 [eD1]	0.8 [abC1]	48.4 [C1]	<0.001
21:00–5:00	1.9 [bC2]	2.8 [cA3]	0.0 [aA]	0.0 [aA]	87.7 [eE2]	87.7 [C1]	0.0 [aA]	0.0 [aA]	7.5 [dA1]	0.0 [aA]	7.5 [A1]	<0.001
Prob.	<0.001	<0.001	<0.001	<0.001	<0.001	<0.001	<0.001	<0.001	<0.001	<0.001	<0.001	-
Paternal a₂												
5:00–9:00	4.0 [cD]	2.4 [bC1]	8.0 [dC2]	1.0 [aB]	38.7 [gB1]	47.8 [B]	3.4 [bcB1]	18.3 [eC3]	23.2 [fC2]	0.7 [aC]	45.8 [C2]	<0.001
9:00–13:00	1.5 [bC1]	1.9 [bC1]	18.8 [eD2]	3.2 [cD2]	34.8 [fA1]	56.8 [C2]	12.0 [dD3]	13.7 [dB1]	13.8 [dA1]	0.2 [aB]	39.7 [B]	<0.001
13:00–17:00	0.2 [aB2]	1.7 [cC1]	2.3 [cB1]	1.8 [cBC1]	46.4 [fC2]	50.4 [B1]	15.0 [dE3]	13.9 [dB]	18.0 [eB1]	0.8 [bC2]	47.7 [C3]	<0.001
17:00–21:00	0.1 [aB1]	1.3 [bB1]	2.3 [cB2]	1.9 [bcC]	40.0 [fB]	44.2 [A]	5.2 [dC]	23.4 [eD2]	24.0 [eC12]	1.8 [bcD2]	54.4 [D2]	<0.001
21:00–5:00	0.0 [aA1]	0.0 [aA1]	0.0 [aA]	0.0 [aA]	87.4 [cD]	87.4 [D2]	0.0 [aA]	0.0 [aA]	12.6 [bA2]	0.0 [aA]	12.6 [A2]	<0.001
Prob.	<0.001	<0.001	<0.001	<0.001	<0.001	<0.001	<0.001	<0.001	<0.001	<0.001	<0.001	-
Maternal a₃												
5:00–9:00	5.1 [bC]	5.9 [cC2]	3.9 [bC1]	0.7 [aB]	43.3 [eB2]	47.9 [AB]	6.0 [bC2]	9.8 [cB1]	24.9 [dB2]	0.4 [aB]	41.1 [AB1]	<0.001
9:00–13:00	4.1 [cC3]	5.2 [cdC2]	10.7 [eE1]	1.5 [bC1]	37.9 [hA2]	50.1 [B1]	3.5 [cB1]	14.9 [fD12]	21.9 [gA2]	0.2 [aB]	40.5 [AB]	<0.001
13:00–17:00	1.6 [bAB3]	4.8 [cC2]	5.5 [cD2]	1.9 [bC1]	42.8 [fB1]	50.2 [B1]	6.1 [cC1]	12.3 [dC]	24.6 [eAB2]	0.4 [aB12]	43.4 [B2]	<0.001
17:00–21:00	1.0 [aA2]	3.6 [cB2]	2.7 [bcB2]	2.1 [abC]	41.5 [gB2]	46.3 [A]	5.9 [dC]	16.8 [eD1]	25.1 [fB2]	1.3 [aC12]	49.2 [C1]	<0.001
21:00–5:00	1.9 [bB2]	1.9 [bA2]	0.0 [aA]	0.0 [aA]	56.3 [dC1]	56.3 [C1]	0.0 [aA]	0.0 [aA]	39.8 [cC3]	0.0 [aA]	39.8 [A3]	<0.001
Prob.	<0.001	<0.001	<0.001	<0.001	<0.001	<0.001	<0.001	<0.001	<0.001	<0.001	<0.001	-

a, b, c, d, e, f, g indicate significant differences among the different behavioral forms ($p < 0.05$); A, B, C, D, E indicate significant differences among the different periods of the day ($p < 0.05$); 1, 2, 3 indicate significant differences among the different genotypes ($p < 0.05$).

At that time, only 0.2–0.7% of the flock dealt with the beak-abrasive material on average. The peak of dealing with this fell on the time of the day from 17:00 to 21:00. From the data, it can be clearly seen that compared to the morning hours, the rate of usage of beak-abrasives increased four times in the case of the commercial hybrid, two and a half times in the case of the paternal genotype, and more than three times in the case of the maternal genotype. In the case of the paternal offspring group, the extent of dealing with the beak abrasion material had already exceeded the extent of staying in the laying nest during the last light hours of the day.

The paternal genotype dealt with the beak-abrasive material to a significantly greater extent during the days from 13:00 to 17:00 and from 17:00 to 21:00 than the commercial genotype, which also coincided with the data of the weight loss of the beak-abrasive materials. It can also be seen from Table 3 that the proportion of eating behavior was the lowest in the period from 17:00 to 21:00, so that the proportion of laying hens engaged in scratching and using beak-abrasive materials was increased during this period.

The use of beak-abrasive material showed daily fluctuations for all genotypes. Researching the background of the phenomenon, we examined in more depth the development of the use of the abrasive in the pen where the greatest weight loss (Paternal) was experienced since in this pen we had the opportunity to observe this activity the most (Figure 3).

The deal with the beak-abrasive material was observed first in the morning. It can be said that the abrasive block aroused the interest of mainly one individual at a time, less often two, and even less often three or four hens.

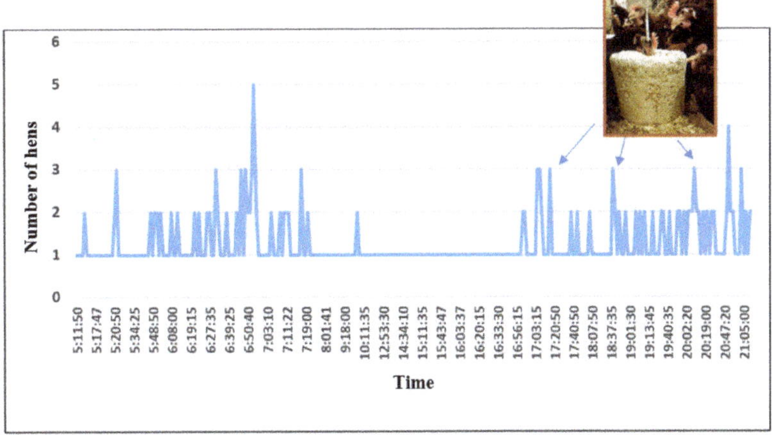

Figure 3. Changes in the number of laying hens using beak-abrasives at different times of the day for the paternal genotype.

The duration of beak-abrasive material showed daily fluctuations also (Figure 4). In the morning, before 8:00 a.m., the feeders are usually empty, in which case the laying hens are hungry and, in addition to scratching, the beak-abrasive materials are also pecked. According to the work schedule, feeding takes place around 8:00 a.m.; when the animals concentrate on feed intake with rather high intensity, their activity is mainly determined by this. In this case, the number of individuals dealing with the abrasive material decreases significantly, and next time their number increased only after 17:00.

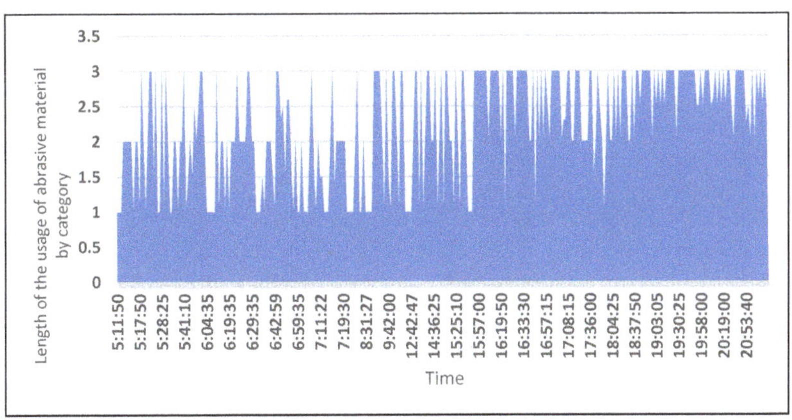

Figure 4. Changes in the duration of dealing with abrasive material at different times of the day in the case of paternal genotype.

Mainly in the early morning hours before feeding, categories 1, 2 were typical, i.e., the hens showed a lower level of interest in the abrasive, which means that they pecked up to one or two and used them for about 10 s. Similar to Figure 3, not only did the number of individuals dealing with abrasive material increase after 16:00, but also the duration of treatment with abrasive material. At that time, the hens used abrasive materials more frequently and longer. In the evening hours, i.e., after 19:00, category 3 was predominantly observed.

3.3. Mortality of Laying Hens Depending on the Use of the Beak Abrasion

In terms of mortality, a significant difference was observed between the genotypes, both in pens with and without beak abrasion material (Table 4). In the case of the commercial hybrid and the paternal genotype, mortality was significantly lower in pens using beak-abrasives, presumably because this enriching element was used by the animals, which could reduce aggression toward each other and cannibalism, and thus, the resulting mortality was also reduced. This enrichment had a positive effect on laying hens, as noted by some researchers [23,24]. We came to similar results as another study [25], where, as an environmental enrichment, brightly colored bottles, balls, and rattles were placed among the laying hens, the hens began pecking them with their beaks, an activity that reduced the stress accumulated in the animals and helped to reduce their excess energy, aggression toward each other.

Table 4. The change in the mortality rate in pens with or without abrasive materials in the case of different genotypes during the 13-week study period (%).

	Commercial Hybrid a_1	Maternal a_2	Paternal a_3	Prob.	Total
Pen with abrasive material	0.00 [a]	15.38 [b]	2.91 [a]	<0.001	6.07
Pen without abrasive material	9.43	5.71	12.38	0.246	9.18
Prob.	0.001	0.023	0.011	-	0.142
Total	4.72	10.53	7.69	0.081	7.63

[a, b] indicate significant differences among the different genotypes ($p < 0.05$).

In contrast, the maternal genotype had three times the mortality rate in pens equipped with abrasive material than in the control groups (C2). The commercial hybrid had no mortality in pens with beak-abrasive material, and the paternal genotype had only a quarter of what was experienced in the control flock (C3).

4. Conclusions

Based on the results of the study, it can be concluded that the insertion of beak-abrasive materials increased the behavioral repertoire of hens, which is particularly beneficial from an animal welfare point of view. This enriching element has been used extensively by animals, probably because the beak-abrasive material has aroused the interest of the birds, occupied them while consuming its ingredients. During the study period, the weight loss of the abrasive material was the highest in the paternal genotype (56.6%), which was significantly higher than in the maternal line (29.6%) and the commercial hybrid (27.4%). Presumably, the reason for this is to be found in the different genetic backgrounds of the studied herd, in their different blood temperatures. The scratching area, where the beak-abrasive material was suspended, was most used by laying hens in the time interval from 17:00 to 21:00. In the period from 17:00 to 21:00, the hens used the beak-abrasive materials the most, because the proportion of other behaviors has already decreased by this period, as the laying of eggs were mainly in the morning, the feed runs out of feeders, that is why the eating and drinking are greatly reduced. The duration of the dealing with the abrasive material showed fluctuations during the day. In the morning, there was less interest in the abrasive before feeding, while after 16:00, the animals were more and more often used the abrasive material, even for periods longer than 15 s. Based on the phenomenon, it can be stated that within a day, over time, the length of use of the beak wearer increased. Insertion of the beak-abrasive material reduced mortality in genotypes that used them to a higher extent (commercial hybrid, paternal offspring group) of beak-abrasives because they produced significantly fewer mortality than their counterparts in the control group. Based on the preliminary results, it is clear that a full evaluation of several study days is needed to gain further and deeper conclusions. Consideration should be given to the use of an object of a different composition or enrichment that provides a better understanding of the background of poultry behavior during alternative housing of hens, as full open-air housing solutions are not available due to the risk of avian influenza and the cost of free-range. Further and longer-term research is needed to determine whether the insertion of the beak-abrasive block into the scratching area has a beneficial effect on the mortality data of the experimental groups through enrichment, either through physical abrasion of the beak or both.

Author Contributions: Data curation, A.R.; Formal analysis, T.P.F. and Z.S.; Investigation, T.P.F., E.G. and Z.S.; Resources, A.O.; Supervision, S.S. and Z.S.; Validation, T.P.F.; Visualization, T.P.F.; Writing—original draft, T.P.F. and Z.S.; Writing—review and editing, T.P.F. and S.S. All authors have read and agreed to the published version of the manuscript.

Funding: This research was funded by the Hungarian State, grant number 2018-1.3.1-VKE-2018-00042. The publication is supported by the EFOP-3.6.3-VEKOP-16-2017-00008 project.

Institutional Review Board Statement: All animals were handled according to the principles stated in the Directive 2010/63/EU regarding the protection of animals used for experimental and other scientific purposes [26].

Informed Consent Statement: Not applicable.

Data Availability Statement: The data presented in this study are available on request from the corresponding author. The data are not publicly available due to sensitive data about some pedigree lines.

Conflicts of Interest: A.O. is the employee of the Bábolna TETRA Ltd., which is the owner of the examined genotypes and the provider of the beak-abrasive materials. The remaining authors declare no conflict of interest. The funders had no role in the design of the study, in the collection, analyses, or interpretation of data, in the writing of the manuscript, or in the decision to publish the results.

References

1. Zomborszky, Z.; Budai, Z.; Milisits, G.; Szász, S.; Farkas, T.P.; Ujváriné, J.; Horn, P.; Sütő, Z. Eltérő genetikai hátterű, tojó típusú, csőrkurtítatlan jérce állomány nevelés alatti és tojóházi kiesésének elemző vizsgálata, különös tekintettel az agresszióra. In *XXI. Kaposvári Baromfitenyésztési Szimpózium, Kaposvár, Hungary, 29 September 2018*; Sütő, Z., Ed.; Faculty of Agricultural and Environmental Sciences, Kaposvár University: Kaposvár, Hungary, 2018; pp. 78–87.
2. Banks, E.M. Social organization in Red Jungle fowl hens (*Gallus gallus* subsp.). *Ecology* **1956**, *37*, 240–248. [CrossRef]
3. Horn, P. *A Baromfitenyésztők Kézikönyve*; Mezőgazdasági Kiadó: Budapest, Hungary, 1981; pp. 137–189.
4. Niebuhr, K.; Zaludik, K.; Gruber, B.; Thenmaier, I.; Lugmair, A.; Troxler, J. Epidemiologische Unter-suchungen zum Auftreten von Kannibalismus und Federpicken in alternativen Legehennenhaltungen in Österreich. In *Endbericht Forschungsprojekt*; Nr. 1313 ITT; University Vienna: Vienna, Austria, 2006. [CrossRef]
5. Rodenburg, T.B.; Van Krimpen, M.M.; De Jong, I.C.; De Haas, E.N.; Kops, M.S.; Riedsrta, B.J.; Nicol, C.J. The prevention and control of feather pecking in laying hens: Identifying the underlying principles. *World's Poult. Sci. J.* **2013**, *69*, 361–373. [CrossRef]
6. Damme, K. Der Einfluß der Herkunft und des Schnabelkupierens auf die Leistung, Befiederung und Nestakzeptanz verschiedener Weißlegehybriden in Bodenhaltung. *Arch. Für Geflügelkunde* **1999**, *63*, 93–99.
7. Nicol, C. Feather pecking and cannibalism. *Adv. Poult. Welf.* **2018**, 175–197. [CrossRef]
8. Sepeur, S.; Spindler, B.; Schulze-Bisping, M.; Habig, C.; Andersson, R.; Beyerbach, M.; Kemper, N. Comparison of plumage condition of laying hens with intact and trimmed beaks kept on commercial farms. *Arch Geflugelkd* **2015**, *79*. [CrossRef]
9. Weeks, C.A.; Lambton, S.L.; Williams, A.G. Implications for Welfare, Productivity and Sustainability of the Variation in Reported Levels of Mortality for Laying Hen Flocks Kept in Different Housing Systems: A Meta-Analysis of Ten Studies. *PLoS ONE* **2016**, *11*, e0146394. [CrossRef] [PubMed]
10. Guesdon, V.; Ahmed, A.; Mallet, S.; Faure, J.; Nys, Y. Effects of beak trimming and cage design on laying hen performance and egg quality. *Br. Poult. Sci.* **2006**, *47*, 1–12. [CrossRef] [PubMed]
11. Cheng, H. Morphopathological changes and pain in beak trimmed laying hens. *World's Poult. Sci. J.* **2006**, *62*, 41–52. [CrossRef]
12. Breward, J.; Gentle, M.J. Neuroma formation and abnormal afferent nerve discharges after partial beak amputation (beak trimming) in poultry. *Cell. Mol. Life Sci.* **1985**, *41*, 1132–1134. [CrossRef]
13. Duncan, I.J.H.; Slee, G.S.; Seawright, E.; Breward, J. Behavioural consequences of partial beak amputation (beak trimming) in poultry. *Br. Poult. Sci.* **1989**, *30*, 479–488. [CrossRef]
14. Gentle, M.J. Cutaneous sensory afferents recorded from the nervus intramandibularis ofGallus gallus vardomesticus. *J. Comp. Physiol. A* **1989**, *164*, 763–774. [CrossRef] [PubMed]
15. Bestman, M.; Koene, P.; Wagenaar, J.-P. Influence of farm factors on the occurrence of feather pecking in organic reared hens and their predictability for feather pecking in the laying period. *Appl. Anim. Behav. Sci.* **2009**, *121*, 120–125. [CrossRef]
16. Dennis, R.L.; Cheng, H.W. Effects of Beak Trimming on Pecking Force. *Int. J. Poult. Sci.* **2010**, *9*, 863–866. [CrossRef]

17. Angevaare, M.J.; Prins, S.; van der Staay, F.J.; Nordquist, R.E. The effect of maternal care and infrared beak trimming on development, performance and behavior of Silver Nick hens. *Appl. Anim. Behav. Sci.* **2012**, *140*, 70–84. [CrossRef]
18. Brown, N.S. The Effect of Host Beak Condition on the Size of Menacanthus Stramineus Populations of Domestic Chickens. *Poult. Sci.* **1972**, *51*, 162–164. [CrossRef]
19. Mullens, B.A.; Chen, B.L.; Owen, J. Beak condition and cage density determine abundance and spatial distribution of northern fowl mites, Ornithonyssus sylviarum, and chicken body lice, Menacanthus stramineus, on caged laying hens. *Poult. Sci.* **2010**, *89*, 2565–2572. [CrossRef]
20. Chen, B.L.; Haith, K.L.; Mullens, B.A. Beak condition drives abundance and grooming-mediated competitive asymmetry in a poultry ectoparasite community. *Parasitology* **2011**, *138*, 748–757. [CrossRef]
21. Carruthers, C.; Gabrush, T.; Schwean-Lardner, K.; Knezacek, T.D.; Classen, H.L.; Bennett, C. On-farm survey of beak characteristics in White Leghorns as a result of hot blade trimming or infrared beak treatment. *J. Appl. Poult. Res.* **2012**, *21*, 645–650. [CrossRef]
22. Blokhuis, H.J.; Van Der Haar, J.W. Effects of pecking incentives during rearing on feather pecking of laying hens. *Br. Poult. Sci.* **1992**, *33*, 17–24. [CrossRef]
23. Jones, R.B. Effects of early environmental enrichment upon open-field behavior and timidity in the domestic chick. *Dev. Psychobiol.* **1982**, *15*, 105–111. [CrossRef]
24. Huber-Eicher, B.; Sebö, F. The prevalence of feather pecking and development in commercial flocks of laying hens. *Appl. Anim. Behav. Sci.* **2001**, *74*, 223–231. [CrossRef]
25. Reed, H.; Wilkins, L.; Austin, S.; Gregory, N. The effect of environmental enrichment during rearing on fear reactions and depopulation trauma in adult caged hens. *Appl. Anim. Behav. Sci.* **1993**, *36*, 39–46. [CrossRef]
26. European Union. Directive (EC) no. 2010/63/EU of the European Parliament and of the Council of 22 September 2010 on the protection of animals used for scientific purposes. *Off. J. Eur. Union* **2010**, *276*, 33–79.

Case Report

Effect of Feeding a High Calcium: Phosphorus Ratio, Phosphorous Deficient Diet on Hypophosphatemic Rickets Onset in Broilers

Lei Xu [1], **Ning Li** [1], **Yuhua Z. Farnell** [2], **Xiaoli Wan** [1], **Haiming Yang** [1,*], **Xiangqian Zhong** [3] and **Morgan B. Farnell** [2,*]

[1] College of Animal Science and Technology, Yangzhou University, Yangzhou 225009, China; xlei@yzu.edu.cn (L.X.); Aning0715@gmail.com (N.L.); wanxl@yzu.edu.cn (X.W.)

[2] Department of Poultry Science, Texas A&M AgriLife Research, Texas A&M University, College Station, TX 77843, USA; yfarnell@tamu.edu

[3] Suqian Animal Husbandry and Veterinary Station, Suqian 223800, China; yhmdlp@sina.com

[*] Correspondence: hmyang@yzu.edu.cn (H.Y.); mfarnell@tamu.edu (M.B.F.);
Tel.: +86-514-8797-9045 (H.Y.); +1-979-847-7363 (M.B.F.)

Abstract: Recently, a P-deficient diet caused rickets in commercial chicks within three days. This study aimed to investigate the duration of onset of rickets in chicks. Data were collected from 3–11 day old chicks raised on 88 commercial farms. Male day-old Arbor Acres Plus broilers (n = 450) were studied in three trials, with three to four treatments each. Each treatment used one of the following crumbled feeds: control feed (calcium (Ca): phosphorus (P)-1.41), slightly high Ca:P feed (SHCa:P, Ca:P-2.69), high Ca:P ratio, P deficient feed (HCa:P, Ca:P-3.08), and HCa:P feed plus 1.5% dicalcium phosphate (HCa:P + DP). Each treatment had three replicates with 15 birds each. Rickets was induced by HCa:P, and cured by HCa:P + DP, confirmed by gross anatomy, gait score, serum P concentration and growth performance. Lameness was not found in control groups, whereas, observed in the HCa:P groups as early as day 2.7 on commercial farms and day 3 in experimental farm. Serum P was reduced in HCa:P ($p < 0.01$). Bodyweight and feed intake started decreasing at day 3 on commercial farms and in all trials ($p < 0.01$). The duration of onset of hypophosphatemic rickets in broiler chicks fed HCa:P crumbled feed is approximately three days.

Keywords: lameness; broiler chicken; poultry; hypophosphatemic rickets; Ca:P ratio; phosphorus deficiency; phosphorus requirement; duration of onset; animal nutrition; animal husbandry

Citation: Xu, L.; Li, N.; Farnell, Y.Z.; Wan, X.; Yang, H.; Zhong, X.; Farnell, M.B. Effect of Feeding a High Calcium: Phosphorus Ratio, Phosphorous Deficient Diet on Hypophosphatemic Rickets Onset in Broilers. *Agriculture* **2021**, *11*, 955. https://doi.org/10.3390/agriculture11100955

Academic Editor: István Komlósi

Received: 21 August 2021
Accepted: 26 September 2021
Published: 1 October 2021

Publisher's Note: MDPI stays neutral with regard to jurisdictional claims in published maps and institutional affiliations.

1. Introduction

The word "rickets", derived from the old English "wrikken", to twist or bend, refers to a condition of impaired mineralization of growing bones which ultimately results in their bowing and twisting [1]. In humans, rickets is most commonly attributed to vitamin D deficiency [2]. In growing poultry, rickets most often results from feed mixing errors resulting in changes in dietary vitamin D, calcium, or phosphorus (P) concentrations [3]. Calcium (Ca) and P deficiency are well-known factors affecting bone strength, bone mineralization, and growth performance [4–11]. Vitamin D plays a role in maintaining the homeostatic balance between calcium and phosphorus in birds [12], affecting bone growth and mineral deposition [13]. In addition to absolute deficiencies of Ca or P, an improper Ca:P ratio can affect growth performance [14,15], the incidence of leg abnormalities [14], gait scores, and prevalence of hock burn and footpad lesions in birds [14–17]. When feed lacks inorganic P, rickets could occur two weeks earlier in birds fed a high Ca:P diet compared to those fed a low Ca:P diet [7]. The reason may be that a high dietary Ca:P ratio decreases the bioavailability of Ca and P to sub-optimal levels [18], thereby causing a deficiency of one or both of these elements.

Hypophosphatemic rickets was reported in 33 northern China poultry farms in broiler chicks ranging from three to 11 day-olds (n ≈ 410,400). Lameness (an inability to stand, walk or run freely due to physical problems, poor gait score) is a typical characteristic of rickets. In our investigation on this case, farmers reported that the prevalence of lameness started between two to four days after accidental commencement of feeding of nutritionally deficient rations. Through our survey, we observed that the incidence of lameness rose rapidly to 70% within six days but was ameliorated within two days in birds switched to a complete ration provided by another company (data for 70,800 chicks on 6 farms). The duration of onset of rickets observed was much sooner than in previous reports which mentioned onsets at seven [7,8], seven to ten [9], and 18 days [8]. Few reports concerning an early duration of onset of rickets caused by P deficiency in poultry in recent years could be found. Thus, the conception of a slow onset of hypophosphatemic rickets in poultry may hinder the timely diagnosis of this syndrome by veterinarians and nutritionists. As early identification of a nutrient deficiency offers an opportunity to ameliorate the ill effects of a feed processing error, the early diagnosis and immediate treatment of hypophosphatemic rickets is of utmost importance [19], with irreparable economic losses resulting if this is not possible. We hypothesized that the duration of onset of hypophosphatemic rickets might be around three days after commencement of feeding the chicks a high Ca:P ratio, P deficient (HCa:P) crumbled feed.

We aimed to evaluate rickets' duration of onset in broiler chicks that fed HCa:P crumbled feed via a field survey and three trials, based on our observations concerning gross anatomy, serum P concentration, the incidence of lameness and growth performance. The purpose of this study is to provide information enabling more efficient diagnosis and management of early-onset hypophosphatemic rickets in chicks by nutritionists, quality supervisors, flock supervisors, and veterinarians.

2. Materials and Methods

2.1. Ethical Statement

The procedures followed for the field research and all trials were approved by the Institutional Animal Care and Use Committee of Yangzhou University (ethical protocol code: YZUDWSY 2017-05-09.03) and conducted according to the relevant animal welfare regulations.

2.2. Field Research

We surveyed 88 farms whose chicks had been fed feed from the same mill. Thirty-three farms (problem farms) bought feed produced on the same day and had lameness issues in their bird flocks, whereas 55 farms which bought feed produced on other dates did not have such problems. The scale of these farms varied from 3000 to 12,000 birds per flock, in total including approximately 410,400 fast-growing white feather broilers (Arbor Acres Plus (AA+) or Cobb 500 broilers), male and female sourced from seven hatcheries. All birds had been raised on plastic mesh flooring, 50–80 cm over the ground, at an approximate density of 30–35 birds/m^2 during the first week, and 20–30 birds/m^2 during the second week. Air temperature outside of the chicken houses varied between 22–35 °C, with diurnal temperature ranges of 6–11 °C during our survey in summer. Internal temperatures ranged between 30.2–34.5 °C for chicks aged between 11 and three days old, and humidity ranged between 70–84% in various chicken houses on the days on which we collected field data. Birds had free access to feed and water. We assessed gait scores, lameness incidence, body weights, feed intakes (FI) record, and gross anatomy at these commercial farms. The history of drug use was reported by farmers.

2.3. Trials

2.3.1. Feeds and Treatments

Three farms were selected from the surveyed farms, as representative of healthy, slightly sick, and severely sick chicks, respectively. Thirty bags (40.0 kg each) of crumbled feed were collected from each of these farms. The feeds were classified into the following three categories, according to their nutrient levels and impact on chick's gait score: control feed, which did not cause any lameness in chicks; HCa:P feed, which caused serious lameness in chicks and exhibited a high Ca:P ratio and P deficiency; slightly high Ca:P (SHCa:P) feed, which caused slight lameness in chicks and exhibited a slightly high Ca:P ratio and P deficiency. All the feeds were composed according to the formula and nutrient levels of Table 1. In addition, 1.5% of dicalcium phosphate (DP, $CaHPO_4 \cdot 2H_2O$) was added to 100 kg of HCa:P feed to comprise the fourth (HCa:P + DP) feed. Due to a processing error, the Ca:P ratios, Ca and P concentrations were different among treatments. These values were shown in Table 2.

Table 1. Ingredients and nutrient levels of the commercial feed and the control feed for all trials.

Ingredients and Specifications [1]	Composition	Nutrient Composition	Percent
Corn	55.00	Metabolic energy (MJ/kg)	12.28
Soybean meal(CP, 47%)	25.30	CP	21.00
Wheat flour	6.00	Lysine	1.27
Peanut meal (CP, 46%)	2.00	Methionine	0.52
Dried distilled grain with soubles (DDGS)	5.00	Available total sulfur-containing amino acids (ATSAA)	0.80
Dicalcium phosphate (DP)	1.40	Available lysine	1.15
Lime stone (coarse)	1.06	Available methionine	0.47
Feather powder (CP, 80%)	1.00	Available threonine	0.77
Duck oil	0.68	Available tryptophan	0.19
Lysine (65%)	0.60	Calcium (Ca)	0.85
Threonine (50%)	0.30	Total phosphourus	0.62
Methionine (88%)	0.26	Available phosphourus (AP)	0.38
Salt	0.20	Ca/AP	2.24
Sodium bicarbonate	0.10	Salt	0.23
Choline chloride (75%)	0.10	Sodium	0.14
Premix [2]	1.00	Sodium + potassium-chloride (mEq/kg)	219.5

[1] CP, crude protein. [2] Premix supplied the following nutrients for each kg of feed: vitamin A, 1, 300 IU; vitamin D,4500 IU; vitamin E, 80 IU; vitamin K, 4 mg; vitamin B_1, 4 mg; vitamin B_2, 9 mg; vitamin B_6, 6 mg; vitamin B_{12}, 0.02 mg; niacin,60 mg; pantothenic acid, 19 mg; folic acid, 2.5 mg; biotin 0.30 mg; Fe, 20 mg; Cu, 16 mg; Mn, 120 mg; Zn, 110 mg; I, 1.25 mg; Se, 0.30 mg. Minerals were supplied in the forms of ferrous sulfate, copper sulfate, manganese sulfate, zinc sulfate, potassium iodide, and sodium selenite, respectively, for Fe, Cu, Mn, Zn, I, and Se. Values were calculated values.

Table 2. Calcium (Ca) and phosphorus (P) levels for each treatment in trials one, two and three.

Diets [1]	Nutrients Values (Air-Dry Basis)					Nutrients Values (Dry Matter)				Ca:P	
	Moisture, %	Ca, %	TP [2], %	Phytate P,%	Available P, %	Ca, %	TP, %	Phytate P,%	NPP [2]	Ca: TP	Ca: NPP
Control	11.05	0.87	0.62	0.18	0.44	0.98	0.69	0.20	0.49	1.41	2.00
SHCa:P	10.05	1.13	0.42	0.18	0.24	1.26	0.47	0.20	0.27	2.69	4.71
HCa:P	9.96	1.11	0.36	0.18	0.18	1.23	0.40	0.20	0.20	3.08	6.17
HCa:P + DP	10.27	1.29	0.61	0.18	0.43	1.44	0.68	0.20	0.48	2.11	3.00

[1] Control, Ca and P-balanced feed which did not cause any lameness in chicks. SHCa:P, slightly high Ca:P ratio and P-deficient crumbled feed from farms with slight chick lameness issues; HCa:P, high Ca:P ratio and P-deficient crumbled feed collected from farms with severe chick lameness issues. HCa:P + DP, feed made by adding 1.5% dicalcium phosphate (DP) powder to HCa:P crumbled feed. All the feeds were composed with reference to the same formula, with phytase at 500 FTU/kg feed (air-dry basis). [2] TP: total P; NPP: non-phytate P. The NPP and Ca:P ratios were calculated values, other values were the means of three measurements (n = 3).

As AA+ broiler was more popular in this area, we selected AA+ broiler as the experimental object. Healthy male AA+ broilers from the day of hatch (135 in trials one and two, 180 in trial three) with similar body weight (BW, 41.88 ± 1.37 g) were randomly divided into three treatments, with three replicates per treatment and 15 chicks per replicate. The birds in trial one were fed a control diet, HCa:P diet or HCa:P + DP diet. One week later, birds in trial two were fed with a control diet, SHCa:P diet, or HCa:P diet. After another one week, the treatments used in both earlier trials were repeated in trial three, which included groups of birds fed control diet, HCa:P diet, SHCa:P diet, and HCa:P + DP diet.

As lameness incidence had reached over 80% by day 7 of trial one, the birds were kept only until day 7 in trials two and three.

2.3.2. Animals and Management

Chickens in all three trials were raised under the same room, with the same equipment and environmental controls, and following the AA+ broiler management guide [20]. Broiler chicks were raised in cages of dimensions $0.8 \times 0.6 \times 0.6$ m (31 birds per m^2). Water and feed were provided ad libitum unless otherwise stated. The temperature inside the house was 33 °C on the first day and was reduced by 2 °C per week, with humidity ranging from 60–70%.

2.3.3. Measurements

• Determination of Ca, P, and non-phytate P (NPP) in feeds

Control (n = 19), SHCa:P (n = 10), and HCa:P (n = 11) feed samples were randomly collected from 19 out of 55 normal farms and 21 out of 33 trouble farms. Limestone and DP samples were collected from the feed mill. A total of 2.00 kg samples were taken using a multipoint method, further reduced to 0.25 kg via quartering, before being ground to a fineness that permitted sample particles to pass through a 0.42 mm sieve. Calcium was measured via the potassium permanganate method (GB/T 6436-2018) [21]. Phytate P was measured according to Haugh and Lantzsch [22]. Total P was determined via spectrophotometry [23]. The NPP content was calculated by subtracting phytate P from total P (TP) in each feed [24]. Each sample was measured with three replicates, and their mean was taken as the final value.

• Growth performance

The bodyweight (BW) and FI were recorded on days 0, 3, and 9 in trial one, and on days 0, 3, and 7 in trials two and three. Birds and feed were weighed at 8:00 am on those days, with each cage treated as a replicate. Bodyweight gain (BWG), average daily gain (ADG), and average daily feed intake (ADFI) were calculated. The feed:gain (F:G) ratio was corrected according to mortality and calculated as FI/BWG. The growth performance of commercial farms was measured on the day of our visit. Historical growth performance was reported by farmers.

• Gait score and lameness incidence observations

Broiler's walking ability was evaluated using a three-point gait scoring scale based on that of Farhadi et al. [25]. Briefly, broilers were classified to one of three gait scores, based on the following criteria: easy, well-balanced gait without any irregularity (scored: one, normal gait); irregular, uneven strides and unbalanced gait (scored: two, slight lameness); reluctance to move, takes only a few strides before sitting (scored: three, severe lameness). These observations were made every day between 8:00 to 9:00 am in all trials, after creating a noisy environment by striking the cage with two iron bars ($1 \times 1 \times 100$ cm) at a rate of four times per second. After striking, the birds' gait was observed and scored over a five-second period. The striking and observations were repeated three times for each cage of birds, and counts of lame (score two or three) birds were recorded. Lameness incidence was calculated as the average count of lame birds/total count of observed birds $\times$ 100%. In field research, we took observations of chicken gait scores at the commercial farms based on three randomly selected pens at each farm during our visit. Lameness incidence was estimated on these farms.

• Gross anatomy

Gross anatomy was done on chicks in the field survey and all trials. Ten healthy chicks and lame chicks were randomly selected from commercial farms on the day of the survey (at 3–11 days old). Two healthy chicks from control cages and lame chickens from treatment cages were randomly selected on day 4 and the last day in all trials. A panel of veterinarians (n = 5) from three independent companies inspected the gross anatomy of the lame chickens. Chickens were necropsied after sacrificing by cervical dislocation. The beak, femur, and metatarsus were forcefully pushed (or bent) aside using finger pressure to test for bone hardness. Diagnosis of hypophosphatemic rickets was performed according to Dinev [8] via a longitudinal tibial cut at the proximal tibiotarsus. However, inspection for lesions was performed immediately after necropsy instead of histochemical analysis due to a time pressure to judge who should compensate for the economic loss of farmers. Lesions were checked in visceral organs (heart, liver, spleen, kidney, bursal, gizzard, gut, thymus, and abdominal fat) were also performed.

- Determination of serum P

On day 9 in trial one, six severely lame birds and six healthy birds were randomly selected from the cages from control group and HCa:P group, respectively. Blood samples were collected via cardiac puncture by sterile syringe after CO_2 (30% in air) aspiration for one minute. Then birds were sacrificed by cervical dislocation. The blood samples were transferred into tubes containing a coagulation promoting agent and maintained at room temperature for two hours, before centrifuging at $1000\times g$ at 4 °C for 10 min for serum collection. Serum P was determined in duplicate via phosphomolybdic acid methodology using commercial kits (Zhongsheng Beikong Bio-technology and Science Inc., Beijing, China) [26] analyzed using a Beckman Coulter AU5800 unit (Beckman Coulter Inc.; Brea, CA, USA).

2.4. Statistical Analysis

Figures were edited via OriginPro 8 SR1 (OriginLab Corporation, Northampton, MA, USA) and the mspaint program of accessories for Windows 10.0 (Microsoft corp., Beijing, China). Other data were summarized with Microsoft Excel 2016 (Microsoft corp., Beijing, China). Lameness incidence was a relative value, and it was taken as the mean data of three trials. The growth performance and nutrient levels were analyzed using SPSS statistical software, version 17.0 (SPSS Inc., Chicago, IL, USA). Homogeneity of variance and Kolmogorov-Smirnov (K-S) tests were performed to confirm the assumption of normally distributed data. Data were statistically analyzed via one-way ANOVA. The significance of differences between treatments was determined at $p < 0.05$ based on Duncan's multiple range tests. Welch's ANOVA was conducted where the data did not pass homogeneity testing.

3. Results and Discussion

3.1. Verification of Hypophosphatemic Rickets in Lame Chicks

Birds fed normal rations were clean and could walk and run freely (Figure 1A). Chicks on farms with problematic HCa:P feeds appeared lame, displaying characteristics of rickets including dirty feathers, reluctance to walk, shaking or falling during walking, or walking on their shanks/hocks, unable to reach feed and water, etc. (Figure 1A'). These gait characteristics of lame chicks were consistent with previous reports of hypophosphatemic rickets [8]. Low P diets are known to be capable of reducing tibia ash and tibia P [10,11], likely one of the main reasons for the birds' poor gait scores.

Hypophosphatemic rickets was further identified in the gross anatomy of chicks (Figure 1B–F,B'–F'). The bone structure of chicks on commercial farms was affected by problematic feed as early as the 4th day. Healthy chicks' bones were strong, well calcified, and well formed (Figure 1B–F); Bones in lame chicks fed the HCa:P feeds were pliable, elastic, and tended to be bent (Figure 1B'–F'). Lame chicks had "S"-shape soft ribs (Figure 1B'), swollen and cylinder-like costochondral junctions (Figure 1C'), elastic bones in metatarsus (Figure 1D') and beak; poorly calcified tibia proximal tibiotarsus with visible hollows (dark red parts) (Figure 1E'), and malleable femur (no fractures after pushing outwards (Figure 1F). Consistent with the P-deficient chicks in the case reports of Gröne et al. and Dinev [3,8] and the experiment of Shao et al. [7], these chicks had rickets lesions in the form of pliable bones, bent ribs, swollen costochondral junctions, and non-calcified tibia proximal tibiotarsus growth plates. The visible hollows at the tibia proximal tibiotarsus were consistent with the lesions observed in P-deficient chicks in Shao et al. [7].

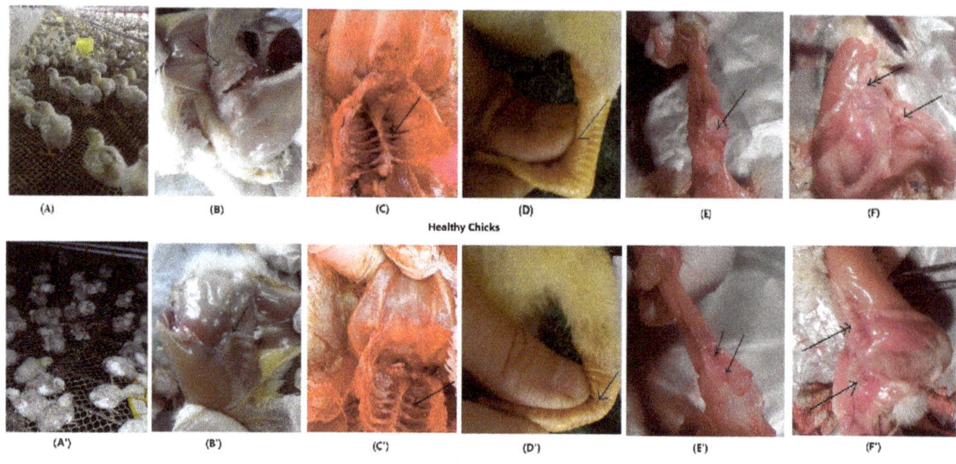

Figure 1. Rickets characteristics apparent in gait features and gross anatomy of 4–11 day-old broiler chicks caused by high-calcium (Ca): phosphorus (P) ratio, P-deficient (HCa:P) crumbled feed. Pictures show changes in walking characteristics (**A**/**A'**), ribs (**B**/**B'**), costochondral junctions (**C**/**C'**), metatarsus (**D**/**D'**), tibia proximal tibiotarsus (**E**/**E'**), and femur (**F**/**F'**).

The commercial feeds which caused the severe lameness in chicks (HCa:P feeds) contained higher (all $p < 0.01$) levels of Ca, Ca:P and Ca:NPP ratios, and lower ($p < 0.01$) levels of TP and NPP compared to the normal feeds that did not cause lameness issues (Table 3). We discovered that during the production of the defective feed, limestone was erroneously substituted for a proportion of the DP in the DP bin. Consistent with the feed processing error and low dietary P level, the concentration of serum P in severely lame chicks (gait score = 3) was only 38% of that of healthy chicks (gait score = 1) on day 9 ($p < 0.01$, Figure 2) in trial one. Low blood P concentration is an indication of hypophosphatemic rickets [8]. Consistent with our study, Li et al. [26] and Zhang et al. [10] reported that the serum P was decreased by dietary P deficiency. The primary reason for low serum P may be due to the high Ca:NPP ratio, as serum P levels were not affected when a balanced Ca:NPP ratio was maintained in a low-P diet [26].

Table 3. Calcium (Ca), phosphorus (P) levels and Ca:P ratios in commercial feeds from field research (air-dry basis).

Variables [1]	Defective Feeds from Commercial Farms [2]			Pooled SEM	*p* Values
	Normal Feeds	**SHCa:P**	**HCa:P**		
Ca, %	0.89 [a]	1.03 [b]	1.07 [b]	0.02	<0.01
TP, %	0.59 [a]	0.54 [b]	0.39 [c]	0.02	<0.01
Ca: TP	1.52 [a]	1.94 [b]	2.72 [c]	0.09	<0.01
NPP, %	0.41[a]	0.36 [b]	0.21 [c]	0.02	<0.01
Ca: NPP	2.19 [a]	2.99 [b]	5.03 [c]	0.21	<0.01

[1] TP: total P; NPP: non-phytate P. [2] Feed samples from commercial farms were classified into normal feeds (n = 19), high-Ca:P ratio, P-deficient (HCa:P, n = 10) feeds, and slightly high Ca:P (SHCa:P, n = 11) feeds, based on chicks fed these feeds being healthy, severely lame or slightly lame. The average moisture content of the feeds was 10.33%. Each sample value given is the mean of three measurements.
[a–c] Means within a row without common superscripts differ significantly ($p < 0.01$).

Based on our studies in both commercial and the experimental farm, we confirmed that a high Ca:P ratio, P deficient crumbled feed was the cause of hypophosphatemic rickets identified in the broilers in this case, as attested by feed processing records, extremely low dietary P levels in feeds and serum, high Ca:P ratios in feeds, and the poor gait characteristics (lameness) and bone development observed.

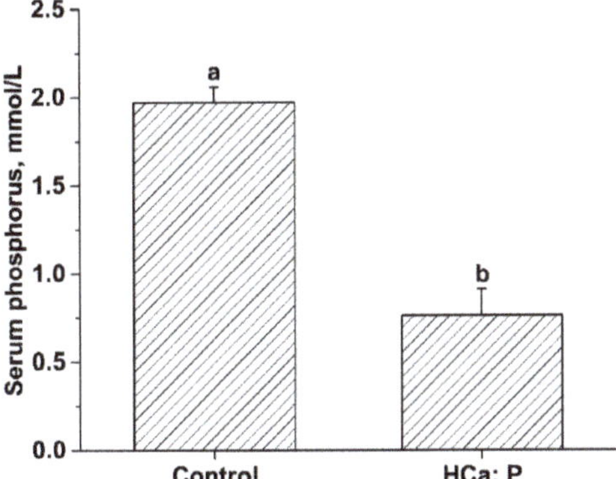

Figure 2. Effect of high-calcium (Ca): phosphorus (P) ratio and P-deficient crumble (HCa:P) feed on serum P of broiler chicks in trial one. Control, Ca and P-balanced feed which did not cause chick lameness. HCa:P, high Ca:P ratio, P-deficient crumbled feed from farms with severe chick lameness issues. [a-b] Means on the different columns without common superscript, differ significantly ($p \leq 0.01$). Data were showed in the form of means + SD.

3.2. Determination of the Duration of Onset of Hypophosphatemic Rickets

The changes in bone structure began to be affected by the feed as early as the 4th day in all trials, similar to the observations on the commercial farms. Low dietary NPP levels is a cause of bone abnormalities, leading to significantly decreased tibia ash, mineral, Ca and P content [5,24], and reduced bone density and bone strength in broilers [24]. Dietary P levels and the ratio of available P (AP) are important factors affecting bone mineralization, bone strength, gait score, and rickets in broiler chickens [14–18]. The rapidity of impact reflected the greater impact of a Ca:P imbalanced diet on tibia bone mineral density and ash content, and therefore, bone-breaking strength, compared with a Ca:P balanced, P-deficient diet [26]. This may be because a high dietary Ca:AP ratio causes the formation of insoluble Ca-P complexes in the intestines [27], reducing the true digestibility of Ca [17], increasing Ca excretion, and decreasing plasma P concentration [16], thus leading to poor tibial mineralization [10,11]. In our study, no consistent difference was observed in lesions of visceral organs (heart, liver, spleen, kidney, bursal, gizzard, gut, thymus, and abdominal fat) between healthy chicks and lame chicks. Data collected from our survey also showed that various drugs (antibiotics, antiviral drugs, Chinese herbal medicines, mycotoxin adsorbent, etc.) were used to treat this problem. However, no curative effect was observed. Thus, infectious disease was excluded from the factors that affect P digestion and absorption. As we did not make anatomical observations on day 3, whether significant differences would be observable at that point remains to be seen.

There was a strong correlation between lameness (gait score) and rickets based on our anatomical observations at commercial and experimental farms. From our survey on farmers, the duration of onset of rickets was 2.70 ± 0.69 day in chicks based on lameness (gait scores, 2 and 3) surveyed at ages of 6.93 ± 2.73 d. On commercial farms feeding HCa:P feeds, the average incidence of lameness was 74.11% after the 6th day. Lameness was not found in the control groups, whereas, it was observed in the HCa:P groups as early as day 4, day 3, and day 4 in trials one, two and three, respectively. The average lameness incidence in the HCa:P group rose to be about 71% and 79% on the 6th and 7th day in three trials; and reached 93% and 100% on the 8th and 9th day in trial one (Figure 3). Adding DP in HCa:P feed resulted in a lameness incidence between 0 to 11% in trials one and three. However, as we visually identified lameness, the possibility that some birds had developed pathological changes but were nonetheless continuing to move normally cannot be excluded. We recommend that more sensitive methods, such as PCR, histochemical methods or immunohistochemical methods, be adopted in future for further study of the early course of rickets in chicks. Compared with commercial farms, the onset was later (4 or 3 days vs. 2.7 days) in all three trials. The higher stocking density of intensive commercial farms might represent another factor aggravating leg problems [28]. In addition, on the commercial farms, the temperatures were between 30 and 34 °C, and the humidity was 70–84%; these are high for 6- to 11-day-old birds and may have aggravated problems stemming from P deficiency syndrome. Based on gait score and lameness incidence in all trials, the duration of onset of hypophosphatemic rickets was as early as three days after commencement of feeding HCa:P diets.

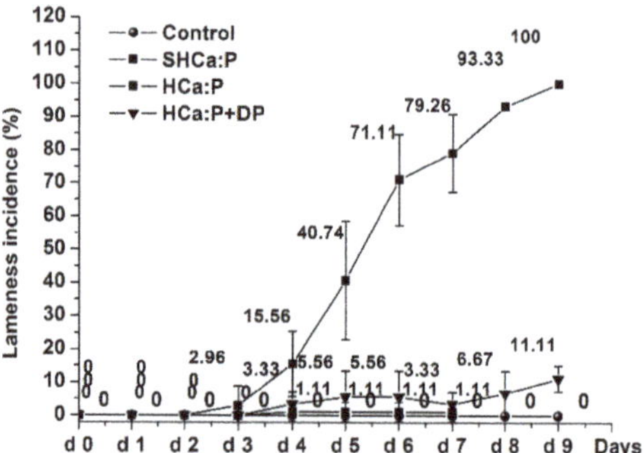

Figure 3. Lameness incidence in broiler chicks caused by high-calcium (Ca): phosphorus (P) ratio and P-deficient crumbled feeds in trials one to three (C). Control, Ca and P-balanced feed which did not cause chick lameness (n = 9 at days 0 to 7; n = 3 at days 8 and 9); HCa:P, high Ca:P ratio, P-deficient crumbled feed from farms with severe chick lameness issues (n = 9). SHCa:P, slightly high Ca:P ratio, P-deficient crumbled feed from farms with slight chick lameness issues (n = 6); HCa:P + DP, feed made by mixing 1.5% dicalcium phosphate (DP) powder into HCa:P crumbled feed (n = 6 at days 0 to 7; n = 3 at days 8 and 9). Lameness incidence is calculated from mixed data in trials one to three and given as mean ± SD.

Growth performance is one of the variables most sensitive to P deficiency, by comparison with Ca deficiency alone or simultaneous Ca and P deficiency [7,10]. Farmers responded that feed intake was lower than broiler performance objectives [29,30] at two days after feeding the P deficient commercial feed, and the growth rate was obviously lower since the third day. The BW, BWG, ADG, FI, and ADFI were decreased by HCa:P feed as early as day 3 in trial one ($p \leq 0.05$, Table 4), trial two ($p < 0.01$, Table 5) and trial three ($p < 0.01$, Table 6). This result is consistent with a previous study which demonstrated that both BW and BWG were sensitive to dietary Ca and P deficiency [4]. However, effects on the F:G ratio were not apparent until day 9 in trial one ($p = 0.05$), day 7 ($p = 0.03$) in trial two, and day 3 ($p \leq 0.01$) in trial three. Similarly, the onset time of lameness had a difference of one day among trials. The inconsistency may be due to different batches of chicks and the change of environmental conditions caused by different experiment times. Consistent with our result, F:G is not very sensitive to low dietary phosphorous by comparison with FI and ADG [11]. The BW and FI at day 3 worsened increasingly as the Ca:TP ratio increased from 1.41 to 2.69 and 3.08 in our trials. Consistent with our study, FI and BW or ADG reduced in broilers fed with low P [7,15] diets or both low P and high Ca:P ratio diet [7]. Adding DP improved the growth performance harmed by HCa:P feed in both trial one (e.g., $ADG_{day\ 0-9}$ and $ADFI_{day\ 3-9}$, all $p \leq 0.05$) and trial three (e.g., $ADG_{day\ 0-3}$, $ADFI_{day\ 0-3}$, F: G day_{0-7}, all $p \leq 0.01$). Data of growth performance suggested that the day of onset of P deficiency was three days, earlier than noted in the previous reports [4,11,15]. The earlier changes in FI and growth rates in our study may reflect the more severe lack of NPP (0.18–0.24%), high Ca (about 1.23%) and high Ca:TP ratio (>2.69) resulted from a replacement of DP by limestone. Hu et al. suggested that a dietary Ca:TP ratio of 1.00–1.25 is optimal in terms of maximizing digestion of Ca and P while maintaining tibia strength and growth performance characteristics [18]. Based on growth performance in all trials, the duration of onset of hypophosphatemic rickets was as early as three days after commencement of feeding HCa:P diets.

Table 4. Effect of high-calcium (Ca): phosphorus (P) ratio, P-deficient (HCa:P) crumbled feed on growth performance in 0–9 day-old broiler chicks during trial one.

Variables [1]	Items [1]	Treatments [2]			Pooled SEM	*p* Value
		Control	HCa:P	HCa:P + DP		
BW, g/bird	$BW_{day\ 0}$	40.33	40.60	40.57	0.65	0.99
	$BW_{day\ 3}$	81.93 [a]	73.07 [b]	77.93 [ab]	1.52	0.03
	$BW_{day\ 9}$	181.37 [b]	164.23 [b]	182.90 [a]	3.29	<0.01
BWG, g/bird	$BWG_{day\ 0-3}$	41.60 [a]	32.47 [b]	37.37 [ab]	1.51	0.01
	$BWG_{day\ 3-9}$	99.43 [ab]	91.17 [b]	104.97 [a]	2.39	0.03
	$BWG_{day\ 0-9}$	141.03 [b]	123.63 [b]	142.33 [a]	3.13	<0.01
ADG, g/bird/d	$ADG_{day\ 0-3}$	13.87 [a]	10.82 [b]	12.46 [ab]	0.50	0.02
	$ADG_{d\ 3-9}$	16.57 [ab]	15.20 [b]	17.50 [a]	0.40	0.03
	$ADG_{day\ 0-9}$	15.67 [a]	13.74 [b]	15.81 [a]	0.35	<0.01
FI, g/bird	$FI_{day\ 0-3}$	45.57 [a]	39.30 [b]	44.43 [ab]	1.23	0.05
	$FI_{d\ 3-9}$	164.40 [a]	139.50 [b]	169.53 [a]	5.06	<0.01
	$FI_{d\ 0-9}$	209.83 [a]	178.73 [b]	213.40 [a]	6.11	<0.01
ADFI, g/bird/d	$ADFI_{day\ 0-3}$	15.19 [a]	13.10 [b]	14.81 [ab]	0.41	0.05
	$ADFI_{day\ 3-9}$	27.40 [a]	23.25 [b]	28.27 [a]	0.84	<0.01
	$ADFI_{day\ 0-9}$	23.32 [a]	19.86 [b]	23.71 [a]	0.68	<0.01
F:G	$F{:}G_{day\ 0-3}$	1.10	1.21	1.20	0.04	0.48
	$F{:}G_{day\ 3-9}$	1.65 [a]	1.53 [b]	1.61 [ab]	0.02	0.05
	$F{:}G_{day\ 0-9}$	1.49	1.45	1.50	0.02	0.47

[1] BW, body weight, BWG, body weight gain; ADG, average daily gain, FI, feed intake; ADFI, average daily feed intake. F:G, feed:gain ratio.
[2] Control, Ca and P-balanced feed which did not cause chick lameness. HCa:P, high Ca:P ratio, P-deficient crumbled feed collected from farms whose chicks had severe lameness issues. SHCa:P, slightly high Ca:P ratio, P-deficient crumbled feed from farms whose chicks had slight lameness issues; HCa:P + DP, feed made by mixing 1.5% dicalcium phosphate (DP) powder into HCa:P crumbled feed. [a–b] Values within a row without common superscripts differ significantly ($p \leq 0.05$).

Table 5. Effect of high-calcium (Ca): phosphorus (P) ratio, P-deficient (HCa:P) crumbled feed on growth performance in 0- to 7- day-old broiler chicks during trial two.

Variables	Items [1]	Treatments			Pooled SEM	*p* Value
		Control	SHCa:P	HCa:P		
BW, g/bird	$BW_{day\ 0}$	42.33	42.67	42.7	0.09	0.19
	$BW_{day\ 3}$	91.13 [a]	84.20 [bc]	80.67 [c]	1.71	<0.01
	$BW_{day\ 7}$	194.60 [a]	171.27 [b]	142.37 [c]	7.65	<0.01
BWG, g/bird	$BWG_{day\ 0-3}$	48.80 [a]	41.53 [b]	37.97 [b]	1.75	<0.01
	$BWG_{day\ 3-7}$	102.87 [a]	87.07 [b]	61.70 [c]	6.29	<0.01
	$BWG_{day\ 0-7}$	102.87 [a]	87.07 [b]	61.7 [c]	6.30	<0.01
ADG, g/bird/d	$ADG_{day\ 0-3}$	16.27 [a]	13.84 [b]	12.66 [b]	0.58	<0.01
	$ADG_{day\ 3-7}$	14.70 [a]	12.44 [b]	8.81 [c]	0.90	<0.01
	$ADG_{day\ 0-7}$	11.43 [a]	9.67 [b]	6.86 [c]	0.70	<0.01
FI, g/bird	$FI_{day\ 0-3}$	45.70 [a]	41.03 [b]	38.17 [c]	1.12	<0.01
	$FI_{d\ 4-7}$	136.270 [a]	123.97 [b]	99.00 [c]	5.56	<0.01
	$FI_{d\ 0-7}$	181.63 [a]	165.00 [b]	137.13 [c]	6.58	<0.01
ADFI, g/bird/d	$ADFI_{day\ 0-3}$	15.23 [a]	13.68 [b]	12.72 [c]	0.37	<0.01
	$ADFI_{day\ 3-7}$	34.07 [a]	31.00 [b]	24.75 [c]	1.39	<0.01
	$ADFI_{day\ 0-7}$	25.95 [a]	23.57 [b]	19.59 [c]	0.94	<0.01

Table 5. *Cont.*

Variables	Items [1]	Treatments			Pooled SEM	*p* Value
		Control	SHCa:P	HCa:P		
	$F:G_{day\ 0-3}$	0.94	0.99	1.01	0.02	0.32
F:G	$F:G_{day\ 3-7}$	1.32	1.43	1.62	0.06	0.07
	$F:G_{day\ 0-7}$	1.20 [a]	1.29 [ab]	1.38 [b]	0.03	0.03

[1] Control, Ca and P-balanced feed which did not cause chick lameness. HCa:P, high Ca:P ratio, P-deficient crumbled feed collected from farms whose chicks had severe lameness issues. SHCa:P, slightly high Ca:P ratio, P-deficient crumbled feed from farms whose chicks had slight lameness issues; HCa:P + DP, feed made by mixing 1.5% dicalcium phosphate (DP) powder into HCa:P crumbled feed. [a–c] Values within a row without common superscripts differ significantly ($p \leq 0.05$).

Table 6. Effect of high-calcium (Ca): phosphorus (P) ratio, P-deficient (HCa:P) crumbled feed on growth performance in 0- to 7-day-old broiler chicks during trial three.

Variables [1]	Items	Treatments				Pooled SEM	*p* Value
		Control	SHCa:P	HCa:P	HCa:P + DP		
BW, g/bird	$BW_{day\ 0}$	42.13	42.3	42.7	42.47	0.09	0.16
	$BW_{day\ 3}$	89.67 [a]	89.63 [a]	81.33 [b]	89.47 [a]	1.14	<0.01
	$BW_{day\ 7}$	191.3 [a]	181.77 [b]	155.27 [c]	187.4 [a]	4.31	<0.01
BWG, g/bird	$BWG_{day\ 0-3}$	47.53 [a]	47.33 [a]	38.63 [b]	47.00 [a]	1.18	<0.01
	$BWG_{day\ 3-7}$	101.63 [a]	92.13 [c]	73.93 [d]	97.93 [b]	3.29	<0.01
	$BWG_{day\ 0-7}$	149.17 [a]	139.47 [b]	112.57 [c]	144.93 [ab]	4.36	<0.01
ADG, g/bird/d	$ADG_{day\ 0-3}$	15.84 [a]	15.78 [a]	12.88 [b]	15.67 [a]	0.39	<0.01
	$ADG_{day\ 3-7}$	25.41 [a]	23.03 [b]	18.48 [c]	24.48 [a]	0.49	<0.01
	$ADG_{day\ 0-7}$	21.31 [a]	19.92 [b]	16.08 [c]	20.71 [ab]	0.62	<0.01
FI, g/bird	$FI_{day\ 0-3}$	45.07 [a]	45.8 [a]	39.93 [b]	44.23 [a]	0.74	<0.01
	$FI_{day\ 3-7}$	138.4 [a]	125 [b]	105.1 [c]	130.53 [b]	5.06	<0.01
	$FI_{day\ 0-7}$	183.47 [a]	170.8 [b]	145.03 [c]	174.77 [b]	6.11	<0.01
ADFI, g/bird/d	$ADFI_{day\ 0-3}$	15.02 [a]	15.27 [a]	13.31 [b]	14.74 [a]	0.25	<0.01
	$ADFI_{day\ 3-7}$	34.6 [a]	31.25 [b]	26.28 [c]	32.63 [b]	0.95	<0.01
	$ADFI_{day\ 0-7}$	26.21 [a]	24.40 [b]	20.72 [c]	24.97 [b]	0.63	<0.01
F:G	$F:G_{day\ 0-3}$	0.97 [a]	0.97 [a]	1.03 [b]	0.94 [a]	0.01	0.01
	$F:G_{day\ 3-7}$	1.37	1.36	1.42	1.33	0.01	0.13
	$F:G_{day\ 0-7}$	1.23 [a]	1.22 [a]	1.29 [b]	1.21 [a]	0.01	<0.01

[1] Control, Ca and P-balanced feed which did not cause chick lameness. HCa:P, high Ca:P ratio, P-deficient crumbled feed collected from farms whose chicks had severe lameness issues. SHCa:P, slightly high Ca:P ratio, P-deficient crumbled feed from farms whose chicks had slight lameness issues; HCa:P + DP, feed made by mixing 1.5% dicalcium phosphate (DP) powder into HCa:P crumbled feed. [a–c] Values within a row without common superscripts differ significantly ($p \leq 0.05$).

This duration of onset of rickets was earlier than in the previous reports, which identified onset durations such as 7 [7,8], 7–10 [9], and 18 days [8]. One reason for this may be a difference in diagnosis methods: the authors of those reports based their judgements on the incidence of rickets on observations of anatomy and serum P rather than gait score changes or growth performance. In recent years, developments in poultry breeding have produced broilers with faster growth rates and higher nutrient requirements, which may contribute to more rapid symptom onset under conditions of P deficiency [26].

The dietary level of AP was 0.28–0.34% in feed (assuming a 0.1% improvement due to phytase). The minimal AP recommendation for broilers chicks is 0.45% [29] and 0.48% [30] of feed, respectively, by breeding companies. The level of dietary NPP (0.21% in feed) was much lower than the recommendation of 0.45 by NRC standard [31] or 0.39% by Liu et al. [24]. A Ca:AP ratio at 2.0 or Ca:NPP ratio at 2.22 is recommended for broiler chicks by breeding companies [29,30] or NRC [31]. In the present case, Ca:NPP ratios were over 2.22, at 4.71 and 6.17, respectively, in the SHCa:P and HCa:P feeds. Consistent with our observations, high Ca:TP ratios caused rickets in growing rheas in a case reported by Gröne et al. [3]. Conversely, when P concentration is low, a feed with a balanced Ca:P ratio resulted in better growth performance and bone structure than feed with high Ca:TP ratios [7,15]. In future, the use of more accurate methods to determine phytic acid and other inositol phosphate contents will be necessary to provide more exact recommendations for safe dietary doses of NPP in broiler feeds [32]. The P is largely mobilized by the yolk sac membrane during 12–17 days of incubation, and its content is very low as the chicken's embryo prepares to hatch [33]. The P leftover in ducks' yolk sac at 3 days before the hatch is only about 13.1% of the first incubation day [34]. During fasting, newly hatched broilers had normal activities during the first 58 h, absorbing about 70% of the nutrients in the yolk sac, but the birds appear dull and lack vitality from the third day [35]. We observed that severe lame birds could not reach feed and water freely. The thirsty and starvation further aggravated rickets. Therefore, a shortage of P in the yolk sac, a dietary P deficiency and an imbalanced Ca:P ratio might be the main reasons for the early onset of hypophosphatemic rickets in chicks in our study.

Physiological reasons for the early onset of rickets may involve the following factors. An unbalanced Ca:P and P deficiency in feed led to a low P concentration in blood circulation (Figure 2), which in turn caused a poor development of bone structures (Figure 1B'–F'). The leg bones were too soft (Figure 1B'–F') to support the standing and walking of birds, leading to early onset of rickets and high lameness incidence (Figure 3). The poorly developed bones, e.g., sternum (Figure 1C') was soft and small, and not enough to support the attachment of heavy muscles, leading to poor growth performance (Tables 4–6). The P is a component of creatine phosphate and adenosine triphosphate (ATP). Depletion of high-energy phosphate may be one reason for the hypotonia of rickets [36]. Dietary P deficiency slows down the ATP synthesis in muscles, leading to myopathy at the early stage [37]. So, a lack of energy (e.g., ATP) may obstacle movement and muscle deposit in chicks. Phosphorus is also a critical component of nucleotides (DNA and RNA), phospholipids in membranes, and phosphorylated intermediates in cellular signaling [38]. Thus, P deficiency may lead to early rickets in broilers partly via impairing these molecules.

4. Conclusions

In conclusion, crumbled feed with a high Ca:P ratio and deficient in P can cause hypophosphatemic rickets in broiler chicks as early as 3 days. Lameness incidence rose to over 70% after chicks feed a high Ca:P ratio and P deficient feed for six days. Several variables, including the dietary P level and Ca:P ratio, and chicks' growth performance, the incidence of lameness, and bone development are helpful in the early diagnosis of hypophosphatemic rickets in broilers. The current multifactorial nature of diagnosis complicates this task, and in future, the development of more specific and sensitive single methods allowing early, rapid, and accurate diagnosis of hypophosphatemic rickets caused by high Ca:P ratios and P deficiency of various degrees, would be of great practical value to poultry breeders.

Author Contributions: Data collection, animal trials, L.X.; writing—original draft preparation, N.L. and L.X.; data analysis, X.W. and X.Z.; writing—review and editing, M.B.F. and Y.Z.F.; funding acquisition, H.Y. All authors have read and agreed to the published version of the manuscript.

Funding: This study was supported by the earmarked fund for Jiangsu Agricultural Industry (Broiler) Technology System (JATS[2021]183), the Priority Academic Program Development of Jiangsu Higher Education Institutions (PAPD), and the 2018 Deputy General Manager Program for Science and Technology of Jiangsu Province's "innovation and entrepreneurship plan" (FZ20180271).

Institutional Review Board Statement: The study was conducted according to the guidelines of the Experimental Animal. The animal trials were approved by the Animal Care and Use Committee of Yangzhou University (protocol code YZUDWSY 2017-05-09.03, approved on 9 May 2017).

Informed Consent Statement: Not applicable.

Data Availability Statement: Data are available on reasonable request from the corresponding author.

Acknowledgments: Special thanks are given to Lu S., Shi W., Liu H., Liu G., Li Z., and Yan W for their support and assistance. The authors are grateful to the faculties from QA, QC, feed production, and R and Day in the anonymous company for their help in the study. The authors thank the A.W.E.S.O.M.E. faculty group in the College of Agriculture and Life Sciences at Texas A and M University for assistance with editing the manuscript. Initial group financial support was through NSF ADVANCE Institutional Transformation Award 1008385, with continued support from the college.

Conflicts of Interest: The authors declare no conflict of interest.

References

1. Imel, E.A.; Carpenter, T.O. Rickets: The skeletal disorders of impaired calcium or phosphate availability. In *Pediatric Endocrinology*; Humana Press: Totowa, NJ, USA, 2013; pp. 357–378.
2. Novi, B.D.; Shahfar, K.; Zachary, K. Rickets in a 12 month old Boy. *Isual J. Emerg. Med.* **2020**, *20*, 100752. [CrossRef]
3. Gröne, A.; Swayne, D.E.; Nagode, L.A. Hypophosphatemic rickets in rheas (rhea americana). *Vet. Pathol.* **1995**, *32*, 324–327. [CrossRef]
4. Yan, F.; Angel, R.; Ashwell, C.; Mitchell, A.; Christman, M. Evaluation of the broiler's ability to adapt to an early moderate deficiency of phosphorus and calcium. *Poult. Sci.* **2005**, *84*, 1232–1241. [CrossRef]
5. Venäläinen, E.; Valaja, J.; Jalava, T. Effects of dietary metabolisable energy, calcium and phosphorus on bone mineralisation, leg weakness and performance of broiler chickens. *Brit. Poult. Sci.* **2006**, *47*, 301–310. [CrossRef]
6. Valable, A.S.; Narcy, A.; Duclos, M.J.; Pomar, C.; Page, G.; Nasir, Z.; Magnin, M.; Létourneau-Montminy, M.P. Effects of dietary calcium and phosphorus deficiency and subsequent recovery on broiler chicken growth performance and bone characteristics. *Animal* **2018**, *12*, 1555–1563. [CrossRef] [PubMed]
7. Shao, Y.; Xing, G.; Zhang, L.; Lyu, L.; Li, S.; Liao, X.; Luo, X. Effects of dietary calcium and phosphorus deficiency on growth performance, rickets incidence characters and tibia histological structure of broilers during 1 to 21 days of age. *Chin. J. Anim. Nutr.* **2019**, *31*, 2107–2118.
8. Dinev, I. Comparative pathomorphological study of rickets types in broiler chickens. *Iran. J. Vet. Sci. Technol.* **2011**, *3*, 1–10.
9. Zhu, L.Q.; Zhu, F.H.; Zhang, Z.M.; Wang, Q.J.; Gong, Q.J. Effect of dietary available phosphorus levels on growth and tissue phosphorous and calcium contents in chicks. *Chinese J. Anim. Nutr.* **2001**, *37*, 14–16.
10. Li, T.; Xing, G.; Shao, Y.; Zhang, L.; Luo, X. Dietary calcium or phosphorus deficiency impairs the bone development by regulating related calcium or phosphorus metabolic utilization parameters of broilers. *Poult. Sci.* **2020**, *99*, 3207–3214. [CrossRef]
11. Nari, N.; Ghasemi, H.A. Growth performance, nutrient digestibility, bone mineralization, and hormone profile in broilers fed with phosphorus-deficient diets supplemented with butyric acid and Saccharomyces boulardii. *Poult. Sci.* **2020**, *99*, 926–935. [CrossRef]
12. Proszkowiec-Weglarz, M.; Angel, R. Calcium and phosphorus metabolism in broilers: Effect of homeostatic mechanism on calcium and phosphorus digestibility. *J. Appl. Poult. Res.* **2013**, *22*, 609–627. [CrossRef]
13. Kim, W.K.; Bloomfield, S.A.; Ricke, S.C. Effects of age, vitamin D3, and fructo oligosaccharides on bone growth and skeletal integrity of broiler chicks. *Poult. Sci.* **2011**, *90*, 2425–2432. [CrossRef] [PubMed]
14. Hulan, H.W.; De Groote, G.; Fontaine, G.; De Munter, G.; McRae, K.B.; Proudfoot, F.G. The effect of different totals and ratios of dietary calcium and phosphorus on the performance and incidence of leg abnormalities of male and female broiler chickens. *Poult. Sci.* **1985**, *64*, 1157–1169. [CrossRef] [PubMed]
15. Delezie, E.; Bierman, K.; Nollet, L.; Maertens, L. Impacts of calcium and phosphorus concentration, their ratio, and phytase supplementation level on growth performance, foot pad lesions, and hock burn of broiler chickens. *J. Appl. Poult. Res.* **2015**, *24*, 115–126. [CrossRef]
16. Naves, L.D.P.; Rodrigues, P.B.; Teixeira, L.D.V.; de Oliveira, E.C.; Saldanha, M.M.; Alvarenga, R.R.; Corrêa, A.D.; Lima, R.R. Efficiency of microbial phytase supplementation in diets formulated with different calcium: Phosphorus ratios, supplied to broilers from 22 to 33 days old. *J. Anim. Physiol. Anim. Nutr.* **2015**, *99*, 139–149. [CrossRef]
17. Anwar, M.N.; Ravindran, V.; Morel, P.C.H.; Ravindran, G.; Cowieson, A.J. Effect of limestone particle size and calcium to non-phytate phosphorus ratio on true ileal calcium digestibility of limestone for broiler chickens. *Brit. Poult. Sci.* **2016**, *57*, 707–713. [CrossRef]
18. Hu, Y.X.; Bikker, P.; Duijster, M.; Hendriks, W.H.; Krimpen, M.M.V. Coarse limestone does not alleviate the negative effect of a low Ca:P ratio diet on characteristics of tibia strength and growth performance in broilers. *Poult. Sci.* **2020**, *99*, 4978–4989. [CrossRef] [PubMed]
19. Maia, M.; Abreu, A.; Nogueira, P.; Val, M.; Carvalhaes, J.; Andrade, M.C. Hypophosphatemic rickets: Case report. *Rev. Paul. Pediatr.* **2018**, *36*, 242–247. [CrossRef]
20. Arbor Acres. *Broiler Management Guide*; Aviagen Limited: Midlothian, UK, 2009; pp. 9–59.
21. Gao, L.H.; Wang, S.S.; Huang, T.; Wang, J.; He, X.K. *Determination of Calcium in Feed. National Standard of the People's Republic of China*; GB/T 6436-2018; China Standards Press: Beijing, China, 2015.
22. Haugh, W.; Lantzsch, H.J. Sensitive method for the rapid determination of phytate in cereals and cereal products. *J. Sci. Food Agric.* **1983**, *34*, 1423–1426. [CrossRef]

23. Shang, J.; Hua, X.H.; Huang, S.X.; Lu, C.; Cao, Y.; Tian, K.; Zhang, H.R.; Sun, B.Q.G. *Determination of Phosphorus in Feeds—Spectrophotometry. National Standard of the People's Republic of China*; GB/T 6437–2018; China Standards Press: Beijing, China, 2018.
24. Liu, S.B.; Liao, X.D.; Lu, L.; Li, S.F.; Wang, L.; Zhang, L.Y.; Jiang, Y.; Luo, X.G. Dietary non-phytate phosphorus requirement of broilers fed a conventional corn-soybean meal diet from 1 to 21 d of age. *Poult. Sci.* **2017**, *96*, 151–159. [CrossRef]
25. Farhadi, D.; Karimi, A.; Sadeghi, G.; Rostamzadeh, J.; Bedford, M.R. Effects of a high dose of microbial phytase and myo-inositol supplementation on growth performance, tibia mineralization, nutrient digestibility, litter moisture content, and foot problems in broiler chickens fed phosphorus-deficient diets. *Poult. Sci.* **2017**, *96*, 3664–3675. [CrossRef]
26. Zhang, L.H.; He, T.F.; Hu, J.X.; Li, M.; Piao, X.S. Effects of normal and low calcium and phosphorus levels and 25-hydroxycholecalciferol supplementation on performance, serum antioxidant status, meat quality, and bone properties of broilers. *Poult. Sci.* **2020**, *99*, 5663–5672. [CrossRef]
27. An, S.H.; Sung, J.Y.; Kong, C. Ileal digestibility and total tract retention of phosphorus in inorganic phosphates fed to broiler chickens using the direct method. *Animals* **2020**, *10*, 2167. [CrossRef]
28. Sanotra, G.S.; Lawson, L.G.; Vestergaard, K.S.; Thomsen, M.G. Influence of stocking density on tonic immobility, lameness, and tibial dyschondroplasia in broilers. *J. Appl. Anim. Welf. Sci.* **2001**, *4*, 71–87. [CrossRef]
29. Arbor Acres Plus. *Broiler Nutrition Specifications*; Aviagen Limited: Midlothian, UK, 2014; pp. 1–8.
30. Cobb 500. *Broiler Performance and Nutrition Supplement*; Cobb-Vantress Lnc.: Siloam Springs, AR, USA, 2018; pp. 1–14.
31. National Research Council. *Nutrient Requirements of Poultry*, 4th ed.; Nutrient Requirements of Domestic Animals; National Academy Press: Washington, DC, USA, 1994.
32. Marolt, G.; Kolar, M. Analytical methods for determination of phytic acid and other inositol phosphates: A review. *Molecules* **2020**, *26*, 174. [CrossRef]
33. Torres, C.A.; Korver, D.R. Influences of trace mineral nutrition and maternal flock age on broiler embryo bone development. *Poult. Sci.* **2018**, *97*, 2996–3003. [CrossRef] [PubMed]
34. Onbasilar, E.E.; Erdem, E.; Hacan, O.; Yalçin, S. Effects of breeder age on mineral contents and weight of yolk sac, embryo development, and hatchability in Pekin ducks. *Poult. Sci.* **2014**, *93*, 473–478. [CrossRef] [PubMed]
35. Wang, H.; Huo, Q.; Li, S.; Yu, H.; Wang, J.; Yang, S.; Lin, J.; Feng, S.; Yin, R. Nutrient transfer in yolk sac of fasting chicks. *Acta Veterinaria et Zootechnica Sinica* **1994**, *25*, 13–19. (In Chinese)
36. Mize, C.E.; Corbett, R.; Uauy, R.; Nunnally, R.; Williamson, S. In vivo time course of muscle phosphocreatine, phosphorus, and adenosine triphosphate during treatmen of rickets. *Pediat. Res.* **1987**, *4*, 345A. [CrossRef]
37. Hettleman, B.D.; Sabina, R.L.; Drezner, M.K.; Holmes, E.W.; Swain, J.L. Defective adenosine triphosphate synthesis. An explanation for skeletal muscle dysfunction in phosphate-deficient mice. *J. Clin. Investig.* **1983**, *72*, 582–589. [CrossRef] [PubMed]
38. Jan de Beur, S.M.; Levine, M.A. Molecular pathogenesis of hypophosphatemic rickets. *J. Clin. Endocrinol. Metab.* **2002**, *87*, 2467–2473. [CrossRef] [PubMed]

Article

Assessment of Composted Pelletized Poultry Litter as an Alternative to Chemical Fertilizers Based on the Environmental Impact of Their Production

Nikolett Éva Kiss *, János Tamás, Nikolett Szőllősi, Edit Gorliczay and Attila Nagy

Institute of Water and Environmental Management, University of Debrecen, 4032 Debrecen, Hungary;
tamas@agr.unideb.hu (J.T.); nszollosi@agr.unideb.hu (N.S.); edit.gorliczay@agr.unideb.hu (E.G.);
attilanagy@agr.unideb.hu (A.N.)
* Correspondence: kiss.nikoletteva@agr.unideb.hu

Citation: Kiss, N.É.; Tamás, J.; Szőllősi, N.; Gorliczay, E.; Nagy, A. Assessment of Composted Pelletized Poultry Litter as an Alternative to Chemical Fertilizers Based on the Environmental Impact of Their Production. *Agriculture* **2021**, *11*, 1130. https://doi.org/10.3390/agriculture11111130

Academic Editor: Vasileios Antoniadis

Received: 7 October 2021
Accepted: 8 November 2021
Published: 11 November 2021

Publisher's Note: MDPI stays neutral with regard to jurisdictional claims in published maps and institutional affiliations.

Abstract: Reducing the use of chemical fertilizers in agriculture is one of the EU Green Deal's priorities. Since poultry production is increasing worldwide, stabilized poultry litter such as composted pelletized poultry litter (CPPL) is an alternative fertilizer option. On the contrary, compared to chemical fertilizers, the environmental impacts of composted products have not been adequately studied, and no data are currently available for CPPL produced by a closed composting system, such as the Hosoya system. The aim of this research was to assess the role of CPPL as a potential alternative for chemical fertilizer by evaluating the environmental impact of CPPL production via the Hosoya system using common chemical fertilizers. Based on life cycle assessment (LCA), the environmental impact (11 impact categories) was determined for the production of 1 kg of fertilizer, as well as for the production of 1 kg of active substances (nitrogen (N), phosphorus pentoxide (P_2O_5), and potassium chloride (K_2O)) and the theoretical nutrient (NPK) supply of a 100 ha field with CPPL and several chemical fertilizer options. The production of CPPL per kilogram was smaller than that of the chemical fertilizers; however, the environmental impact of chemical fertilizer production per kilogram of active substance (N, P_2O_5, or K_2O) was lower for most impact categories, because the active substance was available at higher concentrations in said chemical fertilizers. In contrast, the NPK supply of a 100 ha field by CPPL was found to possess a smaller environmental impact compared to several combinations of chemical fertilizers. In conclusion, CPPL demonstrated its suitability as an alternative to chemical fertilizers.

Keywords: composted pelletized poultry litter; life cycle assessment; Hosoya composting; chemical fertilizers; EU Green Deal

1. Introduction

Chemical fertilizers provide nutrients to plants quickly and easily. Since relatively low amounts of chemical fertilizers with an increased active substance content are sufficient for productivity [1–3], the introduction of chemical fertilizers has decreased the usage of manure to a low level in intensive farming systems. Chemical fertilizers, on the other hand, can hasten the decomposition of soil organic matter, resulting in the degradation of soil structure. Excess fertilization also has the potential to pollute waterbodies by causing leaching and acidity [4–7]. Furthermore, several studies have shown that the production and use of chemical fertilizers produce high levels of NO_x and N_2O; moreover, the use of fertilizers also increases soil CO_2 emissions [8–20]. With the overarching aim of making Europe climate neutral and sustainable by 2050, the EU introduced the European Green Deal. One of its key targets is to reduce the overall use of chemical fertilizers. The positive effects of the use of manure as a fertilizer for soil–plant systems, particularly on the environment, highlight the importance of organic matter-based fertilizer applications. The European Commission presented the "Farm to Fork Strategy" in the spring of 2020. This strategy is one of the major elements of the European Union's Green Deal aiming at the use of sustainable practices, including carbon management and storage in soil, improved nutrient management, and reductions in chemical fertilizer use in precision and organic farming, in order to improve water and soil quality and to reduce emissions [21].

Manure and other organic matters can be a viable alternative to chemical fertilizers since they play an important role in soil resource replenishment [22–27]. In recent years, one of the rapidly growing livestock sectors is broiler farming [28,29], which is expected to become even more important in the future [30–33] to meet the food demand of a growing population. Due to the growing broiler production, the issue of manure utilization is becoming more important not only from an environmental standpoint, but also from a circular economy aspect in accordance with the Green Deal.

In comparison to other organic matter-containing fertilizers, broiler manure includes a high percentage of readily available micro- and macro-elements for plants and enhances the soil physical characteristics, soil organic matter content, water-holding capacity, nutrient uptake, and, ultimately, plant productivity [34–39]. Raw poultry manure is highly recommended to be treated before use directly as a fertilizer due to its pathogen microorganism content. Composting produces a valuable and environmentally favorable end product [40]; however, the production process is not necessarily environmentally friendly, and therefore, the environmental impact of production must be evaluated. The degree of emissions is influenced by the quantity, quality, and composition, storage, and processing of manure, which includes several types of composting. According to Finstein [41], the main issue is with open composting technology, which pollutes the atmosphere by directly releasing gases, water vapor, and odors. CO_2 loss is the most important and contributes greatly to the greenhouse effect, although there are studies that indicate that the effect of ammonia emissions contributes more to GHG emissions than CO_2 [42]. When organic wastes and byproducts with a high nitrogen content are composted, one of the main compounds that causes pollution is ammonia. Ammonia emissions are an issue, not only because ammonia is hazardous to the environment [43–47], but because it also reduces the nitrogen content of the end product [43,44]. Therefore, potential emerging treatment options involve closed and intensive composting technologies, resulting in a lower ammonia loss and GHG emissions [48–50] compared to open composting systems. One such closed and intensive composting technology is the Hosoya system, which produces composted pelletized granules with heat treatment, thus eliminating toxic ammonia emissions, weed seeds, and pathogenic microorganisms [51,52]. Although the technological process of the Hosoya system is well studied [52], there are no studies related to the environmental impact of production based on life cycle assessment (LCA).

The aim of this research was to assess the role of composted and pelletized poultry litter (CPPL) as a potential alternative to chemical fertilizers by evaluating the environmental impacts of CPPL (53% broiler manure and litter, 27% manure layer and litter, 20% chicken meal (meat and bone meal)) production via the Hosoya system using common chemical fertilizers (ammonium nitrate (AN), calcium ammonium nitrate (CAN), urea, triple superphosphate (TSP), monoammonium phosphate (MAP), and potassium chloride (KCl)). Since CPPL includes all macro-elements, based on a life cycle assessment, the environmental impact of CPPL and chemical fertilizer production was not only determined for 1 kg of the product and 1 kg of the active substance (NPK), but also for the nutrient supply of a field with CPPL and combinations of chemical fertilizers at the same NPK level.

2. Materials and Methods

Environmental impact analysis is a complex issue in agriculture. Therefore, the principles, the framework for life cycle assessment (LCA), and the four main phases of an LCA were based on the ISO14040:2006 standard [53] in this research (Figure 1). Though, the life cycle assessment standard is primarily developed for industry, with less frequent application in agricultural systems and byproducts.

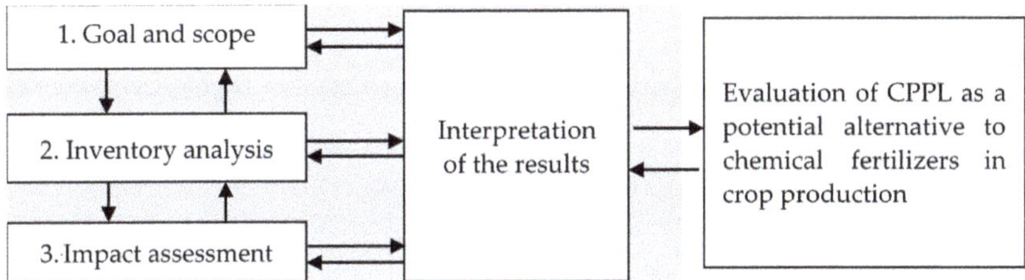

Figure 1. Steps of life cycle assessment (adopted from ISO14040:2006 [53]).

2.1. Definition of the Goal and Scope of LCA

The main objective of this analysis was to assess the role of CPPL as a potential alternative to chemical fertilizers by evaluating the environmental impact of CPPL production via the Hosoya system using common chemical fertilizers. Based on a life cycle assessment, the environmental impact (11 impact categories) was determined for:

- The production of 1 kg of fertilizers: 1 kg of composted pelletized poultry litter (CPPL) and 1 kg of the following chemical fertilizers: ammonium nitrate (AN), calcium ammonium nitrate (CAN), urea, triple superphosphate (TSP), monoammonium phosphate (MAP), and potassium chloride (KCl);
- The production of 1 kg of active substances separately for the N, P_2O_5, and K_2O content of fertilizers to provide comparable inputs to assess fertilizer production per unit of nutrient;
- The nutrient (NPK) supply of a 100 ha field with 1.5 Mg/ha of CPPL (based on Szabó et al.'s [54] method), and with chemical fertilizer combinations with an equivalent NPK supply to analyze the environmental impacts of CPPL as a multi-element fertilizer.

2.2. Life Cycle Inventory Analysis

In the framework of inventory analysis, the input and output materials and energy flows were quantified for the Hosoya composting system using the data of a regionally important poultry producer company in Hungary. In the Hosoya composting plant, deep litter from broiler and layer poultry stock farms and filtered sewage sludge generated by slaughterhouses and hatcheries were collected and treated. The capacity of the plant is 10 mg/day. Poultry houses were littered with heat-treated and grinded straw pellets. Due to the high absorbance capacity of these straw pellets, the deep litter manure also had a low moisture content. The parameters of the broiler and layer manure and litter are shown in Table 1.

Table 1. Parameters of broiler and layer manure and litter.

Parameters	Broiler Manure and Litter (53%)	Layer Manure and Litter (27%)
N content ($w/w\%$)	2.75 ± 0.092	2.14 ± 0.151
P_2O_5 content (mg/kg)	9344 ± 63.692	$20,146 \pm 109.672$
K_2O content (mg/kg)	$26,007 \pm 125.812$	$27,306 \pm 244.178$
Moisture content ($w/w\%$)	27.5 ± 2.750	25 ± 1.944
Organic matter content ($w/w\%$)	64 ± 1.541	56 ± 1.581
Calorific value (J/g)	$12,894 \pm 73.986$	$10,532 \pm 51.088$
C/N ratio	25/1	25/1

The progress of Hosoya composting followed several steps:

Receipt of raw materials—storage, pre-treatment, and mixing: The raw materials were delivered by closed and covered manure transport vehicles. Dehydrated broiler and layer manure and litter were mixed (53% broiler manure and litter, 27% manure layer and litter, and 20% chicken meal). This mixture was stored until use in a closed manure storage building.

Storage of manure to oval tanks: Stored manure was transported to a loading hopper by front loaders. From the loading hopper, manure was transported to the entry points of the Japanese Hosoya-type manure oval tank system by belt feeders with rubber belts. The yearly capacity per a tank is 5000 mg/year.

Moisture content optimization: For optimal composting, the moisture of the raw material must be adjusted to 40–45 $w/w\%$ by adding sewage sludge (50 L/intake) and water (100 L/intake).

Composting: Controlled and monitored composting took place in the tanks. Proper ventilation was provided by a perforated pipe system at the bottom of said tanks, where the air was blown by a compressor. Depending on the technological need, it was possible to change the air temperature from 15 to 70 °C. The moisture content of the raw materials decreased to 22–28 $w/w\%$ by the end of the procedure. Due to the intensive mixing and aeration, very intensive microbiological processes took place in the raw material during the decomposing process. The temperature varied between 60 and 70 °C for several days. At this temperature, weed seeds, which may have come from the litter, already lost their ability to germinate, and the number of colonies of several pathogenic bacteria decreased. The stirring machine with double rotors resulted in continuous mixing of the manure and litter in the tanks. The system completed a full run along the oval tank in approximately 4 h, and the speed of the run was 0.8 m/min. On a daily basis, a maximum of six full runs were able to be completed. One complete run resulted in the displacement of 1.5 m of manure and litter along the tank or a maximum of 9.0 m after six runs completed in 24 h. Otherwise, the raw materials would have cooled down too quickly and this would have hindered the process. The design of the tanks and the applied operational technology ensured a 14-day time period traveling time for fresh manure and litter to reach the exit point as compost (Figure 2). Continuous operation ensured that the same amount of manure and litter entered the tank as the amount of compost leaving it. In the technology, 5 cm thick compost remained at the bottom of the tank as a microbial starter. This layer was mixed with the added amount of fresh manure and litter.

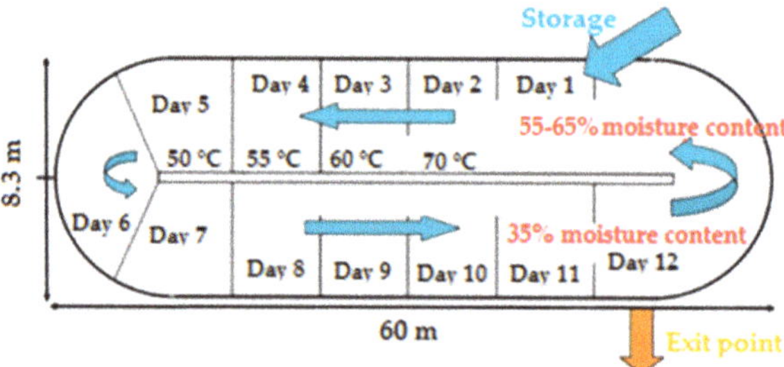

Figure 2. Poultry litter in the Hosoya oval composting tank from storage to the exit point.

Drying: Due to drying progress, it was further decreased from 22–28 to 10–11 $w/w\%$.

Grinding: The dried, heat-treated, and sterile compost raw material was ground into a powder fraction, which became the raw material of the end products.

Pre-storage—nutrient supplementation: The ground compost was supplemented with meat and bone meal as additional nutrients with an 8.6% N content before granulation.

Granulation: Granulation occurred after nutrient supplement.

Cooling: The pellets could reach 80–95 °C temperature after granulation, so it was required to cool down to 20–25 °C.

Aroma coating—packaging: The shaped and cooled pellets were coated with micro-components, fragrances, and biostimulators. Finally, the CPPL was packaged. As a result of the above process, the content of the end product was as follows (Table 2).

Table 2. Parameters of the end product.

Parameters		Parameters	
Moisture content ($w/w\%$)	12 ± 1.189	B content (mg/kg)	31.4 ± 1.155
Organic matter content ($w/w\%$)	69 ± 4.785	Fe content (mg/kg)	545 ± 13.976
Humus content ($w/w\%$)	51.84 ± 1.378	Mn content (mg/kg)	374 ± 14.230
N content ($w/w\%$)	5.5 ± 0.606	Mo content (mg/kg)	3.66 ± 0.482
P_2O_5 content ($w/w\%$)	3 ± 0.707	Zn content (mg/kg)	367 ± 39.438
K_2O content ($w/w\%$)	2.5 ± 0.408	Cu content (mg/kg)	53.3 ± 1.811
Ca content ($w/w\%$)	6 ± 0.770	pH	7.2 ± 0.532
Mg content ($w/w\%$)	0.5 ± 0.264	Calorific value (J/g)	$15,092 \pm 151.391$
S content ($w/w\%$)	1 ± 0.236	C/N ratio	$13/1$

The input flows for the production of 1 kg of CPPL are listed in Table 3. The inputs represent the energy and material flows required for LCA.

Table 3. The flow inputs per kilogram of composted pelletized poultry litter end product.

Flow of Inputs	Amount	Unit
Poultry manure, fresh	1.305	kg
Sludge, 4–6%DM	0.033	kg
Tap water	0.067	kg
Diesel, burned in building machine	0.087	MJ
Electricity, medium voltage	180.12	Wh
Packaging, solid fertilizers or pesticides	1.000	kg

A part of the data was provided by poultry manure treatment plant (manure, sludge, water, and fuel). However, data administration was based on our own calculations (electricity and emissions).

The inputs for the chemical fertilizers (AN, CAN, urea, TSP, MAP, and KCl) were provided by the Agribalyse database [55]. All parameters (e.g., raw materials, such as ammonia for AN, CAN, urea, and MAP, dolomite and nitric acid for CAN, phosphate rock for TSP and MAP, phosphoric acid for TSP, and potash salt for KCl; electricity; heat; steam in the chemical industry; tap water; and packaging) were included in the calculations, except for the transport processes to the application site, since transport is a highly changing variable in terms of distance, type of transport, and vehicle.

2.3. The Life Cycle Impact Assessment

In practice, LCA software is used to carry out life cycle impact assessments. The openLCA software was chosen for this life cycle assessment. Greendelta, a German software development company, created the software in 2006. The software is available for free download and use, and it allows for quick, accurate, and flexible modeling. The openLCA development team ensures that the software is updated on a regular basis.

There are several methods for assessing the impact of a project. The TRACI method, for example, is used in the United States. In Europe, the EcoIndicator, ReCiPe, ILCD, and CML methods are more widely used [56–59]. In this research, the CML 2001 impact assessment method was used. This method was created at the University of Leiden in the Netherlands in 1992, and its name is derived from the acronym Centrum voor Milieukunde (CML) [57]. The most significant influence of CML's methodology is in the field of "impact assessment". The aim of the CML method is to quantitatively explore all direct material and energy exchange relationships between the natural environment and the product system. On the one hand, the method is based on the assumption that emissions with the same effect can be summarized across media and, on the other hand, on the impact-oriented classification of material and energy flows for impact assessment. The method is in line with international standardization efforts, as it covers target definition (goal and scope), life cycle inventory (inventory analysis), impact analysis (impact assessment), and evaluation (interpretation of the results) [57].

Within the openLCA software, the Agribalyse database was used because it provides a large number of LCIs of agricultural products [60–62].

The impact of emissions and consumption on the environment is illustrated with the following impact categories based on other authors [63,64]:

1. Abiotic depletion potential for elements (kg Sb-eq) (ADPe): The 'abiotic depletion potential for elements' refers to the extent of the use of non-renewable sources and minerals. It shows the per capita use of antimony (Sb) and equivalent substances per year.
2. Abiotic depletion potential for fossil fuels (MJ) (ADPf): The 'abiotic depletion potential for fossil fuels' is shown in megajoules, instead of unit antimony equivalents (kg Sb-eq) of the resource.
3. Acidification potential (kg SO_2-eq) (AP): The acidification potential refers to compounds that cause acid rain (SO_2, NO_x, NO, and N_2O), usually denoted by the SO_2 equivalent.
4. Eutrophication potential (kg PO_4-eq) (EP): The eutrophication potential refers to the effects of over-fertilization or an excess supply of nutrients on terrestrial and aquatic environments, with a focus on the two most important nutrients, nitrogen (N) and phosphorus (P). Eutrophication is indicated as the PO_4 equivalent.
5. Global warming potential (kg CO_2-eq) (GWP): The global warming potential is an index improved by the impact of the comparison of different gases on the atmosphere. A higher value of the GWP means a more negative impact on the environment. The basis of the GWP is usually a period of 100 years as the CO_2 equivalent by its measurement.
6. Ozone layer depletion potential (kg CFC-11-eq) (ODP): To determine the ozone depleting potential, the CFC-11 equivalent is used to describe the emissions of all ozone-depleting substances.
7. Photochemical oxidation potential (kg C_2H_4-eq) (POP): The photochemical oxidation potential describes the ethylene equivalent emissions from photochemical oxidation due to a high NOx concentration.
8. Fresh water aquatic ecotoxicity potential (kg 1.4-DB-eq) (FAETP): This indicates the amount of contaminants in freshwater that have an impact on aquatic life pollution.
9. Human toxicity potential (kg 1.4-DB-eq) (HTP): The maximum concentration of compounds that are hazardous to humans.
10. Marine aquatic ecotoxicity potential (kg 1.4-DB-eq) (MAETP): The marine aquatic ecotoxicity potential shows the effects of different chlorine compounds in the atmosphere on marine life and aquatic environments.
11. Terrestrial ecotoxicity potential (kg 1.4-DB-eq) (TETP): This shows the impact of various chlorine compounds on the environment and on humans.

2.4. Methods for the Interpretation of LCA Results

During the interpretation of the LCA results, comparative analyses were carried out to assess the environmental impacts of CPPL. At first, the environmental impact of 1 kg of CPPL was assessed compared to chemical fertilizers.

Then, the environmental impacts were determined separately for the production of 1 kg of active substances (N, P_2O_5, and K_2O) (Table 4).

Table 4. Active substance content of fertilizers (N%), (P_2O_5%), and K_2O%) and the amount of fertilizer needed to apply 1 kg of the active substance.

Active Substance Content (%)		
Fertilizers	Nitrogen content (N%)	Fertilizer (kg) for 1 kg of N
Composted pelletized poultry litter (CPPL)	5.5	18
Ammonium nitrate (AN)	33.5	2.99
Calcium ammonium nitrate (CAN)	27	3.7
Urea	46	2.17
Monoammonium phosphate (MAP)	12	8.33
	Phosphorus pentoxide content (P_2O_5%)	Fertilizer (kg) for 1 kg of P_2O_5
Composted pelletized poultry litter (CPPL)	3	33.33
Triple superphosphate (TSP)	46	2.17
Monoammonium phosphate (MAP)	52	1.92
	Potassium chloride content (K_2O%)	Fertilizer (kg) for 1 kg of K_2O
Composted pelletized poultry litter (CPPL)	2.5	40
Potassium chloride (KCl)	60	1.66

Finally, the environmental impact of producing the nutrient supply of a 100 ha field was assessed and evaluated. The production of 1.5 Mg/ha of CPPL was compared to the production of the CPPL equivalent macro-element content of N, P, and K fertilizers combined. The application of 1.5 Mg/ha (as an optimum based on Szabó et al. [54]) of CPPL was 82.5 kg/ha of active N content, which is in line with Kátai's [65] recommendation of 80 kg/ha as the minimum N requirement for soils with low and medium nitrogen supplies. First, the active substances of 1.5. Mg/ha of CPPL were calculated, and then the CPPL equivalent quantity of the chemical fertilizers was determined for 100 ha (Table 5).

Table 5. NPK treatments of 100 ha of arable land based on the parameters of composted poultry granules.

	Quantity of Fertilizers (Mg/ha)	Quantity of Fertilizers Per 100 ha (Mg/100 ha)
CPPL	1.5	150
AN	0.246	24.6/21.5 *
CAN	0.305	30.5/26.7 *
Urea	0.18	18/15.7 *
TSP	0.096	9.6
MAP	0.086	8.6
KCl	0.063	6.25

* Quantity of N fertilizers when the P fertilizer was MAP (considering the N content of MAP).

In order to supply the CPPL equivalent N, P, and K dosages on a 100 ha field, six combinations of chemical fertilizers were set, and the overall quantity of the combinations was determined (Table 6).

Table 6. Different treatments of the N, P, and K fertilizers.

Name of Combination	NPK Combination	Mg/100 ha
NPK1	AN + TSP + KCl	40.45
NPK2	AN + MAP + KCl	36.35
NPK3	CAN + TSP + KCl	46.15
NPK4	CAN + MAP + KCl	41.51
NPK5	Urea + TSP + KCl	33.85
NPK6	Urea + MAP + KCl	30.59

The environmental impact of CPPL and the six combinations of chemical fertilizers for fertilization of 100 ha was calculated using the quantity required for NPK fertilization of 100 ha and the previously calculated environmental impacts of 1 kg of CPPL and chemical fertilizers.

The CPPL and NPK combinations were identified to have low, medium, and high environmental impact. Three categories were defined based on dividing the difference between the maximum and minimum environmental impact category values into three equal intervals.

3. Results

3.1. Environmental Impact by Producing 1 kg of CPPL and Chemical Fertilizers

The environmental impact of CPPL production and different chemical fertilizers was evaluated per kilogram of the end product (Table 7). Out of the 11 impact categories, 9 cases (ADPe, ADPf, GWP, ODP, POP, FAETP, HTP, MAETP, and TETP) of CPPL production had the smallest environmental impact.

Table 7. Impact assessment of the production of 1 kg of CPPL and fertilizers.

Impact Categories	CPPL	AN	CAN	Urea	TSP	MAP	KCl
ADPe (kg Sb-eq)	7.57×10^{-8}	6.47×10^{-6}	6.37×10^{-6}	7.43×10^{-6}	4.10×10^{-7}	6.70×10^{-6}	4.76×10^{-6}
ADPf (MJ)	0.269	18.338	14.941	27.107	13.987	8.898	4.121
AP (kg SO_2-eq)	0.024	0.006	0.005	0.005	0.010	0.003	0.002
EP (kg PO_4-eq)	0.005	0.002	0.002	0.002	0.004	0.002	0.001
GWP (kg CO_2-eq)	0.273	1.382	1.137	1.127	0.657	0.826	0.399
ODP (kg CFC-11-eq)	3.48×10^{-8}	1.50×10^{-7}	1.23×10^{-7}	2.25×10^{-7}	1.01×10^{-7}	8.54×10^{-8}	3.73×10^{-8}
POP (kg C_2H_4-eq)	2.87×10^{-5}	1.35×10^{-4}	1.17×10^{-4}	1.95×10^{-4}	4.29×10^{-4}	1.32×10^{-4}	7.97×10^{-5}
FAETP (kg 1.4-DB-eq)	0.028	0.274	0.256	0.314	0.198	0.362	0.188
HTP (kg 1.4-DB-eq)	0.032	0.449	0.429	0.534	0.172	0.502	0.334
MAETP (kg 1.4-DB-eq)	47.419	663.080	616.340	790.531	523.135	833.587	504.535
TETP (kg 1.4-DB-eq)	3.14×10^{-4}	1.51×10^{-3}	1.46×10^{-3}	1.82×10^{-3}	5.08×10^{-3}	6.48×10^{-3}	8.61×10^{-4}

In the case of the abiotic depletion potential for elements, the best performing chemical fertilizer was TSP, but it was still five times higher than that of CPPL. For the abiotic depletion potential for fossil fuels, the environmental impact of producing 1 kg of CPPL was 93–99% smaller than the chemical fertilizer production.

Only the acidification and eutrophication potentials were the highest in the production of CPPL. The environmental impact of 1 kg of any chemical fertilizer production was 58–93% smaller in the case of the acidification potential and 24–88% smaller in the case of the eutrophication potential compared to CPPL production.

Among the chemical fertilizers, the GWP was the smallest in the production of KCl fertilizer. The highest emissions were found in the N fertilizers, especially in AN. The production of AN produced a five times higher GWP than CPPL.

The ozone depletion potential was the lowest in CPPL production and in KCl (7% higher than CPPL), while urea had the highest (85% higher than CPPL).

In comparison, the environmental impact of CPPL production was 64–93% smaller in the case of the photochemical oxidation potential. The smallest emission value was calculated for KCl production, while the highest was for TSP amongst the chemical fertilizers.

In addition, of the chemical fertilizers, KCl fertilizer production generated the smallest emissions in the fresh water aquatic ecotoxicity potential (seven times higher than CPPL), the marine aquatic ecotoxicity potential (11 times higher than CPPL), and the terrestrial ecotoxicity potential (three times higher than CPPL). The highest emissions were produced by MAP production. The emission values of CPPL production were 92–95% smaller than the MAP production in the case of these impact categories.

The human toxicity potential was the smallest for TSP, while urea had the highest. The emissions from CPPL production were 81–94% smaller than that of the chemical fertilizers.

In summary, for the production of 1 kg of product, CPPL had the lowest environmental impact in 9 out of the 11 impact categories (ADPe, ADPf, GWP, ODP, POP, FAETP, HTP, MAETP, and TETP), while only 2 impact categories (AP and EP) had a higher environmental impact than the chemical fertilizer production.

3.2. Environmental Impact by Producing of 1 kg of Active Substance

The environmental impact was determined for 1 kg of active substance (N, P_2O_5, and K_2O) in addition to 1 kg of end product. Accordingly, the AN, CAN, and urea fertilizers were included for 1 kg of the N active substance, while the TSP and MAP fertilizers were included for 1 kg of the P_2O_5 active substance and the KCl fertilizer was included for 1 kg of the K_2O active substance content of CPPL.

3.2.1. Environmental Impact by Producing of 1 kg of the Nitrogen Active Substance

Based on the emissions during production, a comparison of the CPPL product and the most major N fertilizers (AN, CAN, and urea) was carried out (Table 8). First, 1 kg of the N active substance was the functional unit. In 6 out of the 11 impact categories (AP, EP, GWP, ODP, POP, and TETP), the environmental impact was higher for CPPL production than for the N fertilizers.

Table 8. Impact assessment of the production of 1 kg of nitrogen content.

Impact Categories	CPPL (5.5% N)	AN (33.5% N)	CAN (27% N)	Urea (46% N)
ADPe (kg Sb-eq)	1.38×10^{-6}	9.06×10^{-6}	2.36×10^{-5}	1.61×10^{-5}
ADPf (MJ)	4.883	54.831	55.283	58.822
AP (kg SO_2-eq)	0.439	0.019	0.019	0.010
EP (kg PO_4-eq)	0.099	0.007	0.007	0.004
GWP (kg CO_2-eq)	4.955	4.133	4.208	2.445
ODP (kg CFC-11-eq)	6.33×10^{-7}	4.48×10^{-7}	4.57×10^{-7}	4.88×10^{-7}
POP (kg C_2H_4-eq)	5.23×10^{-4}	4.04×10^{-4}	4.32×10^{-4}	4.23×10^{-4}
FAETP (kg 1.4-DB-eq)	0.518	0.819	0.947	0.681
HTP (kg 1.4-DB-eq)	0.586	1.341	1.588	1.158
MAETP (kg 1.4-DB-eq)	862.070	1982.609	2280.459	1715.452
TETP (kg 1.4-DB-eq)	0.006	0.005	0.005	0.004

In the instance of CAN, the highest abiotic depletion potential for elements was estimated (17 times higher than CPPL). The values of AN and urea were 6.5 and 11 times higher than those of CPPL.

Chemical fertilizers demonstrated abiotic depletion potential for fossil fuels values ranging from 54.8 (AN) to 58.8 MJ/kg N (urea), while CPPL had a value of less than a 10th of these.

The acidification potential of the AN and CAN fertilizers was approximately equal, and the emissions from urea production were the smallest. The acidification potential of CPPL was 96–98% higher than that of the N fertilizers.

CPPL's eutrophication potential was considerably higher than that of the nitrogen fertilizers. The estimated emission values of AN and CAN were equal, while urea had the smallest EP. The emissions show a 15–26 times difference between CPPL and the N fertilizers.

CPPL had the highest global warming potential (nearly 5 kg CO_2/kg of the N active substance). CAN and AN were close to 4 kg (on average, 16% less emissions than CPPL). Urea had the lowest global warming potential value, which was nearly half that of CPPL.

In comparison to the nitrogen fertilizers, the ozone depletion, photochemical oxidation, and terrestrial toxicity potential values were all higher for CPPL. During the production of CPPL, the values of the above-mentioned impact categories were, on average, 30% higher than in the case of AN, CAN, and urea production.

During the production of CPPL, the values for the impact categories such as the freshwater aquatic ecotoxicity and marine aquatic ecotoxicity potentials, as well as the human toxicity potential, were the smallest. The emissions from the production of CAN were the highest of the impact categories: the freshwater aquatic ecotoxicity potential was 45% higher, while the human toxicity and marine aquatic ecotoxicity potentials were 62–63% higher than CPPL production. The emissions from the production of urea were the lowest of the three N fertilizers in these three impact categories. The freshwater aquatic ecotoxicity potential was 24% higher, while the human toxicity and marine aquatic ecotoxicity potentials were 49–50% higher than CPPL production.

Although only five impact categories (ADPe, ADPf, FAETP, HTP, and MAETP) had lower environmental impacts for CPPL, it should be taken into account that N fertilizers have a much higher N content and were much more concentrated.

3.2.2. Environmental Impact by Producing of 1 kg of the Phosphate Active Substance

The impact assessment of emissions was carried out during the production of phosphate fertilizers in the same way as it was for the N fertilizers. The functional unit used in the comparison of CPPL, TSP, and MAP was 1 kg of the P_2O_5 active substance (Table 9). The environmental impact of CPPL was highest for 6 out of the 11 impact categories (AP, EP, GWP, POP, FAETP, and HTP), while for ADPe and MAETP, CPPL was the second largest emitter.

Table 9. Impact assessment of the production of 1 kg of phosphate content.

Impact Category	CPPL (3% P_2O_5)	TSP (46% P_2O_5)	MAP (52% P_2O_5)
ADPe (kg Sb-eq)	2.52×10^{-6}	8.90×10^{-7}	1.29×10^{-5}
ADPf (MJ)	8.952	30.352	17.085
AP (kg SO_2-eq)	0.804	0.022	0.007
EP (kg PO_4-eq)	0.181	0.009	0.003
GWP (kg CO_2-eq)	9.084	1.426	1.587
ODP (kg CFC-11-eq)	1.16×10^{-6}	2.20×10^{-7}	1.64×10^{-7}
POP (kg C_2H_4-eq)	0.0010	0.0009	0.0003
FAETP (kg 1.4-DB-eq)	0.949	0.429	0.694
HTP (kg 1.4-DB-eq)	1.074	0.372	0.965
MAETP (kg 1.4-DB-eq)	1580.462	1135.203	1600.487
TETP (kg 1.4-DB-eq)	0.010	0.011	0.012

The value of the abiotic depletion potential for elements was the highest in the production of MAP, being 5 times higher than CPPL and 14 times higher than TSP.

The abiotic depletion potential for fossil fuels was smallest for CPPL, being roughly half that of the value of MAP and one-third that of TSP.

In terms of the acidification potential, the emissions during the production of CPPL were, on average, 98% higher than the acidification potential of TSP and MAP.

The highest emissions based on the eutrophication potential were calculated for the production of CPPL. In comparison to the emissions of TSP and MAP production, P_2O_5 emissions per kilogram were 20 and 56 times higher, respectively.

In the case of the P fertilizers, the values of GWP were similar. The production of CPPL, on the other hand, emitted 83–84% greater CO_2 than the P fertilizers.

Similarly to GWP, the ozone-depleting potential value for CPPL was the highest. CPPL produced emissions that were more than 80% higher than that of the P fertilizers.

The value of the photochemical oxidation potential was the lowest for MAP production. The emission rates for CPPL and TSP were nearly similar. These results were 73–74% higher than the emissions produced by the MAP production process.

The freshwater aquatic ecotoxicity and human toxicity potential values were the highest in the production of CPPL. In comparison to CPPL, the freshwater aquatic ecotoxicity potential was 55% smaller during TSP production and 27% smaller during the production of MAP. In the case of the human toxicity potential, the TSP emissions were the lowest, whereas the MAP production emissions were only 10% lower than in the case of CPPL.

The greatest emissions in the production of MAP were observed in both the marine aquatic and terrestrial ecotoxicity potentials. In terms of the marine aquatic ecotoxicity potential, the MAP and CPPL emissions were nearly similar. TSP production had a 28% lower emission rate than CPPL production. For the terrestrial ecotoxicity potential, the emission value of CPPL production was the lowest. TSP production was 5% higher than the emissions of CPPL, while MAP production was 19% higher.

It can be concluded that the emissions were clearly lower in just three cases—for ADPf, ODP, and TETP—during the production of CPPL. However, it must be taken into account that the phosphate content of CPPL (approximately 3%) was lower than that of the fertilizers.

3.2.3. Environmental Impact by Producing 1 kg of Potassium Content

A comparison was made based on the emissions of CPPL and KCl fertilizer production. For this, 1 kg of the K_2O active substance served as a functional unit (Table 10). Only ADPe had a lower environmental impact than CPPL (two and a half times lower), while KCl had a lower emission value for the other 10 impact categories.

Table 10. Impact assessment of the production of 1 kg of the potassium substance.

Impact Category	CPPL (2.5% K_2O)	KCl (60% K_2O)
ADPe (kg Sb-eq)	3.03×10^{-6}	7.90×10^{-6}
ADPf (MJ)	10.744	6.840
AP (kg SO_2-eq)	0.965	0.003
EP (kg PO_4-eq)	0.218	0.001
GWP (kg CO_2-eq)	10.901	0.663
ODP (kg CFC-11-eq)	1.39×10^{-6}	6.19×10^{-8}
POP (kg C_2H_4-eq)	0.0011	0.0001
FAETP (kg 1.4-DB-eq)	1.139	0.313
HTP (kg 1.4-DB-eq)	1.289	0.554
MAETP (kg 1.4-DB-eq)	1896.744	837.528
TETP (kg 1.4-DB-eq)	0.013	0.001

In terms of the abiotic depletion potential for fossil fuels, the production of KCl fertilizer emitted 36% less than CPPL.

The amount of acidification and eutrophication potentials during the production of CPPL was 99% higher than that of the production of KCl.

In comparison to KCl fertilizer production, the global warming and ozone depleting potential values were 94% and 96% higher during the production of CPPL, respectively.

The emission rates were similar according to the results of the photochemical oxidation and terrestrial ecotoxicity potentials. In both impact categories, CPPL production had an 89% higher environmental impact.

In the case of the human toxicity and marine aquatic ecotoxicity potential impact categories, similar rates were obtained. The emission rate of CPPL production was 56–57% higher per 1 kg of the K_2O active substance. The freshwater aquatic ecotoxicity potential values for the two products were different: the KCl fertilizer production emitted 73% lower emissions than CPPL production.

However, it should be noted that (as with the N and P_2O_5 substances) CPPL contained much less K_2O (2.5% K_2O) than the more concentrated KCl fertilizer (60% K_2O).

3.3. Environmental Impact of a Medium-Sized Farm's Nutrient Replenishment

The environmental impact of producing the nutrient supply of a 100 ha field was assessed and evaluated. The production of 1.5 Mg/ha of CPPL (150 Mg/100 ha) was compared to the production of the equivalent macro-element contents of the N, P, and K fertilizers combined (Table 11).

Table 11. Environmental emissions generated by the production of the applied CPPL and NPK treatments on 100 ha of arable land.

Impact Category	CPPL	NPK1	NPK2	NPK3	NPK4	NPK5	NPK6
ADPe (kg Sb-eq)	0.011	0.193	0.227	0.228	0.257	0.167	0.205
ADPf (MJ)	40,290	614,640	496,928	618,834	500,097	648,453	529,320
AP (kg SO_2-eq)	3620	262.9	173.3	265.3	175.3	196.0	115.6
EP (kg PO_4-eq)	816.1	98.7	65.8	101.0	67.7	75.8	46.1
GWP (kg CO_2-eq)	40,880	43,005	39,357	43,654	39,886	29,113	27,372
ODP (kg CFC-11-eq)	0.0052	0.0049	0.0042	0.0050	0.0043	0.0053	0.0045
POP (kg C_2H_4-eq)	4.31	8.02	4.54	8.25	4.74	8.18	4.71
FAETP (kg 1.4-DB-eq)	4,270	9862	10,192	10,923	11,109	8731	9241
HTP (kg 1.4-DB-eq)	4833	14818	16,069	16,868	17,841	13,323	14,823
MAETP (kg 1.4-DB-eq)	7112,789	24607,329	24610,646	27084,044	26749,074	22428,863	22797,379
TETP (kg 1.4-DB-eq)	47.06	92.30	93.98	99.47	100.18	87.60	90.08

Green color = low environmental impact; yellow color = medium environmental impact; red color = high environmental impact.

With the production of 150 Mg of CPPL, the abiotic depletion potential for elements and abiotic depletion potential for fossil fuels indicators were the smallest. Meanwhile, the production of various fertilizer combinations produced 93–95% higher emissions than CPPL on average.

However, compared to the chemical fertilizer treatments, the environmental impact of 150 Mg of CPPL production was higher in terms of the acidification potential (94% on average) and the eutrophication potential (90% on average). The possible reason for this, based on de Vries et al.'s [66] research, is that the main contributor to the high acidification potential of CPPL is ammonia, while the high eutrophication potential is due to nitrate emissions.

In comparison to the NPK1 and NPK3 combinations, CPPL produced a smaller global warming potential by 5.5% on average. CPPL's GWP values were similar to those of the NPK2 and NPK4 combinations. The GWP of those combinations where the N fertilizer was urea (NPK5 and NPK6) was 29–33% lower than CPPL due to low environmental impact of urea production, because urea is the most concentrated nitrogen fertilizer (46% N) and smaller amounts of it cover the desired quantity.

There was no substantial difference between the NPK combinations and CPPL in terms of the ozone depletion potential (the values varied between 0.0042 and 0.0053 kg CFC-11-eq).

In those combinations where TSP was used for the P fertilizer (NPK1, NPK3, and NPK5), the photochemical oxidation potential value was, on average, 47% higher than the during the production of CPPL. Meanwhile, the value of this category was 7.5% higher when MAP was used.

The emissions were similar for the impact categories of the freshwater aquatic eco-toxicity potential, the marine aquatic ecotoxicity potential, the human toxicity potential, and the terrestrial ecotoxicity potential. When comparing the production of NPK to the production of CPPL, the freshwater ecotoxicity potential values were around 57% higher, on average, for the NPK fertilizer combinations. The emission value during the production of the chemical fertilizer combinations was, on average, 69–71% higher than the production of CPPL in terms of the human toxicity and marine aquatic ecotoxicity potentials. The terrestrial ecotoxicity potential was also higher in the production of chemical fertilizers, with an average of 50%.

Overall, Table 11 shows that the production of CPPL has a lower environmental impact than the production of equivalent macro-nutrient chemical fertilizers (7 out of the 11 impact categories were of "low environmental impact"). Out of all the chemical fertilizer combinations, NPK5 had the most favorable environmental impact. The NPK1, NPK2, NPK3, and NPK4 fertilizer combinations had the highest environmental impact, since AN or CAN fertilizers were the N source.

Additional calculations were carried out to determine the transportation-related environmental emissions associated with the CPPL and NPK fertilizer combinations mentioned above. To estimate distances of 10, 20, 50, and 100 km, a vehicle with a carrying capacity of 15 t was used. The results of CPPL and NPK fertilizer production were added to the emission data. As a result, transportation had no significant effect on the production-induced changes in emission rates.

4. Discussion

Although this study evaluated 11 environmental impact categories, the most extensively used and calculated impact category, the global warming potential, was used in order to understand the relevance of the calculated environmental impacts of CPPL and chemical fertilizers.

Since there is a lack of scientific knowledge in the field of environmental impact assessment that includes Hosoya technology's composted and pelletized poultry litter products, the CPPL product was compared with other organic matter treated with semi-closed and closed composting methods (Table 12).

Fresh laying hen manure and carcasses were composted by Zhu et al. [67]. When compared to CPPL based on Hosoya technology, the investigated composting technology emitted 3–6 times less CO_2. Although, during the 11-week experiment, these compost piles were only remixed and reconstructed once compared to the continuously mixed Hosoya compost. Fresh air was introduced into each compost bin via an air distribution plate to provide ventilation.

Table 12. Global warming potential (kg CO_2-eq) values based on a comparison between different composting technologies by scientific publications.

	kg CO_2-eq/ kg of Product	Country/Region	Reference
Hen carcasses and manure	0.045–0.082	China	[67]
Sludge	0.089–0.298	Europe	[68]
Chicken and cow manure	0.147	Egypt	[69]
Poultry manure	0.27	Europe	This study
Livestock waste	0.475–2.307	Europe	[68]

The emissions were studied during the sewage sludge composting process in the frame of the scientific research program by ADAME [68]. According to their results, the observed emissions ranged between 0.089 and 0.298 kg CO_2-eq. In the framework of the ADAME program, the emissions from composting livestock waste were also evaluated. The measured emission rate in this investigation was five times higher than the emission value from the Hosoya composting technology.

Luske [69] examined the composting of chicken and cattle manure. The emissions generated from this composting plant were approximately 50% less (0.147 kg CO_2-eq/kg of the product) compared to the emissions of Hosoya composting technology. The study also demonstrated that the composition and proportion of the input components have a major impact on emissions. Furthermore, composting technology (mixing, aeration, etc.) plays an important role in GWP production. The continuously mixed Hosoya composting technology investigated in this study was moderate compared to the results of other studies.

The environmental effects of chemical fertilizer production have been widely studied. The CO_2-equivalent gas emissions calculated for 1 kg of active substances were summarized to evaluate the results of the present study with other research works (Table 13).

The emissions from the production of AN varied between 3.5 and 7.2 kg CO_2 eq/kg of N at the European level. The highest emission value was measured in China, where 10 kg of CO_2-equivalent emissions per 1 kg of N-substance were detected [70].

The emission factor of CAN production was also smaller in Europe. The rate of the N substance was 3.7 kg CO_2-eq [70], while it was 4.2 kg CO_2-eq for 1 kg of the N substance in this study. The emission factors in Russia, the USA, and China varied between 7.7 and 10.6 kg CO_2-eq.

In the case of urea production-related emissions, the European emission factors were as follows: 1.6 [71], 1.9 [69], and 3.5 kg CO_2-eq/kg of N [72]. The value observed in this study was approximately the average of these three factors (2.4 kg CO_2-eq/kg of N). Higher emission values were calculated in China [70].

Among the P fertilizers, the global warming potential values generated by the production of TSP range from 0.4 to 1.6 kg CO_2-eq/kg of P_2O_5 in Europe and throughout the world (1.42 kg CO_2-eq/kg of P_2O_5 in this study) [70–72].

The emissions from the production of MAP are already much more variable. According to Brentrup et al. [70], the average emission factor is 1.4 in Europe, 1.7 in Russia and the USA, and 2.89 kg CO_2-eq/kg of P_2O_5 in China. The factor calculated in present study is between the former two values (1.6 kg CO_2-eq/kg of P_2O_5). Albaugh et al. [73] recorded a much higher factor in the USA, which was 6.4 kg CO_2-eq/kg of P_2O_5. Based on Zhang et al.'s work [74], it varies between 7.8 and 8.9 kg CO_2-eq/kg of P_2O_5 in China.

Table 13. Global warming potential (kg CO_2-eq) values generated by the production of different chemical fertilizers.

Chemical Fertilizers	kg CO_2-eq/ kg of Active Substance	Country/Region	Reference
Ammonium nitrate (kg CO_2-eq/kg of N)	4.1	Europe	This study
	6.2	Europe	[71]
	7.2	United Kingdom	[72]
	3.5/8/10.3	Europe/Russia, USA/China	[70]
Calcium ammonium nitrate (kg CO_2-eq/kg of N)	3.7/7.7/8.7/10.6	Europe/Russia/USA/China	[70]
	4.2	Europe	This study
Urea (kg CO_2-eq/kg of N)	1.6	Europe	[71]
	1.9/2.7/5.5	Europe/Russia, USA/China	[70]
	2.4	Europe	This study
	3.1	Southeastern USA	[73]
	3.5	United Kingdom	[72]
Triple superphosphate (kg CO_2-eq/kg of P_2O_5)	0.4–0.54	Russia, USA, China	[70]
	1.2	United Kingdom	[72]
	1.43	Europe	This study
	1.6	Europe	[71]
Monoammonium phosphate (kg CO_2-eq/kg of P_2O_5)	1.4/1.7/2.89	Europe/Russia, USA/China	[70]
	1.6	Europe	This study
	6.4	Southeastern USA	[73]
	7.8–8.9	China	[74]
Potassium chloride (kg CO_2-eq/kg of K_2O)	0.14–0.25	China	[75]
	0.23	Europe	[73]
	0.36	New Zealand	[76]
	0.55	China	[74]
	0.66	Europe	this study

The global warming potential of the N and P fertilizer production in this study is similar to that of other studies in Europe. In general, it was also found that China has the highest values.

The emission factor (0.66 kg CO_2-eq/kg of K_2O) in the present study for the production of the KCl fertilizer was close to the highest calculated value in China (0.55 kg CO_2 equivalent) [74]. Based on other studies, the rate of this emission factor is between 0.14 and 0.36 kg CO_2-eq/kg of K_2O [73,75,76].

Since Hungary is in the region of Central Eastern Europe, the global warming potential of CPPL production for utilization of 100 ha was also assessed in Europe, including references for Russia. NPK combinations were calculated based on the relevant GWP references for Europe and Russia (listed in Table 13) using the method applied in this study to calculate NPK fertilizer combinations, which is described in Tables 5 and 6. The GWPs were calculated from the average of the European and Russian global warming potential values (Table 14).

Table 14. Global warming potential values from other scientific publications compared to this study.

Impact Category	CPPL	NPK1	NPK2	NPK3	NPK4	NPK5	NPK6
GWP (kg CO_2-eq)	40,880	55,693	50,449	50,717	46,106	27,933	26,197

The global warming potential for producing 150 Mg of CPPL was 40,880 kg CO_2-eq. In comparison, the average CO_2 equivalent emissions of NPK1 were 27% higher than those of CPPL. The GWP of NPK2 and NPK3 was 19% higher than that of CPPL, while NPK4 had an 11% higher environmental impact. However, for the combinations of NPK5 and NPK6, where the N fertilizer was urea, the environmental impact was 32–36% smaller.

5. Conclusions

As a final statement, considering the environmental impact by producing 1 kg of active substances, CPPL has a higher environmental impact compared to individual chemical fertilizers. On the contrary, considering that CPPL provides nutrients as a complex fertilizer, the CPPL equivalent combinations of chemical fertilizers have a higher impact in the case of the abiotic depletion potential for elements and abiotic depletion potential for fossil fuels, the photochemical oxidation potential, the human toxicity potential, the freshwater and marine aquatic ecotoxicity potentials, and the terrestrial ecotoxicity potential.

Considering the results, the nutrient (NPK) supply of a 100 ha field with 1.5 Mg/ha of CPPL, as well as combinations of chemical fertilizers with an equivalent NPK supply, CPPL is a potential alternative for the complex fertilization of arable lands. The only exemption is in those cases when urea was used in the NPK combinations (NPK5 and NPK6), due to its low environmental impact. Thus, CPPL can be used as a substitute for chemical fertilizer combinations where N replenishment is not provided by urea. Nevertheless, CPPL provides organic components; a high micro-element content; a number of other beneficial effects on soil fertility, structure, and organic matter content; and water management properties. Therefore, in further research, not only NPK but the micro-element content of CPPL too shall be included in further investigation to assess CPPL as a potential macro- and micro-element complex fertilizer alternative for sets of chemical fertilizers.

Author Contributions: Conceptualization, A.N. and J.T.; methodology, N.É.K. and A.N.; software, N.É.K.; investigation, N.É.K., E.G. and N.S.; writing—original draft preparation, N.É.K. and A.N.; All authors have read and agreed to the published version of the manuscript.

Funding: The APC was funded by EFOP-3.6.3-VEKOP-16-2017-00008 project. The project is cofinanced by the European Union and the European Social Fund.

Institutional Review Board Statement: Not applicable.

Informed Consent Statement: Not applicable.

Data Availability Statement: The datasets generated and/or analyzed during the current study are available from the corresponding author on reasonable request.

Acknowledgments: This research was supported by EU grant to Hungary GINOP 2.2.1.-15-2017-00043. The publication is supported by the EFOP-3.6.3-VEKOP-16-2017-00008 project. The project is co-financed by the European Union and the European Social Fund.

Conflicts of Interest: The authors declare that they have no conflict of interest.

References

1. Scholl, L.; Nieuwenhuis, R. *Soil Fertility Management*; Agromisa Foundation: Wageningen, The Netherlands, 2004; pp. 48–55.
2. Chen, J.-H.; Wu, J.-T.; Young, C. The Combined Use of Chemical and Organic Fertilizers and/or Biofertilizer for Crop Growth and Soil Fertility. *Environ. Sci.* **2007**, *10*, 1–12.
3. Han, S.H.; Young, J.; Hwang, J.; Kima, S.B.; Parka, B. The Effects of Organic Manure and Chemical Fertilizer on the Growth and Nutrient Concentrations of Yellow Poplar (Liriodendron tulipifera Lin.) in a Nursery System. *For. Sci. Technol.* **2016**, *12*, 137–143.
4. Alimi, T.; Ajewole, O.C.; Awosola, O.; Idowu, E.O. Organic and Inorganic Fertilizer for Vegetable Production under Tropical Conditions. *J. Agric. Rural. Dev.* **2007**, *1*, 120–136.
5. Savci, S. Investigation of Effect of Chemical Fertilizers on Environment. *APCBEE Procedia* **2012**, *1*, 287–292. [CrossRef]
6. Adediran, J.A.; Taiwa, L.B.; Akande, M.O.; Sobulo, R.A.; Idown, O.J. Application of Organic and Inorganic Fertilizer for Sustainable Maize and Cowpea Yield in Nigeria. *J. Plant Nutr.* **2004**, *27*, 1163–1181. [CrossRef]
7. Bíró, T.; Tamás, J.; Thyll, S. Risk assessment of nitrate pollution in lower watershed of the Berettyó River. In *Soil Water Environment Relationships*; Wageningen–Debrecen; Filep, G., Ed.; Wageningen University and Research: Wageningen, The Netherland; University of Debrecen: Debrcen, Hungary, 1998; pp. 239–247.
8. Dhadli, H.S.; Brar, B.S. Effect of long-term differential application of inorganic fertilizers and manure on soil CO_2 emissions. *Plant Soil Environ.* **2016**, *62*, 195–201.
9. Adviento-Borbe, M.A.A.; Kaye, J.P.; Bruns, M.A.; McDaniel, M.D.; McCoy, M.; Harkcom, S. Soil Greenhouse Gas and Ammonia Emissions in Long-Term Maize-Based Cropping Systems. *Soil Biol. Biochem.* **2010**, *74*, 1623–1634. [CrossRef]
10. Mbonimpa, E.G.; Hong, C.O.; Owens, V.N.; Lehman, M.R.; Osborne, S.L.; Schumacher, T.E.; Clay, D.E.; Kumar, S. Nitrogen fertilizer and landscape position impacts on CO_2 and CH_4 fluxes from a landscape seeded to switchgrass. *Glob. Chang. Biol.-Bioenergy* **2014**, *7*, 836–849. [CrossRef]
11. Dhadli, H.S.; Brar, B.S.; Kingra, P.K. Temporal Variations in N_2O Emissions in Maize and Wheat Crop Seasons: Impact of N-Fertilization, Crop Growth, and Weather Variables. *J. Crop Improv.* **2016**, *30*, 17–31. [CrossRef]
12. Peng, Q.; Dong, Y.S.; Qi, Y.C.; Xiao, S.S.; He, Y.T.; Ma, T. Effects of nitrogen fertilization on soil respiration in temper-ate grassland in Inner Mongolia, China. *Environ. Earth Sci.* **2011**, *62*, 1163–1171. [CrossRef]
13. Zhang, J.J.; Li, Y.F.; Jiang, P.K.; Zhou, G.M.; Shen, Z.M.; Liu, J.; Wang, Z.L. Effects of fertilization on soil CO_2 flux in Castanea mollissima stand. *J. Appl. Ecol.* **2013**, *24*, 2431–2439.
14. Nyamadzawo, G.; Wuta, M.; Nyamangara, J.; Smith, J.L.; Rees, R.M. Nitrous oxide and methane emissions from cultivated seasonal wetland (dambo) soils with inorganic, organic and integrated nutrient management. *Nutr. Cycl. Agroecosystems* **2014**, *100*, 161–175. [CrossRef]
15. Khan, A.; Tan, D.K.Y.; Munsif, F.; Afridi, M.Z.; Shah, F.; Wei, F.; Fahad, S.; Zhou, R. Nitrogen nutrition in cotton and control strategies for greenhouse gas emissions: A review. *Environ. Sci. Pollut. Res.* **2017**, *24*, 23471–23487. [CrossRef] [PubMed]
16. Shao, R.; Deng, L.; Yang, Q.H.; Shangguan, Z.P. Nitrogen fertilization increase soil carbon dioxide efflux of winter wheat field: A case study in Northwest China. *Soil Tillage Res.* **2014**, *143*, 164–171. [CrossRef]
17. Zamanian, K.; Zarebanadkouki, M.; Kuzyakov, Y. Nitrogen fertilization raises CO_2 efflux from inorganic carbon: A global assessment. *Glob. Chang. Biol.* **2018**, *24*, 2810–2817. [CrossRef]
18. Zhang, L.H.; Shao, H.B.; Wang, B.C.; Zhang, L.W.; Qin, X.C. Effects of nitrogen and phosphorus on the production of carbon dioxide and nitrous oxide in salt-affected soils under different vegetation communities. *Atmos. Environ.* **2019**, *204*, 78–88. [CrossRef]
19. Kong, D.L.; Li, S.Q.; Jin, Y.G.; Wu, S.; Chen, J.; Hu, T.; Wang, H.; Liu, S.; Zou, J. Linking methane emissions to methanogenic and methanotrophic communities under different fertilization strategies in rice paddies. *Geoderma* **2019**, *347*, 233–243. [CrossRef]
20. Lin, S.; Zhang, A.; Shen, G.; Shaaban, M.; Ju, W.; Cui, Y.; Duan, C.; Fang, L. Effects of inorganic and organic fertilizers on CO_2 and CH_4 fluxes from tea plantation soil. *Sci. Anthr.* **2021**, *9*, 090. [CrossRef]
21. COM. *Communication from the Commission to the European Parliament, the European Council, the Council, the European Economic and Social Committee and the Committee of the Regions*; The European Green Deal: Brussels, Belgium, 2019.
22. Moyo, S.; Swanepoel, F.J.C. Multifunctionality of livestock in developing communities. In *The Role of Livestock in Developing Communities: Enhancing Multifunctionality*, 1st ed.; Swanepoel, F.J.C., Stroebel, A., Moyo, S., Eds.; University of Free State (UFS) and the technical Centre for Agricultural and Rural Cooperation (CTA): Cape Town, South Africa; Wageningen, The Netherlands, 2010; pp. 1–11.
23. Mézes, L.; Nagy, A.; Gálya, B.; Tamás, J. Poultry feather wastes recycling possibility as soil nutrient. *Eurasian J. Soil Sci.* **2015**, *4*, 244–252. [CrossRef]

24. Magnusson, U. *Sustainable Global Livestock Development for Food Security and Nutrition Including Roles for Sweden*; Ministry of Enterprise and Innovation: Stockholm, Sweden; Swedish FAO Committee: Stockholm, Sweden, 2016.

25. Gorliczay, E.; Boczonádi, I.; Kiss, N.É.; Tóth, F.A.; Pabar, S.A.; Biró, B.; Kovács, L.R.; Tamás, J. Microbiological Effectivity Evaluation of New Poultry Farming Organic Waste Recycling. *Agriculture* **2021**, *11*, 683. [CrossRef]

26. He, Z. Organic Animal Farming and Comparative Studies of Conventional and Organic Manures. In *Animal Manure: Production, Characteristics, Environmental Concerns, and Management*, 1st ed.; Waldrip, H.M., Pagliari, P.H., He, Z., Eds.; American Society of Agronomy: Madison, WI, USA, 2020; Volume 67, pp. 165–182.

27. He, Z.; Pagliari, P.H.; Waldrip, H.M. Applied and Environmental Chemistry of Animal Manure: A Review. *Pedosphere* **2016**, *26*, 779–816. [CrossRef]

28. Chia, S.Y.; Tanga, C.M.; van Loon, J.J.; Dicke, M. Insects for sustainable animal feed: Inclusive business models involving smallholder farmers. *Curr. Opin. Environ. Sustain.* **2019**, *41*, 23–30. [CrossRef]

29. Nalunga, A.; Komakech, A.J.; Jjagwe, J.; Magala, H.; Lederer. J. Growth characteristics and meat quality of broiler chickens fed earthworm meal from Eudrilus eugeniae as a protein source. *Livest. Sci.* **2021**, *245*, 104394. [CrossRef]

30. Enahoro, D.; Lannerstad, M.; Pfeifer, C.; Dominguez-Salas, P. Contributions of livestock-derived foods to nutrient supply under changing demand in low- and middle-income countries. *Glob. Food Secur.* **2018**, *19*, 1–10. [CrossRef]

31. Janković, L.J.; Petrujkić, B.; Aleksić, N.; Vučinić, M.; Teodorović, R.; Karabasil, N.; Relić, R.; Drašković, V.; Nenadović, K. Carcass characteristics and meat quality of broilers fed on earthworm (*Lumbricus rubellus*) meal. *J. Hell. Vet. Med Soc.* **2020**, *71*, 2031–2040. [CrossRef]

32. Kasule, L.; Katongole, C.; Nambi-Kasozi, J.; Lumu, R.; Bareeba, F.; Presto, M.; Ivarsson, E.; Lindberg, J.E. Low nutritive quality of own-mixed chicken rations in Kampala City, Uganda. *Agron. Sustain. Dev.* **2014**, *34*, 921–926. [CrossRef]

33. Van Harn, J.; Dijkslag, M.A.; Van Krimpen, M.M. Effect of low protein diets supplemented with free amino acids on growth performance, slaughter yield, litter quality, and footpad lesions of male broilers. *Poult. Sci.* **2019**, *98*, 4868–4877. [CrossRef] [PubMed]

34. Mohamed, A.M.; Sekar, S.; Muthukrishnan, P. Prospects and potential of poultry manure. *Asian J. Plant Sci.* **2010**, *9*, 172–182. [CrossRef]

35. Garg, S.; Bahla, G.S. Phosphorus availability to maize as influenced by organic manures and fertilizer P associated phosphatase activity in soils. *Bioresour. Technol.* **2008**, *99*, 5773–5777. [CrossRef]

36. Bauer, A.; Black, A.L. Organic carbon effects on available water capacity of three soil textural groups. *Soil Sci. Soc. Am. J.* **1992**, *56*, 248–254. [CrossRef]

37. Tisdall, J.M.; Oades, J.M. Organic matter and water-stable aggregates in soils. *J. Soil Sci.* **1993**, *33*, 141–163. [CrossRef]

38. Mbah, C.N.; Nnej, I.R. Effect of different crop residue management technique on selected soil properties and grain production of maize. *Afr. J. Agric. Res.* **2011**, *6*, 4149–4152.

39. Ojeniyi, S.O.; Amusan, O.A.; Adekiya, A.O. Effect of poultry manure on soil physical properties, nutrient uptake and yield of cocoyam (*Xanthosoma saggitifolium*) in southwest Nigeria. *Am.-Eurasian J. Agric. Environ. Sci.* **2013**, *13*, 121–125.

40. Modderman, C. Composting with or without Additives. In *Animal Manure: Production, Characteristics, Environmental Concerns, and Management*; Waldrip, H.M., Pagliari, P.H., He, Z., Eds.; American Society of Agronomy: Madison, WI, USA, 2020; Volume 67, pp. 245–254.

41. Finstein, M.S. Composting in the context of municipal solid waste management. In *Environmental Microbiology*; Mitchell, R., Ed.; Wiley-Liss, Inc.: New York, NY, USA, 1992; pp. 355–374.

42. Kirchmann, H.; Witter, E. Ammonia volatilization during aerobic and anaerobic manure decomposition. *Plant Soil* **1989**, *115*, 35–41. [CrossRef]

43. Martins, O.; Dewes, T. Loss of nitrogenous compounds during composting of animal wastes. *Bioresour. Technol.* **1992**, *42*, 103–111. [CrossRef]

44. Beck-Friis, B.; Smårs, S.; Jönsson, H.; Kirchmann, H. Gaseous emissions of carbon dioxide, ammonia, and nitrous oxide from organic household waste in a compost reactor under different temperature regimes. *J. Agric. Eng. Res.* **2001**, *78*, 423–430. [CrossRef]

45. Zhang, W.; Lau, A. Reducing ammonia emission from poultry manure composting via struvite formation. *J. Chem. Technol. Biotechnol.* **2007**, *82*, 598–602. [CrossRef]

46. Eghball, B.; Power, J.F.; Gilley, J.E.; Doran, J.W. Nutrient, carbon, and mass loss during composting of beef cattle feedlot manure. *J. Environ. Qual.* **1997**, *26*, 189–193. [CrossRef]

47. Eklind, Y.; Kirchmann, H. Composting and storage of organic household waste with different litter amendments. II: Nitrogen turnover and losses. *Bioresour. Technol.* **2000**, *74*, 125–133. [CrossRef]

48. García, A.; Fox, J.G.; Besser, T.E. Zoonotic enterohemorrhagic Escherichia coli: A One Health perspective. *ILAR J.* **2010**, *51*, 221–232. [CrossRef] [PubMed]

49. García-Heredia, A.; Pohane, A.A.; Melzer, E.S.; Carr, C.R.; Fiolek, T.J.; Rundell, S.R.; Lim, H.C.; Wagner, J.C.; Morita, Y.S.; Swarts, B.M.; et al. Peptidoglycan precursor synthesis along the sidewall of pole-growing mycobacteria. *eLife* **2018**, *7*, 37243. [CrossRef]

50. Ke, W.; Chao, H.; Shijie, Y.; Weijie, L.; Wei, W.; Ruijun, Z.; Huanhuan, Q.; Nanqi, R. Transformation of organic matters in animal wastes during composting. *J. Hazard. Mater.* **2015**, *300*, 745–753.

51. Georgakakis, D.; Krintas, T. Optimal use of the Hosoya system composting poultry manure. *Bioresour. Technol.* **2000**, *72*, 227–233. [CrossRef]

52. Hosoya & Co. *Hosoya Manure Fermentation, System*; Hoyosa & Co.: Kanagawa, Japan, 1996.

53. ISO. *ISO14040 Environmental Management—Life Cycle Assessment—Principles and Framework*; International Organization for Standardization (ISO); Standards Policy and Strategy Committee: London, UK, 2006.

54. Szabó, A.; Tamás, J.; Nagy, A. Spectral evaluation of the effect of poultry manure pellets on pigment content of maize (*Zea mays* L.) and wheat (*Triticum aestivum* L.) seedlings. *Nat. Resour. Sustain. Dev.* **2019**, *9*, 70–79. [CrossRef]

55. OpenLCA Nexus, Databases, Agribalyse. Available online: https://nexus.openlca.org/database/Agribalyse (accessed on 23 April 2021).

56. Guinee, J.B.; Gorree, M.; Heijungs, R.; Huppes, G.; Kleijn, R.; de Koning, A.; van Oers, L.; Sleeswijk, A.W.; Suh, S.; Udo de Haes, H.A.; et al. *Handbook on Life Cycle Assessment—Operational Guide to the ISO Standards*; Kluwer Academic Publisher: New York, NY, USA; Boston, MA, USA; Dordrecht, The Netherlands; London, UK; Moscow, Russia, 2002.

57. Gabathuler, H. The CML Story: How Environmental Sciences Entered the Debate on LCA. *Int. J. Life Cycle Assess.* **2006**, *11*, 127–132. [CrossRef]

58. Kabakian, V.; McManus, M.; Harajli, H. Attributional life cycle assessment of mounted 1.8 kWp monocrystalline photovoltaic system with batteries and comparison with fossil energy production system. *Appl. Energy* **2015**, *154*, 428–437. [CrossRef]

59. Lamnatou, C.; Chemisana, D. Evaluation of photovoltaic-green and other roofing systems by means of ReCiPe and multiple life cycle–based environmental indicators. *Build. Environ.* **2015**, *93*, 376–384. [CrossRef]

60. Colomb, V.; Amar, S.A.; Mens, C.B.; Gac, A.; Gaillard, G.; Koch, P.; Mousset, J.; Salou, T.; Tailleur, A.; van der Werf, H.M.G. AGRIBALYSE®, the French LCI Database for agricultural products: High quality data for producers and environmental labelling. *Oilseeds Fats Crop. Lipids* **2015**, *22*, D104. [CrossRef]

61. Koch, P.; Salou, T. *AGRIBALYSE®: Methodology, Agricultural Stage—Version 3.0*; Ed ADAME: Angers, France, 2020.

62. Asselin-Balençon, A.; Broekema, R.; Teulon, H.; Gastaldi, G.; Houssier, J.; Moutia, A.; Rousseau, V.; Wermeille, A.; Colomb, V. *AGRIBALYSE v3.0: The French Agricultural and Food LCI Database. Methodology for the Food Products*; ADEME: Angers, France, 2020.

63. Gaidajis, G.; Kakanis, I. Life Cycle Assessment of Nitrate and Compound Fertilizers Production—A Case Study. *Sustainability* **2021**, *13*, 148. [CrossRef]

64. Baldini, C.; Bava, L.; Zucali, M.; Guarino, M. Milk production Life Cycle Assessment: A comparison between estimated and measured emission inventory for manure handling. *Sci. Total Environ.* **2018**, *625*, 209–219. [CrossRef]

65. Kátai, J. Determination of nitrogen doses (In Hungarian: Nitrogén dózisok megállapítása). In *Soil ecology (In Hungarian: Talajökológia)*, 1st ed.; Kátai, J., Ed.; University Press, University of Debrecen: Debrecen, Hungary; University of West Hungary: Sopron, Hungary; University of Pannonia: Veszprém, Hungary, 2011; pp. 59–61.

66. de Vries, J.W.; Corré, W.J.; van Dooren, H.J.C. *Environmental Assessment of Untreated Manure Use, Manure Digestion and Co-Digestion with Silage Maize*; Report 372; Wageningen UR Livestock Research, part of Stichting Dienst Landbouwkundig Onderzoek (DLO Foundation): Wageningen, The Netherlands, 2010.

67. Zhu, Z.; Dong, H.; Xi, J.; Xin, H. Ammonia and greenhouse gas emissions from co-composting of dead hens with manure as affected by forced aeration rate. *Trans. ASABE Am. Soc. Agric. Biol. Eng.* **2014**, *57*, 211–217.

68. ADEME. *Programme de Recherche de l'ADEME sur les Émissions Atmosphériques du Compostage. Connaissances Acquises et Synthèse Bibliographique*; Isabelle, D., Ed.; Agence de l'Environnement et de la Maîtrise de l'Energie, Waste Prevention and Management Department—Substainable Consumption and Waste Division: Angers, France, 2012.

69. Luske, B. *Reduced GHG Emissions due to Compost Production and Compost Use in Egypt. Comparing Two Scenarios*; 2010-016 LbD.; Louis Bolk Instituut: Bunnik, The Netherlands, 2010.

70. Brentrup, F.; Hoxha, A.; Christensen, B. Carbon footprint analysis of mineral fertilizer production in Europe and other world regions. In Proceedings of the 10th International Conference on Life Cycle Assessment of Food (LCA Food 2016), Dublin, Ireland, 19–21 October 2016.

71. Skowrońska, M.; Filipek, T. Life cycle assessment of fertilizers: A review. *Int. Agrophysics* **2014**, *28*, 101–110. [CrossRef]

72. Williams, A.G.; Audsley, E.; Sandars, D.L. Environmental burdens of producing bread wheat, oilseed rape and potatoes in England and Wales using simulation and system modelling. *Int. J. Life Cycle Assess.* **2010**, *15*, 855–868. [CrossRef]

73. Albaugh, T.J.; Vance, E.D.; Gaudreault, C.; Fox, T.R.; Allen, H.L.; Stape, J.L.; Rubilar, R.A. Carbon Emissions and Sequestration from Fertilization of Pine in the Southeastern United States. *For. Sci.* **2012**, *58*, 419–429. [CrossRef]

74. Zhang, W.-F.; Dou, Z.-X.; He, P.; Ju, X.-T.; Powlson, D.S.; Chadwick, D.R.; Norse, D.; Lu, Y.-L.; Zhang, Y.; Wu, L. New technologies reduce greenhouse gas emissions from nitrogenous fertilizer in China. *Proc. Natl. Acad. Sci. USA* **2013**, *110*, 8375–8380. [CrossRef]

75. Chen, W.; Geng, Y.; Hong, J.; Yang, D.; Ma, X. Life cycle assessment of potash fertilizer production in China. *Resour. Conserv. Recycl.* **2018**, *138*, 238–245. [CrossRef]

76. Ledgard, S.F.; Lieffering, M.; Zonderland-Thomassen, M.A.; Boyes, M. Life cycle assessmenta—A tool for evaluating resource and environmental efficiency of agricultural products and systems from pasture to plate. *Proc. New Zealand Soc. Anim. Prod.* **2011**, *71*, 139–148.

MDPI
St. Alban-Anlage 66
4052 Basel
Switzerland
Tel. +41 61 683 77 34
Fax +41 61 302 89 18
www.mdpi.com

Agriculture Editorial Office
E-mail: agriculture@mdpi.com
www.mdpi.com/journal/agriculture